E. Gröller
H. Löffelmann
W. Ribarsky (eds.)

Data Visualization '99

Proceedings of the Joint
EUROGRAPHICS and IEEE TCVG
Symposium on Visualization
in Vienna, Austria, May 26–28, 1999

Eurographics

SpringerWienNewYork

Ao. Univ.-Prof. DI Dr. techn. Eduard Gröller
DI Dr. techn. Helwig Löffelmann
Institute of Computer Graphics,
Technical University, Vienna, Austria

Dr. William Ribarsky
College of Computing, Georgia Institute of Technology,
Atlanta, GA, USA

Typesetting: Camera-ready by authors

Printed on acid-free and chlorine-free bleached paper

SPIN: 10728676

With 230 partly coloured Figures

ISSN 0946-2767
ISBN-13:978-3-211-83344-5   e-ISBN-13:978-3-7091-6803-5
DOI:10.1007/978-3-7091-6803-5

Preface

This book contains the proceedings of the Joint EUROGRAPHICS - IEEE TCVG Symposium on Visualization (VisSym '99), which took place from May 26 to May 28, 1999, in Vienna, Austria. Following nine successful EUROGRAPHICS workshops on Visualization in Scientific Computing, the tenth event of this series was organized together with the IEEE Technical Committee on Visualization and Graphics (TCVG).

Due to several facts the symposium has been a unique event. We celebrated the tenth anniversary of this series, which shows the strong continued interest in visualization research. For the first time IEEE TCVG has co-sponsored a conference outside the US. The joint conference has attracted strong participation from both sides of the Atlantic. This year we solicited not only research papers but also, for the first time, case studies to illustrate the increasing impact of visualization research in real-world applications. In previous years the main focus of the workshop has been in scientific visualization (e.g., volume and flow visualization). By broadening the scope of the symposium to include information visualization, we are able to present the latest research results in this emerging, highly active discipline.

Due to the strong interest in the event, authors from 22 countries submitted their latest research work. The 64 paper submissions were carefully evaluated and only 30 papers (21 research papers and 9 case studies) were selected for presentation at the symposium and inclusion in the proceedings. We think you will find that these proceedings are valuable not only for scientists in scientific visualization and information visualization, but also for practitioners developing or using visualization applications.

We want to thank all those involved in organizing this symposium and in producing this book. Special thanks go to the people at the Austrian Academy of Sciences, where the event took place, the people at the Institute of Computer Graphics, Vienna University of Technology, and the staff of Springer-Verlag, Wien. Also, the organizers want to express their thanks to all who helped with reviewing (up to eight) of the submitted papers each.

May 1999   Eduard Gröller,
                                                              Helwig Löffelmann,
                                                              William Ribarsky.

Table of Contents

Visualization of Medical Data & Molecules (Research Papers)

Geometry, Grids, and Systems (Research Papers)

Information Visualization and Systems (Case Studies)

Volume, Medical, & Molecular Visualization (Case Studies)

Authors Index

Color Plates

Chairs, IPC, and Referees

Symposium and Program Co-Chairs

Eduard Gröller, Vienna University of Technology, Vienna, Austria
William Ribarsky, Georgia Institute of Technology, Atlanta, Georgia

International Program Committee

K. Brodlie,	H.-C. Hege,	W. Ribarsky,
R. Crawfis,	L. Hesselink,	L. Rosenblum,
J. David,	A. Kaufman,	M. Rumpf,
J. Dill,	U. Lang,	H. Schumann,
T. Ertl,	W. Lefer,	R. Scopigno,
J. Foley,	H. Löffelmann,	D. Silver,
M. Goebel,	W. Lorensen,	P. Slavik,
M. Grave,	N. Max,	W. Straßer,
E. Gröller,	R. Moorhead,	M. Vannier,
M. Gross,	S. Müller,	E. Wenger,
H. Hagen,	H.-G. Pagendarm,	J. van Wijk
C. Hansen,	F. Post,	

Organizing Co-Chairs

Helwig Löffelmann, Vienna University of Technology, Vienna, Austria
Emanuel Wenger, Austrian Academy of Sciences, Vienna, Austria

Referees

D. Bartz,	J. David,	H. Hagen,
C. Best,	J. Dill,	J.-U. Hahn,
D. Bielser,	L. Dimitrov,	C. Hansen,
M. Bock,	M. Doggett,	P. Hastreiter,
U. Bockholt,	K. Engel,	J. Hladuvka,
K. Brodlie,	J. Foley,	J. Huang,
P. Cignoni,	M. Grave,	A. Hubeli,
B. Colbert,	E. Gröller,	T. Hüttner,
R. Crawfis,	H. Haase,	F. Jiang,

R. D. King,
S. King,
T. King,
C. Knöpfle,
R. Koch,
A. König,
W. Kühn,
E. Kuo,
U. Lang,
M. Lanzagorta,
W. de Leeuw,
W. Lefer,
H. Löffelmann,
C. Lürig,
R. Lütolf,
R. Machiraju,
N. Max,
W. Meyer,

C. Montani,
R. Moorhead,
K. Müller,
F. Nussberger,
H.-G. Pagendarm,
F. Post,
E. Puppo,
F. Reinders,
W. Ribarsky,
S. H. M. Roth,
M. Rumpf,
D. Schmalstieg,
J. Schmidt-Ehrenberg,
F. Schröder,
M. Schulz,
H. Schumann,
R. Scopigno,
H.-P. Seidel,

N. Shareef,
D. Silver,
P. Slavik,
O. Sommer,
H. J. W. Spölder,
T. Sprenger,
D. Stalling,
J. Edward Swan II,
R. F. Tobler,
K. N. Tsoi,
M. Vannier,
A. Vilanova i Bartrolí,
G. Voss,
E. Wenger,
R. Westermann,
J. van Wijk,
M. Zöckler,
J. van der Zwaag

Information Visualization

(Research Papers)

Procedural Shape Generation for Multi-dimensional Data Visualization

David S. Ebert[1], Randall M. Rohrer[2], Christopher D. Shaw[3], Pradyut Panda[1], James M. Kukla[1], and D. Aaron Roberts[4]

[1] CSEE Department, University of Maryland Baltimore County
1000 Hilltop Circle, Baltimore, MD 21250, USA
{ebert,panda,jkukla1}@csee.umbc.edu
[2] Department of EECS, The George Washington University
Washington, DC, 20052, USA
rohrer@seas.gwu.edu
[3] Department of Computer Science, University of Regina
Regina, Saskatchewan, S4S 0A2, Canada
cdshaw@cs.uregina.ca
[4] NASA Goddard Space Flight Center, Mailstop 692.0
Greenbelt, MD, 20771, USA
roberts@vayu.gsfc.nasa.gov

Abstract. Visualization of multi-dimensional data is a challenging task. The goal is not the *display* of multiple data dimensions, but *user comprehension* of the multi-dimensional data. This paper explores several techniques for perceptually motivated procedural generation of shapes to increase the comprehension of multi-dimensional data. Our glyph-based system allows the visualization of both regular and irregular grids of volumetric data. A glyph's location, 3D size, color, and opacity encode up to 8 attributes of scalar data per glyph. We have extended the system's capabilities to explore shape variation as a visualization attribute. We use procedural shape generation techniques because they allow flexibility, data abstraction, and freedom from specification of detailed shapes. We have explored three procedural shape generation techniques: fractal detail generation, superquadrics, and implicit surfaces. These techniques allow from 1 to 14 additional data dimensions to be visualized using glyph shape.

1 Introduction

The simultaneous visualization of multi-dimensional data is a difficult task. The goal is not only the display of multi-dimensional data, but the *comprehensible display* of multi-dimensional data. Glyph, or iconic, visualization is an attempt to encode more information in a comprehensible format, allowing multiple values to be encoded in the parameters of the glyphs [10]. The shape, color, transparency, orientation, etc., of the glyph can be used to visualize data values. Glyph rendering [10, 11] is an extension to the use of glyphs and icons in numerous fields, including cartography, logic, and pictorial information systems.

In previous work, we explored the usefulness of stereo-viewing and two-handed interaction to increase the perceptual cues in glyph-based visualization. The Stereoscopic Field Analyzer (SFA) [4] allows the visualization of both regular and irregular grids of volumetric data. SFA combines glyph-based volume rendering with a two-handed minimally-immersive interaction metaphor to provide interactive visualization, manipulation, and exploration of multivariate, volumetric data. SFA uses a glyph's location, 3D size, color and opacity to encode up to 8 attributes of scalar data per glyph. These attributes are used when a vector visualization is not appropriate, such as when displaying temperature and pressure at each glyph. We are extending this work to combine glyph rendering with other visually salient features to increase the number of data dimensions simultaneously viewable.

2 Background

Our use of glyph attributes for visualization is based on human perceptual abilities. Color, transparency, position, and size are perceptually significant visualization attributes that are commonly used in visualization systems. Shape is a more challenging visualization attribute because three-dimensional shape perception is not as well understood as color, size, and spatialization perception. Most evidence suggests that shape variation is a valuable perceptual attribute that we can harness for visualization. Experiments show that humans can pre-attentively perceive three-dimensional shape [8]. Cleveland [3] cites experimental evidence that shows the most accurate method to visually decode a quantitative variable in 2D is to display position along a scale. This is followed in decreasing order of accuracy by interval length, slope angle, area, volume, and color. Bertin offers a similar hierarchy in his treatise on thematic cartography [2].

Our visualization system already utilizes glyph position in 3D, 3D scale (corresponding to Cleveland's length, area and volume) and color. Slope angle is a difficult dimension to use in an interactive system because of arbitrary orientation of the data volume. Therefore, the next opportunity for encoding a scalar value is shape.

One of the most difficult problems in glyph visualization is the design of meaningful glyphs. Glyph shape variation must be able to convey changes in associated data values in a comprehensible manner [10]. This difficulty is sometimes avoided by adopting a single base shape and scaling it non-uniformly in 3 dimensions. However, the lack of a more general shape interpolation method has precluded the use of shape beyond the signification of categorical values [2]. This paper describes three techniques we have explored for the procedural generation of glyph shapes for glyph-based volumetric visualization [7].

3 Perceptually-based Mapping of Shape Attributes

Glyph shape is a valuable visualization component because of the human visual system's pre-attentive ability to discern shape. Shapes can be distinguished at the pre-attentive stage [8] using curvature information of the 2D silhouette contour and, for 3D objects, curvature information from surface shading. Unlike an arbitrary collection of icons, curvature has a visual order, since a surface of higher curvature looks more jagged than a surface of low curvature. Therefore, generating glyph shapes by maintaining control of their curvature will maintain a visual order. This allows us to generate a range of glyphs which interpolate between extremes of curvature, thereby allowing the user to read scalar values from the glyph's shape. Pre-attentive shape recognition allows quick analysis of shapes and provides useful dimensions for comprehensible visualization.

Our use of glyphs is related to the idea of marks as the most primitive component that can encode useful information [2]. Senay points out that shape, size, texture, orientation, transparency, hue, saturation, brightness, and transparency are retinal properties of marks that can encode information [13, 14]. Bertin has studied the use of marks for two-dimensional thematic maps and gives examples of how shape can be misused in the rendering of these maps [2]. In his examples, shapes are used to represent purely categorical data and, for this reason, he uses a small collection of distinct icons such as star, cross, square, circle, triangle, and so on. Because each individual shape does not have any inherent meaning, the reader is forced to continually look up the shape's meaning in the map legend. The main difficulty is that a collection of arbitrary icons does not have any necessary visual order, and so any assignment of shape to meaning is equivalent.

4   Procedural Shape Visualization

We have explored three different procedural techniques for the generation of glyph shape: superquadrics, fractal detail, and implicit surfaces. All three techniques use a procedural approach for glyph design to solve the complex problem of meaningful glyph design. Procedural shape generation techniques provide flexibility, data abstraction, and freedom from specification of detailed shapes [5]. Procedural techniques allow the shape to be controlled by high-level control parameters. The user changes the glyph shape from a more directorial, indirect aspect, where he or she is unburdened from the full explicit specification of detailed shapes. Our goal for glyph design was to allow the automatic mapping of data to shape in a comprehensible, easily controllable manner.

4.1   Fractal shape detail

One simple procedural technique for shape visualization is to generate distorted versions of a basic shape, such as a cube, where the amount of deviation from the basic shape is proportional to the data dimension being visualized: low data values will map to the base shape and high data values will map to very perturbed shapes. Displacement mapping is used to create the deviation from the base shape. We used a fractional Brownian motion (fBM) [6] turbulence function to create the displacement from the original surface. The main idea with

6

this technique is to add a high frequency component to the shape while not distracting from the perception of the low frequency base shape perception. This addition of high frequency information does not detract from the overall spatial pattern of the data, which can occur from the generation of non-related shapes. Figure 1 shows the results of this technique applied to a simple random test data set where the data values near zero are cuboid and the "fuzzy" cubes have data values near one.

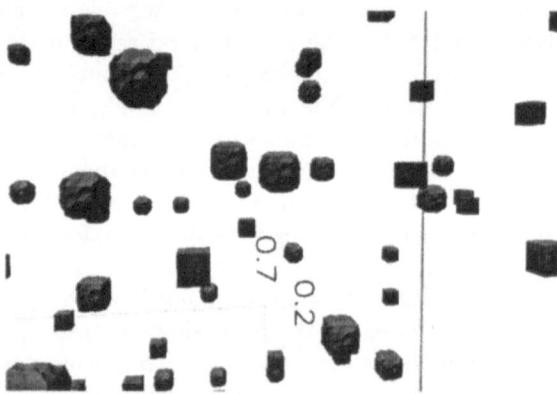

Fig. 1. Example fractal displacement of a base cube shape. The data values are a random data set, with values ranging from 0 (smooth cube) to 1 (fuzzy cube).

4.2   Procedural shape visualization using superquadrics

Superquadrics are a natural tool for automatic shape visualization that can allow from one to two data dimensions to be visualized with shape. Superquadrics, first introduced to computer graphics by Barr [1], are extensions of quadric surfaces where the trigonometric terms are each raised to exponents. Superquadrics come in four main families: hyperboloid of one sheet, hyperboloid of two sheets, ellipsoid, and toroid. For our initial implementation we have chosen superellipses due to their familiarity, but the system can be easily extended to use other types of superquadrics as well as combinations of types. For example, supertoroids could be used for negative values and superellipsoids for positive values.

In the case the of superellipsoids, the trigonometric terms are assigned exponents as follows:

$$x(\eta, \omega) = \begin{bmatrix} a_1 & \cos^{\epsilon 1} \eta & \cos^{\epsilon 2} \omega \\ a_2 & \cos^{\epsilon 1} \eta & \cos^{\epsilon 2} \omega \\ a_3 & \sin^{\epsilon 1} \eta \end{bmatrix}, \quad \begin{matrix} -\pi/2 \le \eta \le \pi/2 \\ -\pi \le \omega < pi \end{matrix}$$

These exponents allow continuous control over the characteristics (the concavity or convexity) of the shape in the two major planes which intersect to form the shape, allowing a very simple, intuitive, abstract schema of shape specification. For example, $\epsilon 1 < 1$ and $\epsilon 2 < 1$ produces cuboid shapes, $\epsilon 1 < 1$ and $\epsilon 2 \sim 1$ produces cylindroid shapes, $\epsilon 1 > 2$ or $\epsilon 2 > 2$ produces pinched shapes while $\epsilon 1 = 2$ or $\epsilon 2 = 2$ produces faceted shapes. As can be seen in Figure 2, varying the exponents achieves smooth, understandable transitions in shape. Therefore, mapping data values to the exponents provides not only a continuous, automatic control over the shape's overall flavor, but a comprehensible shape mapping as well.

To produce understandable, intuitive shapes, we rely on the ability of superquadrics to create graphically distinct [13, 14], yet related shapes. We encode two data dimensions in glyph shape in a manner that allows the easy separation of the shape characteristics.

4.3 Procedural shape visualization with implicit surfaces

Implicit surfaces are a powerful geometric modeling tool which use an underlying density field to represent volumetric information. Implicit techniques provide smooth blending functions for individual implicit field sources [5]. Isosurface generation techniques are then used to create a geometric representation of this volumetric density field. This natural, smooth blending of multiple implicit sources makes implicit surfaces a natural choice for shape visualization.

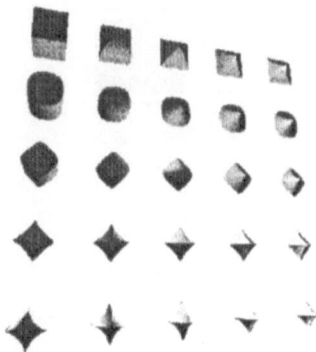
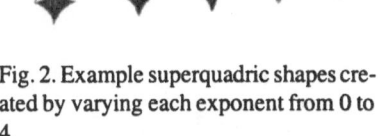
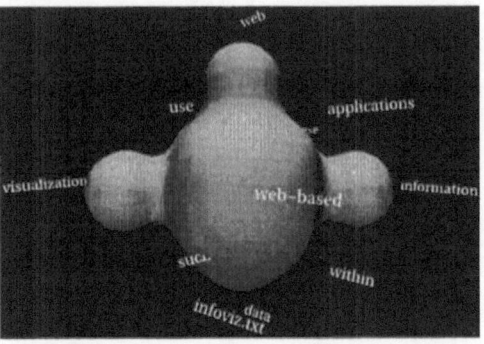

Fig. 2. Example superquadric shapes created by varying each exponent from 0 to 4.

Fig. 3. Example implicit surface shape for a document on "Web-based Information Visualization." The three main bulges correspond to the frequency of the terms "web", "visualization", and "information".

For implicit shape visualization, we map up to 14 data dimensions to uniformly spaced vectors emanating from the center of a sphere. The length of each of these vectors is scaled based on the data value being visualized. At the end of each of these vectors, we place a uniform point source field function. We then generate an isosurface (typically, $F = 0.5$) from the resulting density field to create the glyph shape. This produces "blobby" shapes with bulges in a given direction indicating large data values for that dimension of the data. An example of the a single blobby glyph of this type can be seen in Figure 3.

Since size and spatial location are more significant cues than shape, the importance mapping of data values should be done in a corresponding order. In decreasing order of data importance, data values are mapped to location, size, color, and shape. In our experience, shape is very useful for local area comparisons among glyphs: seeing local patterns, rates of change, outliers, anomalies.

5   Implementation and Results

5.1   Information Visualization Results

We have applied procedurally-generated glyph shapes to the visualization of both scientific and information data. For information visualization, we have chosen an example of the visualization of "thematic" document similarities. Figure 4 shows a visualization of document similarities generated with the Telltale system [9]. The document corpus consists of 1883 articles from *The Wall Street Journal* from September and October 1989. Each glyph in Figure 4 represents a document in the corpus, and the document's X, Y, and Z position, color and shape each represent the similarity of the document to one of the 5 themes.

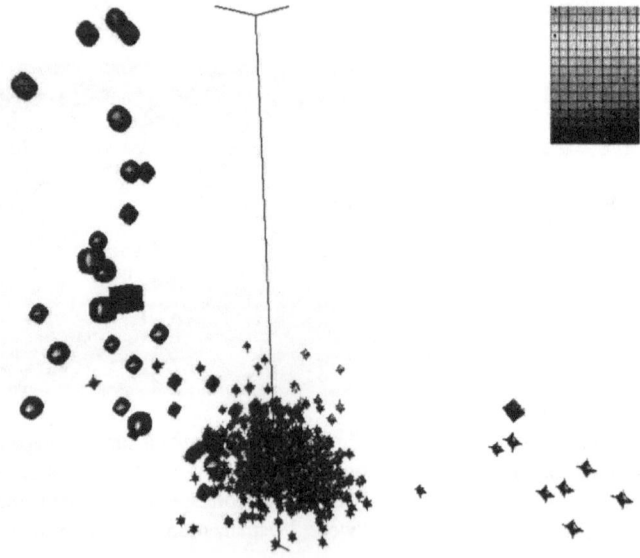

Fig. 4. Three-dimensional visualization of 1833 documents' relationship to gold prices, foreign exchange, the federal reserve, stock prices, and Manuel Noriega.

Document similarity to *gold prices*, the *foreign exchange rate of the U.S. dollar*, and *federal reserve* are respectively mapped to the X, Y, and Z axes. The Y axis is visually indicated in Figure 4 by the vertical line, with the X axis going to the right and the Z axis going to the left. The bulk of the documents have very low similarity to all of these 3 themes, so their glyphs are clustered near the origin at the bottom center.

The documents outside this cluster exhibit two spatial patterns: a cluster of 9 documents to the bottom right and a vertical branch on the left. The right cluster indicates the small number of documents in the corpus that discuss both *gold prices* and the *foreign exchange rate of the U.S. dollar*. The vertical branch depicts a larger collection of documents that discuss both *foreign exchange rate of the U.S. dollar* and the *federal reserve*.

A fourth attribute, similarity to *stock prices*, is inversely mapped to both superquadric exponents of the glyph shape, with highest similarity creating cuboids, then spheres, diamonds, and stars (lowest). Referring to the square array of sample glyphs in Figure 2, the similarity to *stock prices* maps to glyphs on the diagonal from the upper left to the lower right of Figure 2, with upper left indicating high similarity, and lower right indicating low similarity.

In Figure 4 the larger, rounder shapes along the vertical branch exhibit some significant relationship to *stock prices* while the more numerous star-shaped glyphs do not. Clearly the vertical branch contains articles relating *foreign exchange*, *federal reserve* and *stock prices*.

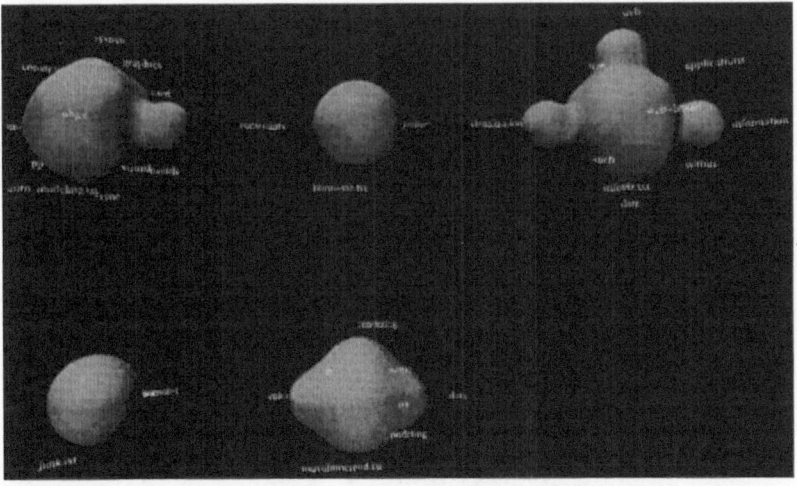

Fig. 5. Multiple documents term frequency visualized as implicit surface shapes. The document in the upper left and the upper right both have a high frequency of the term "information" (bulge to the right). The upper right blob is the same as that in figure 3

Glyph color is mapped inversely to similarity to *Manuel Noriega*. Most of the documents fall in the turquoise and purple range, indicating no significant relationship. However, the documents in the orange, red, and yellow-green range represent documents with a significant relationship to *Manuel Noriega*. Many of these documents mention the effect of the coup attempt against *Manuel Noriega* and its effect on *foreign exchange rate of the U.S. dollar* (vertical axis). The fact that these orange, red, and yellow-green documents are not in either of the branches indicates that these articles did not relate heavily to either *federal reserve* or *gold prices*, and their star shape indicates no relationship with *stock prices*.

We have also used the implicit surface technique for text-based information visualization. The input data for this visualization was word frequency from a collection of text documents [12]. The frequency of the 14 most frequent words were mapped to vector length in positioning the density field sources: documents with common word usage exhibit similar shapes. The results of this process can be seen in Figure 5. From quick examination of this figure, documents with similar topics and word usage are easily distinguishable.

5.2 Scientific Visualization Results

We have used this system to examine several scientific visualization data sets. Figure 6 shows the visualization of a magnetohydrodynamics simulation of the solar wind in the distant heliosphere (20 times the distance of the Earth to the Sun). The simulation data is a $64 \times 64 \times 64$ grid containing the vector vorticity and velocity for the simulation.

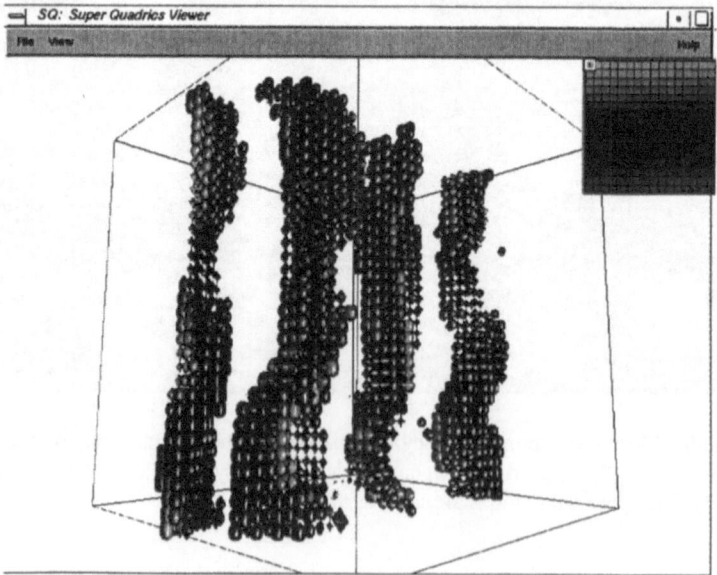

Fig. 6. Visualization of a magnetohydrodynamics simulation of the solar wind in the distant heliosphere showing both velocity components and vorticity components of 6 vortex tubes.

Opacity is used to represent vorticity in the j direction, so that the 6 vortex tubes (only 4 are visible) represent zones in space where this vorticity is somewhat larger than zero. Glyph shape is based inversely on the velocity in the j direction. Positive velocities are displayed as larger, rounder to cuboid shapes and negative velocities are displayed as spiky, star-like shapes. Zero velocity is represented by the diamond shape. The overall columnar pattern of the data is not disturbed by the introduction of the shape mapping, but the velocity variation can still be seen as we traverse the lengths of the tubes. In this case, values close to zero in terms of j vorticity (still fluid) have been masked out.

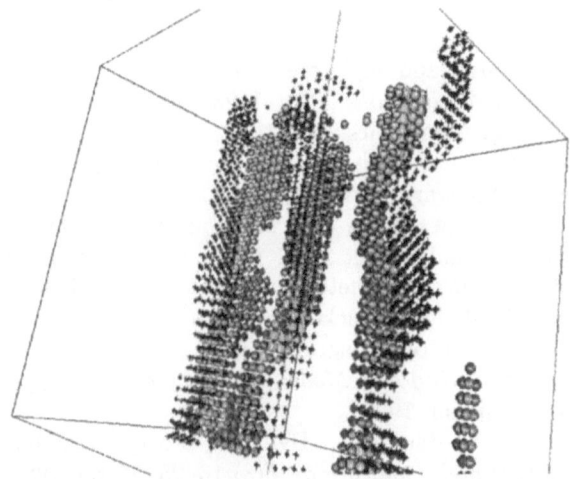

Fig. 7. Visualization of a magnetohydrodynamics simulation of the solar wind in the distant heliosphere displaying 3 vortex tubes with positive j vorticity (cuboids and ellipsoids) and 3 vortex tubes with negative j vorticity (stars).

Figure 7 is a visualization of the same magnetohydrodynamics data, but with the opacity, color, and glyph shape all mapped to the j component of vorticity. Negative vorticity components produce concave shapes (blue stars), while positive values produce convex shapes (orange cuboids and ellipsoids). The use of this data mapping clearly shows three tubes with negative j vorticity and three tubes with positive j vorticity.

6 Conclusions

We have developed several new techniques for intuitive, comprehensible creation of glyph shapes. These techniques are based on procedural shape generation and increase the number of dimensions of data that can be comprehensibly visualized in a glyph-based visualization system. Superquadric functions, fractal surface displacement, and implicit surfaces have been shown to be useful techniques for the automatic generation of glyph shapes and the visualization of

12

multi-dimensional data. Superquadrics are a very natural technique for shape visualization of one to two data dimensions. Fractal surface displacement seems useful for small categorizations (4 to 6 easily discernible shapes) of a single data dimension and can actually be added to the other two techniques as a secondary shape cue. Implicit surface techniques show great promise in visualizing between eight and twenty data dimensions with shape. All of these procedural shapes allow the intuitive understanding of data variation among glyphs, while preserving the global data patterns. We have shown the value of these techniques for both multi-dimensional information and scientific visualization.

References

[1] A. Barr. Superquadrics and angle-preserving transformations. *IEEE Computer Graphics and Applications*, 1(1):11–23, 1981.

[2] J. Bertin. Semiology of graphics, 1983.

[3] William S. Cleveland. *The Elements of Graphing Data.* Wadsworth Advanced Books and Software, Monterey, Ca., 1985.

[4] David Ebert, Chris Shaw, Amen Zwa, and Cindy Starr. Two-handed interactive stereoscopic visualization. *Proc. IEEE Visualization '96*, Oct. 1996.

[5] David S. Ebert. Advanced geometric modeling. *The Computer Science and Engineering Handbook*, Allen Tucker Jr., ed., chap. 56. CRC Press, 1997.

[6] David S. Ebert, F. Kenton Musgrave, Darwyn Peachey, Ken Perlin, and Steven Worley. *Texturing and Modeling: A Procedural Approach, Second Edition.* AP Professional, 1998.

[7] J. D. Foley and C. F. McMath. Dynamic process visualization. *IEEE Computer Graphics and Applications*, 6(3):16–25, March 1986.

[8] Andrew Parker, Chris Cristou, Bruce Cumming, Elizabeth Johnson, Michael Hawken, and Andrew Zisserman. The Analysis of 3D Shape: Psychological principles and neural mechanisms. In Glyn Humphreys, editor, *Understanding Vision*, chapter 8. Blackwell, 1992.

[9] Claudia E. Pearce and Charles Nicholas. TELLTALE: Experiments in a Dynamic Hypertext Environment for Degraded and Multilingual Data. *Journal of the American Society for Information Science (JASIS)*, 47, April 1996.

[10] Frank J. Post, Theo van Walsum, Frits H. Post, and Deborah Silver. Iconic techniques for feature visualization. In *Proceedings Visualization '95*, pages 288–295, October 1995.

[11] W. Ribarsky, E. Ayers, J. Eble, and S. Mukherja. Glyphmaker: creating customized visualizations of complex data. *IEEE Computer*, 27(7):57–64.

[12] Randall M. Rohrer, David S. Ebert, and John L. Sibert. The Shape of Shakespeare: Visualizing Text using Implict Surfaces. In *Proceedings Information Visualization 1998*, pages 121–129. IEEE Press, 1998.

[13] H. Senay and E. Ignatius. A knowledge-based system for visualization design. *IEEE Computer Graphics and Applications*, 14(6):36–47, November 1994.

[14] H. Senay and E. Ignatius. Rules and principles of scientific data visualization. *ACM SIGGRAPH HyperVis Project, 1996*.

Editors' Note: see Appendix, p. 311 for colored figure of this paper

Skeletal Images as Visual Cues in Graph Visualization

I. Herman[1], M.S. Marshall[1], G. Melançon[1], D.J. Duke[2], M. Delest[3], and
J.-P. Domenger[3]

[1] CWI
P.O. Box 94079
1090 GB Amsterdam, The Netherlands
{Ivan.Herman, Scott.Marshall, Guy.Melancon}@cwi.nl
[2] Department of Computer Science
The University of York
Heslington, York, YO10 5DD
duke@cs.york.ac.uk
[3] LaBRI, UMR 5800
351, Cours de la Libération
33405 Talence Cedex, France
{maylis,domenger}@labri.u-bordeaux.fr

Abstract. The problem of graph layout and drawing is fundamental to many
approaches to the visualization of relational information structures. As the data
set grows, the visualization problem is compounded by the need to reconcile the
user's need for orientation cues with the danger of information overload. Put simply: How can we limit the number of visual elements on the screen so as not to
overwhelm the user yet retain enough information that the user is able to navigate
and explore the data set confidently? How can we provide orientational cues so
that a user can understand the location of the current viewpoint in a large data
set? These are problems inherent not only to graph drawing but information visualization in general. We propose a method which extracts the significant features
of a directed acyclic graph as the basis for navigation [1].

1 Introduction

A fundamental challenge for information visualization applications that use graph visualization techniques for relational data sets is the scale and structural complexity of the
data. Beyond the well known and researched problems of graph layout, large-scale data
sets call for new approaches to navigation, and the provision of visual cues to support
the user's awareness of the context or location within the data set. There is a large body
of published research results in this area, which involve the use of zoom [8], pan, visual
cues [2], and focus+context techniques using non-linear filters such as, for example,
fish-eye views [7], hyperbolic geometry [5], and distortion-oriented presentations [9].
This paper contributes a method which can be used to produce a schematic view of a
directed acyclic graph or DAG to the tools and techniques available for viewing graph
structures.

---

[1] Note: The color figures and a demonstration applet are available at the web site:
http://www.cwi.nl/InfoVisu.

The *skeleton* of a graph is the set of nodes and edges that are determined to be significant by a given metric. The skeleton can give the impression of a structural backbone. Because it is a selection of a small subset of important nodes, the skeleton eliminates the problem of information overload while still providing information essential for further exploration. The skeleton also allows the user to characterize a particular graph by providing a simple image which contains the most important features or 'landmarks' of a graph. In this way, the skeleton provides the user with a map for orientation and navigation. The features chosen by the metric may be structurally important or reflect some other measure. By changing the metrics used to extract the skeleton, we may produce different maps for different purposes.

The highlighting of trees according to the underlying Strahler values was proposed as an aid to navigation in [2]. In this paper, we will explain how to apply Strahler and other metrics to trees and DAGs to obtain a skeleton. Obviously, the metric is crucial in the determination of the skeleton. We have looked for metrics which result in a skeleton that is a good indicator of the underlying structure of the graph.

Our current methods require that the graph be acyclic. Although it is possible to extract a DAG from an arbitrary graph, for simplicity we have chosen in this phase of work to assume that the graph is already directed and acyclic. As with other types of graphs, DAGs can be quite overwhelming when the number of nodes is large and are found in many applications. This makes the DAG an excellent candidate for skeleton extraction. Of course, any result derived for DAGs is also applicable for trees.

2 General methodology

The general methodology for extracting skeletons is as follows:

- Choose a metric function for a graph and a threshold value.
- Traverse the graph and extract the nodes whose metric values are equal to or above the threshold value.
- Display the final skeleton. This consists of the nodes which have been extracted and the edges connecting them.
- Display the leftover nodes and edges, possibly merged or simplified.

The metric should reflect the relative significance of each node of the graph. The metric and the resulting skeleton should correspond to a clear mental model which aids the user during navigation. The threshold value determines the level of detail which is represented by the skeleton.

The last step of our method involves representing the nodes and edges not selected by the extraction process. For trees, the subtrees not belonging to the skeleton may be simply replaced by triangles or other shapes, resulting in a schematic view of the tree. Representing the non-skeletal nodes and edges from a DAG calls for more sophisticated techniques using different colours and intensities to distinguish between skeletal and non-skeletal parts of the DAG (See Section 2.3 for details).

Introductory example: Figure 1 shows the original structure of a tree. Figure 2 shows the skeleton which results from selecting the nodes which have metric values equal to or above a threshold value. Our program has replaced the excluded nodes by triangles

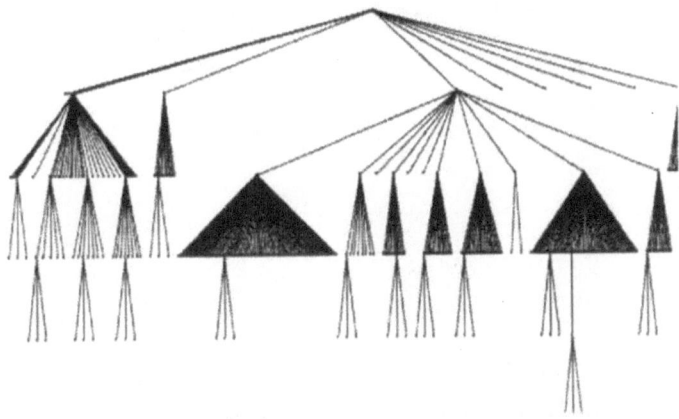

Fig. 1. A fully displayed tree

to create a *schematic view* of the tree. A schematic view is a simplified representation of a graph which makes use of the skeleton and replaces non-skeletal parts of the graph with lines or shapes.

2.1 Metrics

The extraction of a skeleton from a DAG requires the computation of a value for each node of a DAG in the same fashion as for trees. In this section, we will talk about two different metrics and give an impression of the skeletons which they give as a result. These metrics were chosen for two reasons. First, each can be explained in terms of a simple metaphor, which we believe will help users develop an intuition about the effect of the metric, without needing to understand the underlying mathematics. Second, experimental results have shown that these metrics provide an impression of the overall structure of the DAG. Later, we will also indicate how different metrics can be composed. The composition of metrics can produce quite useful results and can be applied much the same way as one might apply several optical filters to a camera lens.

The Strahler metric. We used Strahler numbers for trees as a metric to extract the skeleton in Figure 2. This metric was already presented in [2] and we will recall it here; for a full account on Strahler numbers see [3].

The Strahler value of a leaf is set to 1. For any other node v, a value is computed using the formula:

$$S(v) = \max(S(k_1), \ldots, S(k_p)) + \begin{cases} p - 1 & \text{if all values } S(k_i) \text{ are equal} \\ p - 2 & \text{otherwise} \end{cases} \quad (1)$$

where $k_1, \ldots, k_p$ are the successors of v.

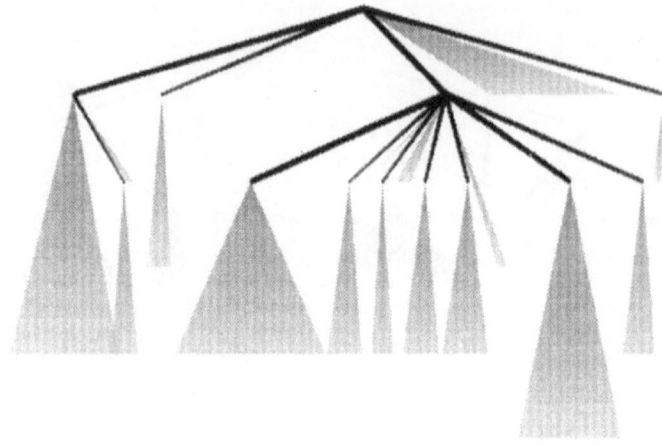

Fig. 2. A schematic view of the tree in Figure 1

Strahler numbers have proven to be a good measure of the branching structure of hierarchical networks (trees). They were also used as the basis of a method for producing realistic images of 2D trees in a paper from Viennot et al. [10]. The results presented there certainly confirm the potential that Strahler numbers bear as a means for describing graphical effects on trees.

In the case of trees, it is easy to check that greater values are attained by balanced trees. Numbers such as the Strahler number for trees are often referred to as *synthetic values*, because of their links with attribute grammars [4] and their use in combinatorial mathematics [6]. Other values can be computed using the same recursive scheme. For example, giving a value of 1 to every leaf of a tree and setting the value of a node to be the sum of the values of its children leads to a synthetic computation of the numbers of leaves for each subtree. This metric can be given more application-specific values, through the use of weights (see [2]). The weight of a node could represent, for example, the number of visits in the case of a web page or the size of a file in case of directory trees.

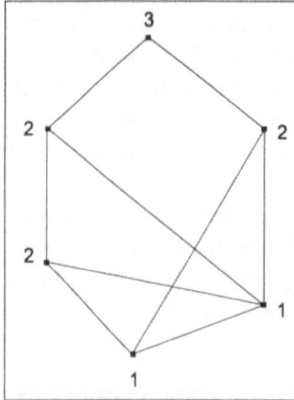

Fig. 3: Strahler for DAGs

The same computation scheme can be applied to any graph without cycles. Indeed, it is the absence of cycles in the structure that makes it possible to define a function depending on the set of *successors* of a node. A DAG has no cycles and provides an explicit direction to traverse its nodes. In a DAG, the subset of nodes having incoming degree zero are called *source nodes*. Similarly, the nodes having outgoing degree zero are called *sink nodes*. Obviously, a full traversal of a DAG may start from the source nodes and end in the sink nodes. The traversal of a DAG may also start from the sink nodes, depending on the desired results.

The Strahler metric can be easily generalized to DAGs by setting $S(b) = 1$ for every sink node, and by applying Eq. (1) to the set of *successors* of a node v. Figure 3 gives an example. Note: There is an implicit downward direction in Figures 3 and 4.

The Flow metric. The second metric we present is based on a water flow metaphor. A downward scan of the DAG emphasizes the distribution or flow of information from a node to its successors.

The DAG could be compared to a set of connected pipes through which water flows

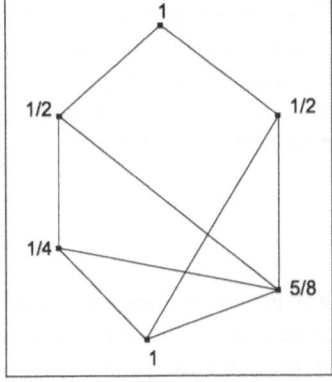

from top to bottom. We will call this metric the *Flow metric* and denote it by M.

Let $M(t) = 1$ for every source node t. Then compute values for every other node the following way: Divide the value at a node by the number of its successors to find its contribution to each of them. A node receiving a set of values from its ancestors sums them up. More precisely, the value $M(v)$ for a node v is obtained by summing contributions over the set of all its ancestors $a_1, \ldots, a_q$ $(q \geq 1)$. That is,

Fig. 4: Flow metric for DAGs

$$M(v) = \sum_j M(a_j)/\text{number of successors of } a_j \qquad (2)$$

The DAG in Figure 4 provides an example. Observe that values produced by this metric do not necessarily decrease (or increase) along a path from source to sink nodes. The value $M(v)$ at a node v evaluates the flow going through that node.

General framework. Of course, both the Strahler and Flow metrics produce values which are proportional to the node's importance. This should be kept in mind when designing any other metric to be applied to a DAG. The pattern used for Strahler, as well as the Flow metric, can easily be extended to a general computation scheme as follows: Suppose arbitrary values $K(b)$ are given to the sink nodes of a DAG. Set

$$K(v) = F((K(k_1), \ldots, K(k_p))) \qquad (3)$$

for any other node v, where $k_1, \ldots, k_p$ are the successors of v and F is a function (or formula) depending on the values $K(k_1), \ldots, K(k_p)$. Hence, values are assigned to nodes of the DAG through an upward search, as with the Strahler metric. We could also define a function computing values through a downward search. In that case, we use a recurrence:

$$K'(v) = F((K'(a_1), \ldots, K'(a_q))) \qquad (4)$$

where $a_1, \ldots, a_q$ are the ancestors of v, and assign starting values $K'(t)$ to source nodes of the DAG. This computation scheme was used to define the Flow metric. Observe that a dual Strahler metric could be defined by applying the opposite computation scheme using the same formula (Eq. (1)), but applying it to ancestors instead of successors. The same observation applies to the Flow metric, yielding a measure for an "upward" flow of information or data.

The actual function to compute is in some sense application dependent. However, the choice or design of a metric should be strongly linked to a clear interpretation of its effect on the extraction process. From this point of view, metrics corresponding to well understood metaphors might have a wider range of uses and applications. This is the case for the Flow metric since it is supported by the water flow metaphor: the nodes in the skeleton are those through which much of the water flows. Also, weights can be used the same way they are used with Strahler to influence values of the nodes. Another possibility could be to give distinct starting values to source nodes of a DAG.

2.2 Skeleton extraction

Given a metric, the simplest approach to extract a skeleton is to collect the nodes with a value greater than or equal to a lower bound. We can compute a lower bound that extracts a specific percentage of the nodes, so we will express it in terms of the percentage.

Figure 5 shows the skeleton which results from selecting the nodes with Flow values in the top 30 %. The square bold-faced nodes are those belonging to the skeleton. The thicker arcs are those joining nodes in the skeleton. This example actually has no need for a skeletal view because it is not very complex but the smaller number of nodes allows us to more easily illustrate the essential concepts. For examples using a larger number of nodes, please see the color plates in the Appendix.

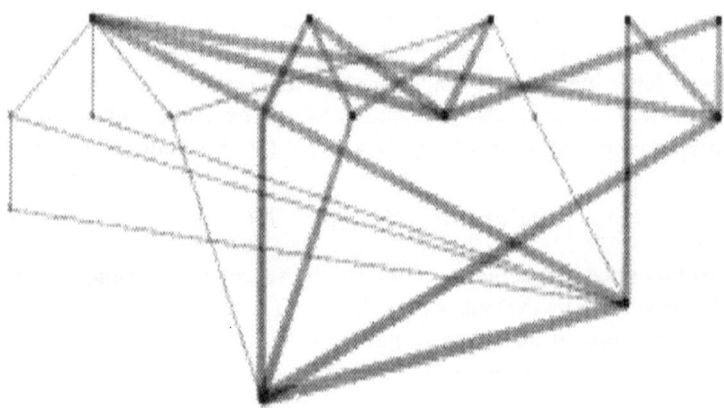

Fig. 5. The nodes selected based on Flow values in the top 30 %

All source nodes are part of the skeleton since they have an assigned values of 1. The water flow metaphor aids in understanding why a given node is present in the skeleton. This is obvious, for example, for the node with incoming degree 4 and outgoing degree 0 in the upper part of the skeleton; the contributions it collects from all sources nodes except one sum up to a value of 1.25, hence its presence in the skeleton. The node at the bottom of the right part on the skeleton bears the same value. Observe that it collects values from six different nodes, only two of which are part of the skeleton. The fact that the value 1.25 makes those nodes part of the skeleton depends on the set of values reached by all nodes in the graph and our choice to display the 30% top nodes.

2.3 Implementing a schematic view

The goal of the schematic view is to emphasize the "backbone", as produced by the skeleton extraction. The schematic view consists of two displayed parts: the skeleton itself, and the leftover nodes and edges.

Instead of lines, very long and thin trapezoids are used to display the skeleton edges. Trapezoids were chosen because they can have different widths at each end as an extra visual cue. The width of the trapezoids at the nodes are proportional to the metric values of the incident nodes. In other words, the sizes of the edges give an indication of the magnitude of the metrics at the incident nodes. Similarly, a continuous visual indication is provided by colour. Skeleton edges and nodes are drawn using a different hue than the leftover nodes (e.g. red on the colour plates in the Appendix). As a further visual cue, the saturation component of the colour along each edge is interpolated from the metric values at the source and destination nodes.

For DAGs, the leftover nodes and edges are simply drawn using a low-contrast hue (light gray on the Appendix colour plates). For trees, the monotonicity of the metrics, as well as the simpler structure of trees, allows for an alternative representation: triangles are used to replace the leftover nodes and edges. The size of the triangle image is proportional to the subtree being represented (see Figure 2). A continuous colour transition similar to the scheme for the skeleton edges is also used on the triangles. The top of the triangles have a saturation proportional to the node's metric value, and the triangle gradually changes colour and saturation toward a shade of the background color. Of course, more complicated representations than triangles could be used for the subtrees such as the type of images used in the Aggregate TreeMaps of Chuah [1]).

In the cases of both trees and DAGs, the use of Alpha blending has also been an effective aid for both trees and DAGs. The transparency provided by alpha blending ensures that the intersections of edges and triangles do not interfere with the clarity of the figure.

3 Metric combination

Any good metric should emphasize a specific aspect of the DAG. *Combining* different metrics into new ones is a way to capture multiple aspects of the graph; some examples will be presented in this section.

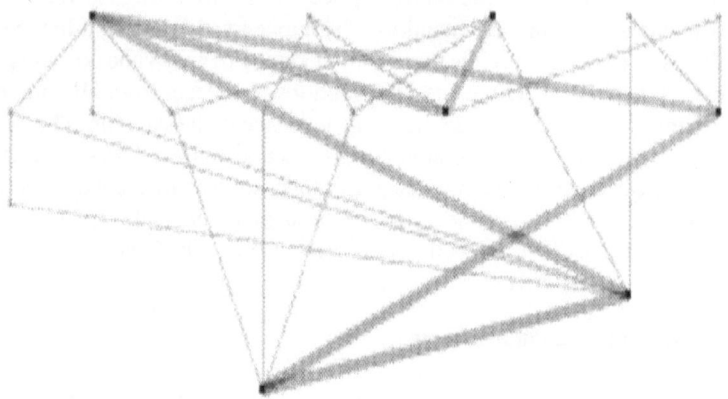

Fig. 6. Skeleton of a DAG based on a combination of Strahler and Flow metrics

Combining Strahler and Flow metrics. We use an example to illustrate how a combination of metrics can be achieved. When looking at the skeleton in Figure 5, one might object to the fact that *all* the source nodes were selected, i.e. that the Flow metric does not allow us to distinguish among the source nodes. Indeed, one may want to use a metric reflecting the fact that a subgraph, starting at a specific source, is more complex than another. Using our water flow metaphor, the metric should provide pipes with different diameters, depending on the complexity of the corresponding subgraph. The Strahler value of a node, which measures the structural complexity of its subgraph, is then a good candidate for measuring this complexity. This, in combination with the Flow metrics, may be used to define the desired new metric. The detailed definition of the new metric is as follows.

The Flow metric is modified so that the node receives a value from its ancestor proportional to its Strahler value. Denote by $\mu(v)$ the sum of the Strahler values of the children of a given node v. That is, $\mu(v) = \sum_j S(k_j)$, where the sum extends over the set of successors $k_1, \ldots, k_p$ of v. The new metric is then defined by $P(t) = 1$ for all source nodes t and by the equation:

$$P(v) = \sum_j P(a_j) \cdot \frac{S(v)}{\mu(a_j)}$$

for all other nodes, where the sum extends over the set of ancestors $a_1, \ldots, a_q$ of v. That is, the value $P(v)$ is obtained by summing contributions obtained from ancestors of v. A specific ancestor will give its children a part of its own value proportional to their Strahler values.

The skeleton based on this modified Flow scheme is shown in Figure 6. It is extracted from the same DAG as in Figure 5 using identical threshold values. Notice how this new computation scheme sorts the source nodes to extract only those playing a more important role in the whole graph (or network of pipes).

Combining directions. A further combination of metrics can be achieved if their directions are also taken into consideration. Indeed, the choice between Eq. (3) or Eq. (4) for the computation scheme priviledges a direction. If both a metric and its "dual" are used on the same graph, each node is assigned two different values, reflecting directional measures. These values can then be combined (for example, by taking their average value), thereby yielding a new metric again. This "average" metric reflects both the "upward" and the "downward" characteristics of the DAG relative to the metric.

As a specific example, the modified Flow metric of the preceding section has its dual metric, too. This dual metric uses the dual Flow metric and the *dual Strahler values*. Finally, the two directional Flow metric can be combined into an average Flow metric. This metric has been used to obtain the color plates in the Appendix.

4   Conclusions and further research

Our skeleton extraction methodology can be applied to any tree or DAG without using domain-specific knowledge, i.e. the semantic information usually associated with nodes or edges in a graph visualization application. However, it is possible to add domain-specific weights to the metric in order to sift the nodes for features of interest. In this way, it is possible to tailor a metric in order to implement a search. We have discussed the first type of metric which extracts interesting features from the graph relations inherent to the data, resulting in a structural view.

In our java application, skeletal views play an important role as navigational aids, complementing techniques such as zoom, pan, and fish-eye views. Although this is the only application of skeletal views which we have discussed, there are others. For example, we have created thumbnail images with skeletal views of a DAG. These thumbnails can then be used as the representation of a folded subtree. Another well known application of thumbnails is as a bird's-eye view of the graph with an indication of the current viewing location.

The primary goal of our future research is to extend the skeleton idea to more general graphs. The definition of metrics for such graphs may be a significant problem because convergence must be ensured. Other metrics for DAGs, as well as other techniques to display the skeleton and the leftover nodes should be explored. A more elaborate usability study on the utility of skeletal views is also a possible future activity.

5   Acknowledgements

Part of this work has been funded by the Franco-Dutch research cooperation programme "van Gogh". We are also grateful to Bèhr de Ruiter (CWI), who developed an interactive interface for tree visualization, including zoom, pan, and fish-eye. Extending this framework to DAGs, instead of developing a completely new framework, allowed us to concentrate on the research issues in a more timely manner.

References

1. Chuah M.C. Dynamic Aggregation with Circular Visual Designs. In Wills G. and Dill J., editors, *Proceedings of the IEEE Symposium on Information Visualization (InfoViz '98), Los Alamitos*, pages 35 – 43. IEEE CS Press, 1998.
2. Delest M., Herman I., and Melançon G. Tree Visualization and Navigation Clues for Information Visualization. *Computer Graphics Forum*, 17(2), 1998.
3. Flajolet P. and Prodinger H. Register allocation for unary-binary trees. *SIAM Journal. of Computing*, 15:629 – 640, 1986.
4. Knuth D.E. Semantics of context-free languages. *Math. Systems Theory*, 2:127–145, 1968.
5. Lamping J., Rao R., and Pirolli P. A focus+context technique based on hyperbolic geometry for viewing large hierarchies. In *ACM CHI'95*. ACM Press, 1995.
6. Delest M. Algebraic languages: a bridge between combinatorics and computer science. In *Actes SFCA '94*, volume 24 of *DIMACS*, pages 71–87. American Mathematical Society, 1994.
7. Sarkar M and Brown M.H. Graphical fisheye views. *Communication of the ACM*, 37(12):73–84, 1994.
8. Schaffer D., Zuo Z., Greenberg S., Bartram L., Dill J., Dubs S., and Roseman M. Navigating hierarchically clustered networks through fisheye and full–zoom methods. *ACM Transactions on Computer-Human Interaction*, 3(2), 1996.
9. Keahey T.A. The Generalized Detail-In-Context Problem. In Wills G. and Dill J., editors, *Proceedings of the IEEE Symposium on Information Visualization (InfoViz '98), Los Alamitos*. IEEE CS Press, 1998.
10. Viennot X.G., Eyrolles G., Janey N., and Arques D. Combinatorial analysis of ramified patterns and computer imagery of trees. *Computer Graphics (SIGGRAPH '89)*, 23:31 – 40, 1989.

Editors' Note: see Appendix, p. 312 for colored figure of this paper

Visualization by Examples: Mapping Data to Visual Representations using Few Correspondences

Marc Alexa and Wolfgang Müller

Darmstadt University of Technology, Department of Computer Science, Interactive Graphics System Group, Rundeturmstr. 6, 64283 Darmstadt, Germany
{alexa,mueller}@gris.informatik.tu-darmstadt.de,
WWW home page: http://www.igd.fhg.de/~{alexa,mueller}

Abstract. In this paper we propose a new approach for the generation of visual scales for the visualization of scalar and multivariate data. Based on the specification of only a few correspondences between the data set and elements of a space of visual representations complex visualization mappings are produced. The foundation of this approach is the introduction of a multidimensional space of visual representations. The mapping between these spaces can be defined by approximating or satisfying the user defined relations between data values and visual atributes.

1 Introduction

The visualization of scalar and multivariate quantitative data involves the mapping of data onto a visual scale. The principles of such a mapping of scalar data to scales of visual attributes are well-known for a number of basic scales. In color mapping, data values are mapped to appropriate hue of lightness values. Scatter plots are based on the principle of mapping data values to positions, or more exactly, to distances from an axis. Other variables often applied for such purposes are scale, form, and texture. Bertin [4] describes a general methodology of how to select an appropriate mapping to these visual variables and how to combine them. Cleveland gives a ranking of their effectiveness [9]. For multivariate data with two upto five dimensions more complex color and texture scales can provide solutions in some cases. For the visualization of local variations higher dimensions data glyphs have been proposed and been applied successfully. Chernoff faces [8] and stick figures [13] are well-known examples for this visualization approach. Tufte gives a good overview of relevant visualization techniques and discusses their effectiveness in certain application areas [17]. Nevertheless, a generally accepted set of visualization rules does not exist.

All the techniques mentioned above represent fundamental approaches to the visualization of scalar and multivariate data. The general understanding is that no single of these visualizations is effective in all possible situations. One visualization may – and should of course – produce new insights and, by this, new questions which again produce the need of different views to the data. A good number of applications have proven that a user can gain knowledge about some unknown data more effectively when provided with highly interactive techniques [15]. Techniques such as Focusing and Linking [7] extend this interactivity even further by connecting different views to the data using interactive feedback.

However, effective visualization still is very, very difficult. There are a number of reasons for this. First, a mapping of application data to these fundamental variables involves an abstraction. If the user is familiar with the idea of this mapping, he may understand the generated visualization pretty good. However, often the application context is lost and the visualization which was applied to simplify the analysis of the data involves an analysis step or experience by its own.

Second, it is not easy to visualize a number of parameters using these fundamental mappings only. Usually, the visualization of multiparameter data involves the generation of application specific models and solutions. General solutions and general scales do not exist for this purpose. Consequently, the user needs methods to define a visual scale for such applications very quickly and easily. Such methods do hardly exist to date.

Last, and may be even most important, it is still difficult to change visualization parameters and to produce a new, appropriate visualization intuitively. For example, to highlight a specific data value one has usually to modify the data filtering or to completely redefine the mapping of the data to visual attributes. A direct and interactive modification of local data mappings is hardly provided by any visualization technique today.

In this paper, we introduce a new approach for the visualization of scalar and multivariate data, which addresses the problems presented above. This approach is based on the direct and interactive specification of local data mappings.

2 Visualization by Examples

The general paradigm of our visualization approach is to enable the user to visualize some data by specifying the mapping of a small number of selected data values. We call this Visualization by Example.

This approach can be explained best with an example. A simple mapping of some scalar quantitative data to color can be defined by linking two arbitrary data values with appropriate color values. This results in a linear mapping. Further links can be supplied to adjust the mapping locally, resulting in a more sophisticated mapping function. The visualization of any local feature and its neighborhood is directly and intuitively controlled by the user and may be easily changed when provided with appropriate visual attributes or objects on which the data may be mapped. Note that this strategy is fundamentally different from the selction of a color scale and applying this scale to all data.

While the direct and interactive linking of data values with visual representations is not new, it has not been combined with appropriate methods for the approximation of data mappings based on the supplied correspondences.

In the example presented above this approach seems simple and easy to follow.

However, the generalization of this approach calls for a mathematical model describing the data objects, flexible visual scales, and the mapping between this values based on a small number of parameters and features.

In addition, the user will have some additional knowledge about the data in many cases, which can be exploited with our approach. Moreover, the user might want to highlight several data values by mapping them to special representations.

There exist approaches to describe data spaces based on mathematical models [6]. Mathematical models have currently been proposed to describe the spaces of visual scales and representations in combination with the use of morphing to construct the corresponding graphical objects for this purpose [2] [12]. This approach allows to define rich sets of useful visual representations from only a few graphical base objects. As such, this method provides the appropriate foundation for a Visualization by Example.

In the following section we will discuss the fundamentals and characteristics in more detail.

3 Space of Visual Representations

For our visualization technique, the visual representations have to be structured as a multidimensional space. That is, a visual object has to be element of an n-dimensional space and represented by a vector $r \in \mathbb{R}^n$.

For many visual scales such a representation is quite natural:

- Color can be represented by real values in $[0, 1]^3$, color scales might be represented by real numbers in $[0, 1]$.
- Size or position is naturally a real number.
- For textures one defines a number of real valued parameters, which control their appearance.
- In general: If the visual representations are used to depict quantitative data there has to be a reasonable understanding in terms of real valued vector spaces.

In [2] and [12] we have proposed a more flexible way to define spaces of visual representations: Given a number of graphical objects of any class (images, polyhedra, etc.) we construct a space by morphing among these objects. Morphing is usually applied to generate animations and, as such, exploited only for blending between two objects. In this approach, a morph among multiple objects by performing several morphing operations between two objects subsequently defines a multidimensional space of objects, whereby each of the morphing operations adds another dimension to the space. Barycentric coordinates can be used to represent elements of that space. Elements are constructed by morphing between two objects recursively. In that case, a space constructed from n graphical base objects is an affine space with dimension $n - 1$ (a more rigorous treatment of the mathematical properties of such spaces can be found in [1]).

Since morphing is applicable to produce scales such as color, position, size, etc. we understand this to be a generalization of these techniques to define spaces of visual representations. For the mapping technique explained in the next section, however, the actual construction of the space is not relevant. All we need are visual representations that can be referred to by n-vectors.

The strength of using morphing techniques to generate visual representations of data becomes evident when applied to multivariate data. As mentioned before, it has been proven difficult to find intuitive visual representations for multivariate data and multidimensional objects. By morphing among multiple objects one could visualize multivariate data as elements of a space of graphical objects.

4 Mapping Data to Coordinates

The main idea of our approach is to let the user define several relations between data values and graphical representations. These correspondences are used to construct a mapping from data to visual representations. We want to allow the user to define any number of correspondences, usually beginning with only a small number of correspondences. Depending on the application the user might decide to generate an affine mapping in any case, or, if no simple affine mapping can be found, to accept a non-linear function to depict the mapping.

To define a single correspondence, the user first chooses a data value and then searches the space of graphical representations for a suitable element. That is, the user gives an example for the intended relation between data and visualization. We denote the space of data values as V^d, i.e. each vector consists of d variates. Assume a number e of data vectors $v_0, v_1, \ldots, v_{e-1}$ should be mapped to a coordinate $r_0, r_1, \ldots, r_{e-1}$ in the space of visual representations. If $e \leq d$ then an affine mapping $a : V^d \to \mathbb{R}^n$ exists that satisfies $a(v_i) = r_i$ for all $i \in 0, \ldots, e-1$. However, if $e > d$ that mapping does not necessarily exist.

We suggest to use a linear mapping whenever possible, i.e. in case $e \leq d$. The reason for this is that visual scales are still meaningful, properties of the data variates are preserved in the visualization, and the order of the data values is not changed.

If e is greater than d we offer three choices:

1. An affine mapping that fits the given correspondences as far as possible.
2. A non-linear mapping that satisfies all relations by locally changing a linear approximation.
3. A non-linear mapping that satisfies all relations by globally interpolating the relations.

The second and third mapping are constructed using the same approach as discussed in section 4.2. Both need an approximation of the affine mapping similar to the one used in the first approach. In the upcoming subsection we explain how to calculate that affine mapping, independent of the number of given relations.

4.1 Finding an affine mapping

We want the mapping a to be represented by a matrix multiplication. Thus, we are searching for a $n \times d$ matrix A that maps from V^d to coordinates in the space of visual representations. In case $e \leq d$, A has to satisfy the simultaneous equations

$$
\begin{aligned}
Av_0 &= r_0 \\
Av_1 &= r_1 \\
\vdots &= \vdots \\
Av_{e-1} &= r_{e-1},
\end{aligned}
\tag{1}
$$

in case $e > d$ we like to minimize the residual

$$
(\|Av_0 - r_0\|, \|Av_1 - r_1\|, \ldots, \|Av_{e-1} - r_{e-1}\|) \ .
$$

We now first solve the first case. The techniques we employ here will automatically produce a solution to the second case.

If we look at the i-th row a_i of A we get the simultaneous equations

$$
\begin{aligned}
a_i v_0 &= r_{0_i} \\
a_i v_1 &= r_{1_i} \\
\vdots &= \vdots \\
a_i v_{e-1} &= r_{e-1_i}.
\end{aligned}
\tag{2}
$$

We define the $d \times e$ matrix

$$
B = \begin{pmatrix} - & v_0 & - \\ - & v_1 & - \\ & \vdots & \\ - & v_{e-1} & - \end{pmatrix}
\tag{3}
$$

to rewrite the simultaneous equations in (2) as a matrix equation:

$$
Ba_i^T = \left(r_{0_i}, r_{1_i}, \vdots, r_{e-1_i}, \right)^T
\tag{4}
$$

The solutions of these n systems of linear equations yield the rows of A. In order to solve one of these systems we use the Singular Value Decomposition (SVD, [14]). The SVD is a decomposition of a matrix into a product of an orthogonal matrix, a diagonal matrix and again an orthogonal matrix. This decomposition is always possible [11]. For most applications the values of the diagonal matrix (singular values) are of particular interest. If any of the singular values is zero, the above matrix equation has no solution. If we replace any zero singular value by infinity, we can invert the diagonal matrix (the orthogonal matrices are invertible anyway). By multiplying both sides of (4) with the three inverted matrices we always find a solution for a_i, no matter what the condition of B is. The SVD has several nice properties that are interesting for our problem:

1. It gives a stable solution in the quadratic case, even in the presence of degeneracies in the matrix.

2. It solves the underspecified case in a reasonable way, i.e. out of the space of solutions it returns the one closest to the origin.
3. It solves the overspecified case by minimizing the quadratic error measure of the residual.

Using the SVD we can compute all rows of A and thus have found the affine mapping we were searching for.

4.2   Non-linear mappings

We want to find a mapping a the satisfies all equations $a(v_i) = r_i$. This could be seen as a scattered data interpolation problem where we try to find a smooth interpolation between the values r_i given at locations v_i. Contrary to some other application domains of scattered data interpolation, we deal with different and high dimensions of the vectors and typically the number of relations is close to the dimension of the input data.

As explained before, it seems desirable to have an affine mapping from the data values to the space of visual representations. Therefore, we always start with a linear approximation of the mapping (as calculated in the previous section) and then fit the relations in the mapping by tiny adjustments. For these adjustments we use radial sums. The idea of combining an affine mapping with radial sums for scattered data interpolation is considered in e.g. [3] and [16] (for two-dimensional vectors, only).

Hence, we define a by

$$a(v) = Ar + \sum_j w_j f(|v - v_j|), v \in V^d, r \in \mathbb{R}^n \tag{5}$$

where $w_i \in \mathbb{R}^n$ are vector weights for a radial function $f : \mathbb{R} \to \mathbb{R}$. We consider only two choices for f:

1. The gaussian $f(x) = e^{-x^2/c^2}$, which is intended for locally fitting the map to the given relations.
2. The shifted log $f(x) = \log \sqrt{(x^2 + c^2)}$, which is a solution to the spline energy minimization problem and, as such, results in more global solutions.

We compute A beforehand as explained in the previous section. Thus, the only unknown in (5) is a pure radial sum, which is solved by constituting the known relations

$$r_i - Ar_i = \sum_j w_j f(|v_i - v_j|) \tag{6}$$

This can be written in matrix form by defining

$$F = \begin{pmatrix} f(0) & f(|v_0 - v_1|) & \cdots & f(|v_0 - v_{e-1}|) \\ f(|v_1 - v_0|) & f(0) & \cdots & f(|v_1 - v_{e-1}|) \\ \vdots & \vdots & \ddots & \vdots \\ f(|v_{e-1} - v_0|) & f(|v_{e-1} - v_1|) & \cdots & f(0) \end{pmatrix}$$

and separating (6) according to the n dimensions of r_i and w_i:

$$F \begin{pmatrix} w_{0_i} \\ w_{1_i} \\ \vdots \\ w_{e-1_i} \end{pmatrix} = \begin{pmatrix} r_{0_i} - a_i r_0 \\ r_{1_i} - a_i r_1 \\ \vdots \\ r_{e-1_i} - a_i r_{e-1} \end{pmatrix} , i \in 0, \ldots, n-1 \qquad (7)$$

Again, we solve these n equations by calculating the SVD of F. This time we are sure that an exact solution exist, because the solvability for the above radial functions f can be proven [10].

5 Results

We will demostrate the techniques at two examples. These examples show two principally different application scenarios:

- The first example shows the mapping from multivariate data onto low-dimensional visual representation. That is, the dimension of the data is much higher than the dimension of the representations.
- The second example shows a mapping from scalar data onto either basic or more complex, multiparameter representations. Here, specific aspects of the scalar data set are mapped to a specific channel of the visual attribute enhancing the expressiveness of the visualization.

5.1 Visualizing city rankings

In this example we visualize an overall (scalar) ranking of cities in the USA. Suppose we want a visual aid for a decision which of the major cities would be nice to live in. In order to quantify the different amenities and drawbacks of these cities we use data from "The places rated almanac" [5]. This data contains values for nine different categories. That means, we need to project nine-valued vectors onto scalar values.

To visualize the ranking of the cities we use a Chernoff-like approach. The faces are generated by morphing among a standard set of facial expressions (see [2]). In this example we make use of only a smile and a grumble, defining a one-dimensional visual scale. Thus, the degree of smiling represents the living quality determined by a combination of the nine data attributes from [5].

One way to find this mapping might be to inspect the nine different categories and try to find some weights for the values. This requires not only to define nine values, also the correlation to the outcome of this mapping does not take into account the user's knowledge about the cities.

A more intuitive approach is to allow the user to supply a ranking based on personal experience. Remark that a ranking of a subset of all cities is sufficient. In figure 1 only three examples were given to generate an affine mapping. Namely, Chicago was thought to be nice to live in and was mapped to a smiling

Fig. 1. Cities of the united states represented by mona lisa faces. The representation is generated from 9-valued ranking vectors. The mapping was defined by mapping Chicago to a smile, Washington to neutral face, and Miami to a grumble.

face, whereas Miami was unacceptable and mapped to a grumble. Additionally, Washington appeared nice but expensive and, therefore, mapped to a neutral face.

5.2 CT scan data

In this example, we inspect CT scan data from the "Visible Human"-project. The data is given as 16-bit data values on a 512 by 512 grid. A standard linear mapping of the relevant CT data to gray values is depicted in figure 2. Note, that this image could be produced by picking the two boundary values to define an affine map.

In figure 2 the soft tissue is display relatively bright. We can adjust this for a better distinction of bones and soft tissue by simply seleting one of the data values from the soft tissue and assigning a dark gray to it. This time an affine mapping is a bad choice, because the three correspondences cannot be satisfied. Instead, we fit the mapping globally to the data value - gray value pairs by using radial basis sums with the shifted logarithm as the radial function. The resulted is depicted in figure 3 and clearly shows the advantage in comparison with figure 2.

If we take a closer look at figure 3 we find a brighter substructure in the stomach. We would like to bring this region of data values to better attention in the visual representation. We do this by mapping a data value of this region to a red color. That is, instead of using gray values in the visualization we now use RGB color. Note, that it is not necessary to use specific two-dimensional color scales: We simply specify which data value maps to which RGB triple. The gray value

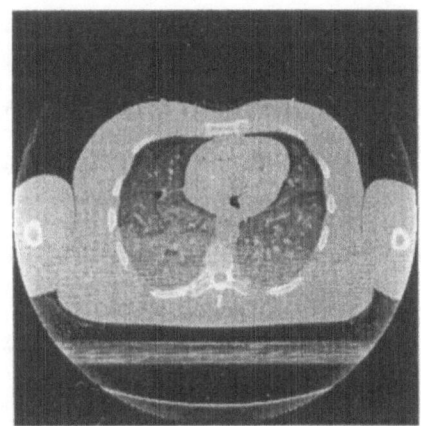 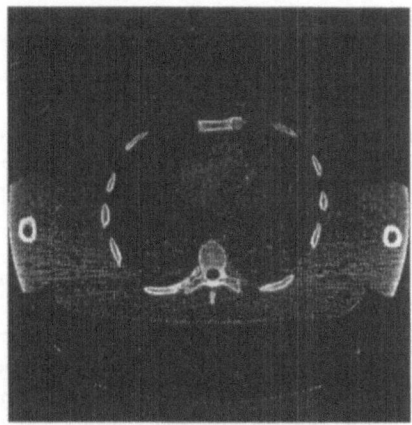

Fig. 2. The CT-scan of the chest of a man. This image is generated from the raw CT-data by linearly mapping the range of usefull CT-data values to a greyscale

Fig. 3. Here, the CT scan was generated by a mapping defined from three correspondences. The background was mapped to black, the bones were mapped to white, and the soft tissues surrounding the lung were mapped to dark grey.

representations of the three correspondences defined earlier are mapped onto corresponding RGB values. The resulting mapping is shown in figure 4 in the color section. Note, how the empty structures are colored in the complementary color of red. This gives a nice distinction of empty spaces and tissues.

6 Conclusion

In this paper we presented a new approach to the construction of visual scales for the visualization of scalar and multivariate data. Based on the specification of only a few correspondences between data values and visual representations, complex visualization mappings are produced, hereby introducing a Visualization by Examples.

This approach exploits the user's knowledge about the data in a more intuitive way. Moreover, the user is enabled to adapt the visualization interactively and easily. The technique of Visualization by Examples can be used in combination with any visual representation.

References

1. Marc Alexa and Wolfgang Müller. The Morphing Space. Proceedings of the WSCG '99, Plzen, Chzech Republic, 1999
2. Marc Alexa and Wolfgang Müller. Visualization by Metamorphosis. IEEE Visualization '98 Late Breaking Hot Topics Proceedings, 33-36, 1998

32

3. Nur Arad and Daniel Reisfeld. Image Warping Using few Anchor Points and Radial Basis Functions. Computer Graphics Forum, 14, 1, 23-29, 1995
4. J. Bertin. Semiology of Graphics, The University of Wisconsin Press, 1983
5. Richard Boyer and David Savageau. Places Rated Almanac. Rand McNally, 1985, also available from http://www.stat.cmu.edu/datasets/places.data
6. Ken Brodlie. A classification scheme for scientific visualization. in: A. Earnshaw and D. Watson: Animation and Scientific Visualization - Tools and Applications. Academic Press Ltd., London, 1993, pp. 125-140
7. Andreas Buja, John Alan McDonald, John Michalak, and Werner Stuetzle. Interactive Data Visualization using Focusing and Linking. Proc. IEEE Visualization '91, Boston, 1991, pp. 156-163
8. H. Chernoff. The use of faces to represent points in k-dimensional space graphically. Journal of American Statistical Association, 63, 361-368, 1973
9. William S. Cleveland. The Elements of Graphing Data, Wadsworth Advanced Book Program, Pacific Grove, 1985
10. Nira Dyn. Interpolation and approximation by radial and related functions. Approximation Theory VI, Vol. 1, Academic Press, 1989, pp. 211-234
11. Gene H. Golub and Charles F. van Loan. Matrix Computations. The Johns Hopkins University Press, Baltimore, 1983
12. Wolfgang Müller and Marc Alexa. Using Morphing for Information Visualization. ACM Workshop on New Paradigms in Information Visualization and Manipulation, Bethesda, 1998
13. Ronald M. Pickett and Georges G. Grinstein. Iconographics display for visualizing multidimensional data. Proceedings IEEE Conference on Systems, Man, and Cybernetics, 514-519, 1988
14. William H. Press, Saul A. Teukolsky, William T. Vetterling, and Brian P. Flannery. Numerical Recipes. Cambridge University Press, Cambridge, 1992
15. Penny Rheingans. Color, Change, and Control for Quantitative Data Display. Proc. IEEE Visualization '92, Boston, 1992, pp. 252-259
16. Detlef Ruprecht and Heinrich Müller. Image Warping with Scattered Data Interpolation. IEEE Computer Graphics and Applications, 15, 2, 37-43, 1995
17. Edward R. Tufte. The Visual Display of Quantitative Information. Graphics Press, Cheshire, Connecticut, 1983

Editors' Note: see Appendix, p. 313 for colored figure of this paper

Flow Visualization

(Research Papers)

2D Vector Field Visualization Using Furlike Texture

Leila Khouas and Christophe Odet and Denis Friboulet

Creatis, CNRS Research Unit (UMR 5515)
affiliated to INSERM, Lyon, France
khouas@creatis.insa-lyon.fr

Abstract. This paper presents a new technique for 2D vector field visualization. Our approach is based on the use of a furlike texture. For this purpose, we have first developed a texture model that allows two dimensional synthesis of 3D furlike texture. The technique is based on a non stationary two dimensional Autoregressive synthesis (2D AR). The texture generator allows local control of orientation and length of the synthesized texture (the orientation and length of filaments). This texture model is then used to represent 2D vector fields. We can use orientation, length, density and color attributes of our furlike texture to visualize local orientation and magnitude of a 2D vector field. The visual representations produced are satisfying since complete information about local orientation is easily perceived. We will show that the technique can also produce LIC-like texture. In addition, due to the AR formulation, the obtained technique is computationally efficient.

1 Introduction

In many areas such as fluid dynamics, electomagnetics, weather or medical imaging, the analysis of studied phenomena produces complex data that consist of large vector fields. The vectors represent some characteristic of each point on the field such as: a fluid vorticity, a wind velocity or a motion speed. The visualization of vector fields is not straightforward because they have no natural representation. However, many advances have been realized in this research area over the past few years. A state of the art of techniques used in fluid dynamics is given in [1][2]. Basic techniques are iconic representation, particles traces (pathlines) and streamlines, flow topology. More suitable techniques for global flow visualization are those based on the use of texture like the *Spot Noise* [3] and the *LIC* [4] techniques.

Here, we propose to build a new representation of vector fields based on furlike texture. We assume that this will provide a natural representation of a dense vector field. Such an idea has already been mentioned by Kajiya and Kay in [5] but using their fur synthesis technique (based on full 3D texture modeling) would be computationally expensive. Our approach consists of the development of a texture model that allows 2D synthesis of furlike texture having a 3D aspect. The model is based on a non stationary two dimensional Autoregressive synthesis

(2D AR). This provides a simple and efficient 2D generator of furlike texture which is then applied to vector fields representation. The work presented in this paper, enables one to encode in a natural way complete information of a vector field (direction and magnitude) in a still image. Recently, a variant of the *LIC* technique that gives similar results, has been proposed in [6]: the *OLIC* technique. A fast implementation of this technique called *FROLIC*[7] has also been described. We will show that with our completely different approach, we can achieve a computation efficiency that can be compared to the *FROLIC* technique.

Section 2 is devoted to a brief description of the AR based texture modeling [8]. Then, results obtained by applying this model for 2D vector fields visualization are shown in section 3.

2   A 2D texture model for furlike texture synthesis

Several approaches were considered to perform 3D synthesis of realistic furlike texture. Some of the proposed solutions are based on a geometrical modeling of each individual filament. Others attempt to reproduce the complex surface appearance through 3D textures and lighting models [5][9]. These techniques yield realistic results but remain too time consuming. As an alternative to the above complex models, we propose to build a 2D texture model for a simple and efficient synthesis of furlike texture. This requires producing furlike texture with arbitrary orientation and length (i.e. length of individual filaments); in addition, the appearance of the generated texture is desired to be quite realistic. We have used the AR approach described below to build such a model.

2.1   AR modeling of Texture

The AR modeling of a texture is based on a supposed linear dependence of each pixel of the texture image on its neighbors. A texture image is then considered as the output of a 2D linear filter in response to a white noise (Figure 1).

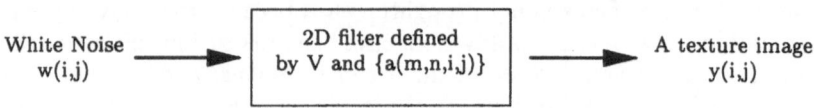

Fig. 1. 2D Autoregressive Filter

Let y be a texture image to be modelized, with a zero mean and $y(i,j)$ the value of a pixel (i,j). The AR modeling of this texture can be described by the following difference equation:

$$y(i,j) = w(i,j) + \sum_{(m,n) \in \mathcal{V}} a(m,n,i,j)y(i-m,j-n) \tag{1}$$

w is a zero mean white noise with variance σ_w^2. $\mathcal{V}$ defines the prediction region, i.e. the set of neighbors on which the pixel $y(i,j)$ depends (Figure 2). The shape and size of the neighborhood respectively define the causality of the model and its order. $a(m,n,i,j)$ are the model parameters which characterize the texture. In the stationary case, the values of the parameters $a(m,n,i,j)$ are independent of the pixel position (i,j).

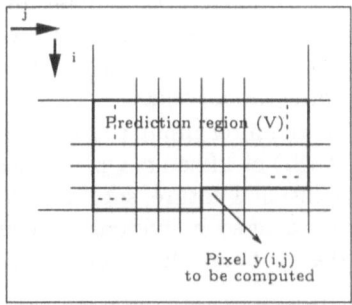

Fig. 2. Prediction region

In our case, we are thus looking for the set of AR parameters that best fits a given stationary furlike texture y. Given an appropriate neighborhood $\mathcal{V}$ (shape and size), the parameters $a(m,n)$ can be estimated using the knowledge of the autocorrelation function (ACF) of the texture image y [10].

Indeed, if we note the ACF by $r(.,.)$, the parameters $a(m,n)$ of the AR model can be obtained by solving the following set of linear equations:

$$r(s,t) - \sum_{(m,n)\in\mathcal{V}} a(m,n)r(s-m,t-n) = 0, \forall (s,t) \in \mathcal{V} \qquad (2)$$

and the input noise variance is computed using:

$$\sigma_w^2 = r(0,0) - \sum_{(m,n)\in\mathcal{V}} a(m,n)r(m,n) \qquad (3)$$

2.2 AR Synthesis of Furlike Texture

In order to build the AR model for furlike texture, it was necessary to have a set of real fur images with different characteristics. Since we needed precise knowledge of length and orientation for each image, we have not used real fur. Instead, we have used as input, images generated by a 3D texture synthesis technique based on the hypertexture modeling developed by K. Perlin and M. Hoffert [9]. This method in which we have introduced orientation, allows a full control of all visual aspects of the texture but, as mentioned earlier, is computationally intensive.

Based on one stationary reference image generated by the oriented hypertexture model and corresponding to one fixed orientation and length, we estimate the ACF $r(m,n)$. Then we compute the AR parameters and the noise variance using equations (2) and (3). The synthesis process consists then of the simple 2D linear filter defined by equation (1) with the corresponding AR parameters.

The neighborhood is chosen so that a causal synthesis can be performed (all pixels in the prediction region are computed before the current pixel). Its shape and size are defined by two parameters jx and jy illustrated in Figure 3. This neighborhood allows a synthesis of orientations in a given range of orientation values (Figure 3). This is due to its causality and its form. We describe in section 2.3 how we deal with any orientation value.

The white noise used as the filter input was chosen to be a 2D Random Sequence of scaled Impulses (RSI) of a fixed magnitude *val* and with an occurrence frequency p. Due to its impulsional nature, such a noise allows one to synthesize images having visually well defined individual filaments. Moreover, this noise enables to easily control the apparent filaments density with the parameter p.

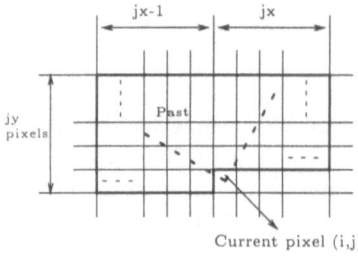

Current pixel (i,j)

Fig. 3. Neighborhood $\mathcal{V}$

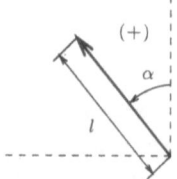

Fig. 4. Coordinates system used to represent the two texture attributes: length l and orientation α

To produce non stationary texture (with varying orientation and length), we use a lookup table of AR parameters, indexed by length and orientation [8] (see Figure 4). This table covers the ranges: $[0, 0.5]$ with a step of 0.05 for length(l) and $[-\frac{\pi}{6}, \frac{\pi}{3}]$ with a step of 1 degree for orientation (α) (see section 2.3).

We present in Figure 5 some results. The first row contains reference images (5.a,5.b) generated with the oriented hypertexture model and images (5.a',5.b') obtained by an AR synthesis with parameters computed from the ACFs of images 5.a and 5.b and using as input noise, a 2D sequence of impulses with parameters *val*=255 and p=0.033. This noise gives strong directional and depth aspects which are of a great interest for the expected application. The second row contains results of non stationary synthesis: image 5.c is a mosaic of some different orientations; and image 5.d is a furlike texture with constant orientation and length increasing from left to right.

The 2D texture model allows correct synthesis of orientations in some values range. This is due to the causality of the neighborhood and its form (Figure 3).

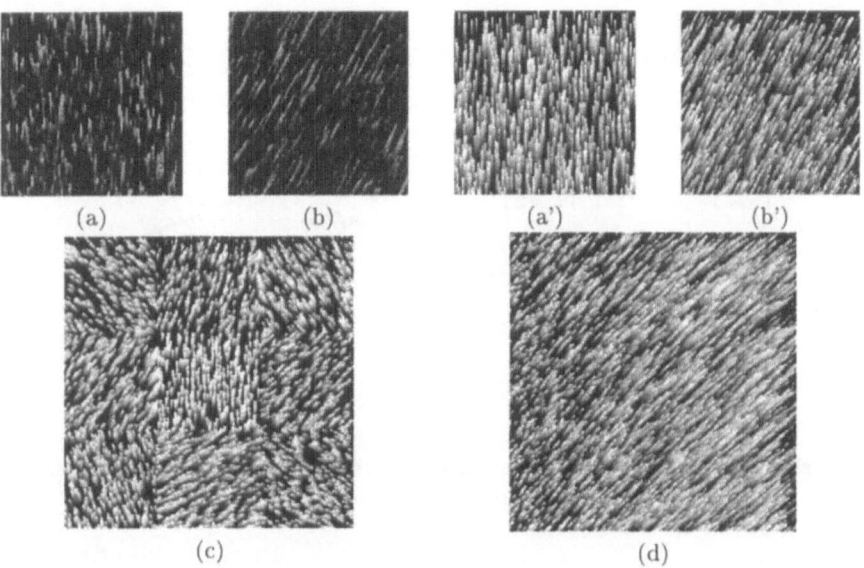

Fig. 5. AR modeling; (a,b) original images. (a',b') corresponding images obtained by AR synthesis using the sizes $(jx,jy)=(3,2)$, $(4,2)$. (c,d) are obtained by non stationary synthesis

2.3 Dealing with any orientation

The model built for furlike texture synthesis allows a correct production of orientations α in a bounded range of values. This range, which is determined by experimentation, is about $\{-\frac{\pi}{6}, \frac{\pi}{3}\}$. This is due to the causality of the neighborhood and its form (Figure 3). To deal with any orientation value, the synthesis procedure consists of four passes through the image with different orders of scanning. Each pass will generate all orientation values fitting in a convenient range as given in Figure 6. To keep a correct synthesis across the boundaries of the passes, the ranges of two successive passes present a small overlapping.

3 2D Vector Field Visualization

3.1 Visualization process

In the previous section, we have built a furlike texture generator allowing a 2D non stationary synthesis of a texture with any orientation value and using length in some available range of values. We will now use this texture model to represent 2D vector fields defined on a 2D support. We assume that the support is a regular grid. A 2D vector given in polar coordinates (α, l) (according to the coordinates system shown in Figure 4) is defined on each point of the grid. The visualization process consists of the synthesis of a furlike texture whose

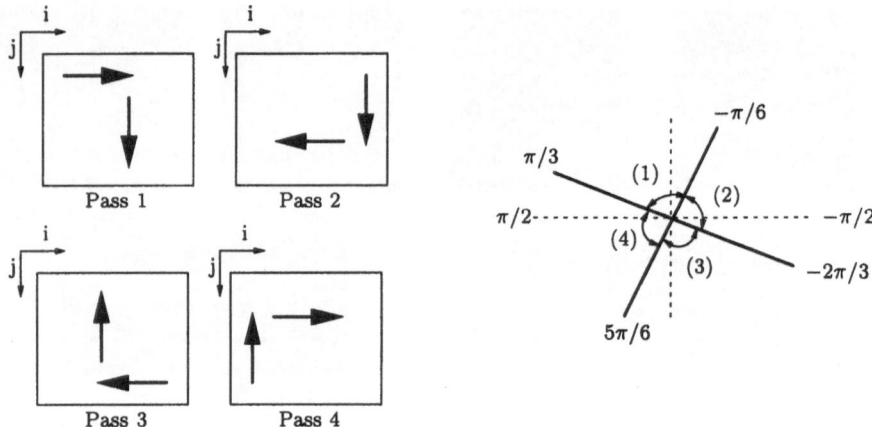

Fig. 6. Orders of scanning the image for the four passes of the synthesis process, with the ranges of orientations produced by each pass

attributes (orientation, length, color, ...) at each point depend on the vector value at the corresponding grid point.

In the following, visualization results obtained using simulated data are presented. We have computed fields similar to those produced around several critical points of different types. Critical points have a zero vector field magnitude. Their study (their location and characteristics) is useful for the analysis of the vector field topology [11][2].

The orientation component (coordinate α) of a vector field is naturally associated to the orientation attribute of the furlike texture. The magnitude component (coordinate l) of the field can be either represented via the length attribute of the texture or encoded using the color or the density.

In all visualizations shown in the following, the RSI noise parameters are set to $p = 0.02$ and $val = 255$.

3.2 Results

Figure 7 presents the result of the visualization process. The critical points are all attracting nodes with different shapes (circle, spiral, ...). The texture length is fixed to $l = 0.20$, the size of the neighborhood is defined by $jx = 5$ and $jy = 2$.

The visual representations obtained seem satisfying since complete information about local orientation is visually perceived. In particular, the depth aspect of the texture makes the direction information non ambiguous even with the dense vector field visualized. This remains true with more complex vector fields such as those presented in Figure 8.

A complete representation of a vector field can be obtained by encoding its magnitude by the length attribute of the texture. For each vector field representation, all the available values of texture length are used in such a way that

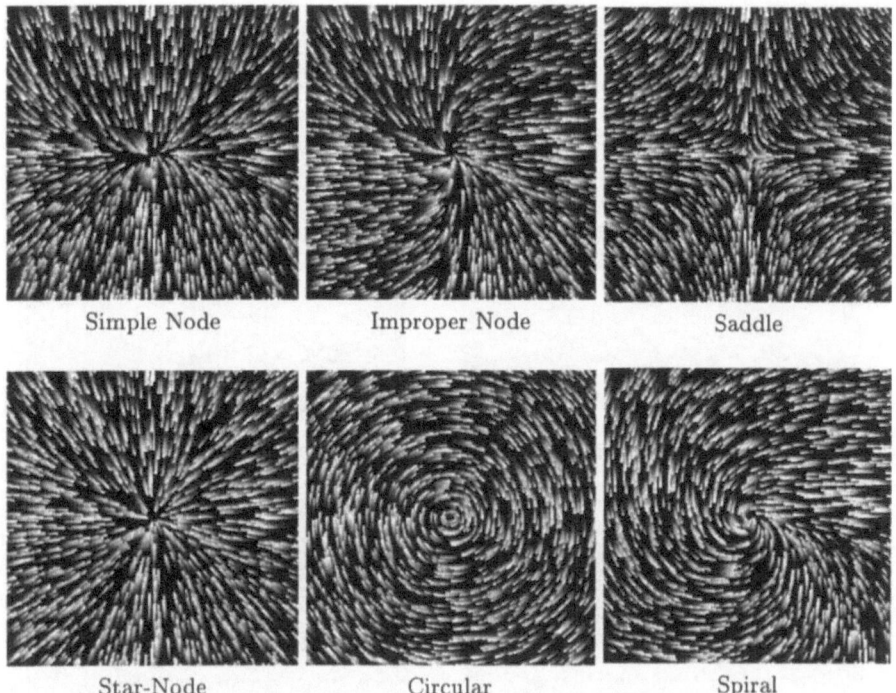

Fig. 7. Visualization of the orientation component of a vector field

greatest values correspond to the regions of highest magnitude. The images obtained are shown in Figure 9. We use the same fields as in Figure 8.

The length attribute gives in a natural way a global information about the spatial variation of the magnitude component. However, since the length is a texture regional attribute, it is difficult to use it to encode a point characteristic (the field magnitude), especially when this latter varies quickly. Note however that a solution could consist of computing the texture at a higher resolution than the original field resolution.

A more precise representation of the magnitude component of a vector field can be achieved by using the color attribute. For this aim, a standard colormap is used in which high values of magnitude are coded by the hottest colors (Red represents the highest value). In Figure 10 (see also the color plate at the end of the book), the same fields as in Figure 8 are visualized using orientation and color attributes of the texture.

Our technique can also produce *LIC*-like texture by using a fine white noise instead of the impulsional one used to generate furlike texture. In Figure 11, we show the result of visualizing the same fields as in Figure 8 by simply using a gaussian white noise as input of the AR filter.

Fig. 8. Fields with several critical points ($p = 0.02$ and $l = 0.5$)

Fig. 9. Length attribute to encode the magnitude ($p = 0.02$)

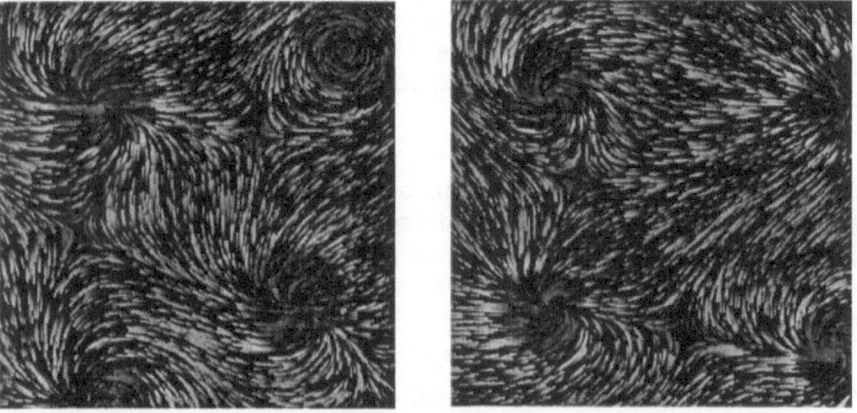

Fig. 10. Color encoding for the magnitude ($p = 0.02$ and $l = 0.5$)

Fig. 11. *LIC*-like images produced by the AR filter with a gaussian white noise ($l = 0.5$)

3.3   Implementation and Complexity

Given a 2D vector field defined on a grid with N data points, the visualization process complexity can be described as follows:

- A noise image (N pixels) (RSI) with the desired density is generated.
- Four passes via the AR filter are performed (the output of one pass is used as the input of the next one). Each pass performs the following operations for each pixel (i, j) (among the N):
  - The vector field value at the corresponding data point is retrieved. It consists in two values (α, l). If the orientation do not belong to the range values of the current pass, the pixel is skipped. Else, α and l are used as input indices to the lookup table of AR coefficients. A set of $M = 2jxjy - jx - jy$ coefficients $a(m, n)$ is then returned. M is the size of the neighborhood which is defined by the two dimensions jx and jy.
  - the value at pixel (i, j) is computed using the filter difference equation (1). The computation consists in M additions and M multiplications.
- Thus, the four passes involve near $2N(2jxjy - jx - jy)$ arithmetic operations. This number depends only on the texture resolution and the neighborhood size (which is fixed to the convenient values $jx = 5$ and $jy = 2$).

The technique is implemented in C language on an SGI station with MIPS R4400 200Mhz processor. As an example, each one of the images presented in Figure 10 is produced in less than 1 *second* (this time includes all the steps of the visualization: noise generation, four passes). The resolution of the images is 400×400. Thus, our technique allows one to achieve a time computation efficiency comparable to the most recently proposed techniques such as the *OLIC*[6] and *FROLIC*[7] which give similar results.

4 Conclusion

In this work, we have presented a new technique for vector fields visualization based on the use of a furlike texture. For an efficient generation of such texture, we have proposed a 2D texture model based on a 2D AR synthesis. We have shown that the use of an AR model gives satisfying visual results, even with very small sizes of neighborhood. This makes the technique fast. In addition, several texture attributes can be locally controlled (orientation, length, density via the noise parameter, color). We have demonstrated the use of this texture model for the visualization of 2D vector fields. We have shown that several representations can be easily generated. In particular, *LIC*-like representations can also be produced. As future work, we expect to animate the visualization and apply the technique developed to real data and to unsteady flow. Another perspective consists of the application of the technique to visualize 3D vector fields on 3D surfaces by introducing appropriate texture mapping procedures.

References

1. V. Watson and P. P. Walatka, "Visual analysis of fluid dynamics," in *State of the Art in Computer Graphics - Visualization and Modeling* (R. David and E. Rao, eds.), pp. 7–18, New York: Springer, 1991.
2. F. H. Post and J. J. Van Wijk, "Visual representation of vector fields: recent developments and research directions," in *Scientific Visualization, Advances and Challenge*, pp. 368–390, San Diego: Academic Press, 1994.
3. J. J. Van Wijk, "Spot noise," *Computer Graphics*, vol. 25, pp. 309–318, July 1991.
4. B. Cabral and L. Leedom, "Imaging vector fields using line integral convolution," *Computer Graphics*, vol. 27, no. Annual Conference Series, pp. 263–270, 1993.
5. J. T. Kajiya and T. L. Kay, "Rendering fur with three dimensional textures," *Computer Graphics*, vol. 23, pp. 271–280, July 1989.
6. R. Wegenkittl and E. Groller, "Animating flowfields: rendering of oriented line integral convolution," in *Proceedings of Computer Animation'97, Geneva, Switzerland, June 5-6, 1997*, pp. 15–21, June 1997.
7. R. Wegenkittl and E. Groller, "Fast oriented line integral convolution for vector field visualization via the internet," in *Proceedings of Visualization '97, Phoenix, Arizona, Oct 19-24, 1997*, pp. 309–315, Oct. 1997.
8. L. Khouas, C. Odet, and D. Friboulet, "3D furlike texture generation by a 2D autoregressive synthesis," in *The sixth International Conference in Central Europe on Computer Graphics and Visualisation*, pp. 171–178, Feb. 1998.
9. K. Perlin and E. M. Hoffert, "Hypertexture," *Computer Graphics*, vol. 23, pp. 253–262, July 1989.
10. S. M. Kay, *Modern spectral estimation: theory and application*, pp. 479–518. Prentice Hall, 1988.
11. A. R. Rao and R. C. Jain, "Computerized flow field analysis: oriented texture fields," *IEEE Trans. on Pattern Anal. and Machine Intel.*, vol. 14, pp. 693–709, July 1992.

Editors' Note: see Appendix, p. 314 for colored figure of this paper

Visualization of Global Flow Structures Using Multiple Levels of Topology

Wim de Leeuw and Robert van Liere

Center for Mathematics and Computer Science CWI,
P.O. Box 94097, 1090 GB Amsterdam, Netherlands
{wimc,robertl}@cwi.nl

Abstract. The technique for visualizing topological information in fluid flows is well known. However, when the technique is used in complex and information rich data sets, the result will be a cluttered image which is difficult to interpet. This paper presents a technique for the visualization of multi-level topology in flow data sets. It provides the user with a mechanism to visualize the topology without excessive cluttering while maintaining the global structure of the flow.
Keywords: multi-level visualization techniques, flow visualization, direct numerical simulation.

1 Introduction

The importance of data visualization is clearly recognized in large scale scientific computing. However, the demands imposed by modern computational fluid dynamics (CFD) simulations severely test the limits of today's visualization techniques. This trend will continue as solutions to more complex problems are desired.

An important aspect of a flow field is its topology [1]. A technique for the visualization of vector field topology in fluid flows was introduced by Helman and Hesselink [2]. It is a technique that extracts and visualizes topological information, and combines simplicity of schematic depictions with the quantitative accuracy of curves computed directly from the data. Visualization of topology is impressive when applied to not too complex flow fields, but in high-resolution turbulent flows problems may arise. As fluid flow computations generate more complex and information rich data sets, the set of computed critical points will become very large. This results in a cluttered image which is difficult to interpet. For example, consider figure 1. The data is a 2D slice of a 3D data set turbulent flow around a square cylinder. A set of 322 critical points has been computed from the data. Small colored icons are used to display the set of critical points: a yellow spiral icon denotes a focus, a blue cross denotes a saddle point, and cyan/magenta disks denote repelling/attracting nodes. Spot noise is used to present the global nature of the flow. Note that most critical points are clustered in regions around the square cylinder. To prevent additional cluttering, streamlines linking critical points have been omitted.

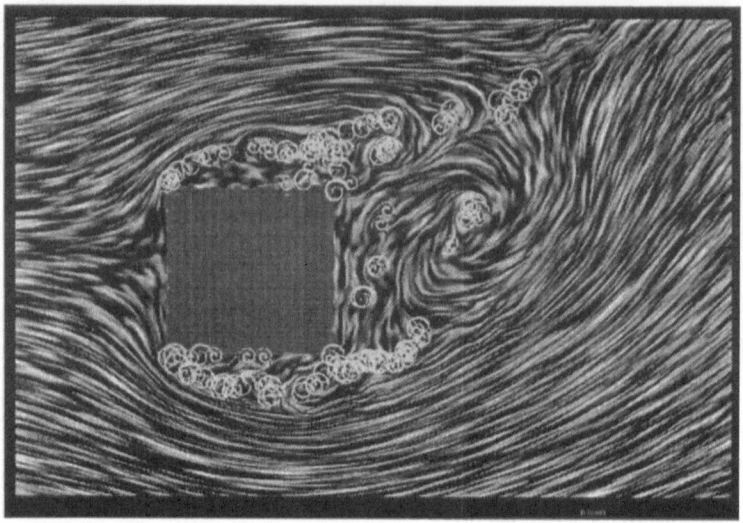

Fig. 1. Topological information of turbulent flow around a square cylinder. Spot noise is used to present the global nature of the flow.

In this paper we present a technique for the visualization of multi-level topology in flow data sets. The multi-level topology technique provides the user with a mechanism to select the set of critical points which define the global flow structure.

The format of this paper is: In the next section we review previous work on vector field topology. In section 3 discusses the multi-level topology technique. Finally, in section 4 we show how the technique has been used to explore a turbulent flow field.

2 Previous work

Vector field topology was introduced by Helman and Hesselink, [2]. It presents essential information by partitioning the flow field in regions using critical points which are linked by streamlines. Critical points are points in the flow where the velocity magnitude is equal to zero. Each critical point is classified based on the behavior of the flow field in the neighborhood of the point. For this classification the velocity gradient tensor is used. The velocity gradient tensor – or Jacobian – is defined as

$$\mathbf{J} = \nabla \mathbf{u} = \begin{pmatrix} u_x \ u_y \\ v_x \ v_y \end{pmatrix} \tag{1}$$

in which subscripts denote partial derivatives. The classification is based on the the two complex eigenvalues ($R1 + i\,I1$, $R2 + i\,I2$). Assuming that the the critical point is hyperbolic, i.e. the real part of the eigenvalues is non zero, five different cases are distinguished (see figure 2) :

1. *Saddle point*, the imaginary parts are zero and the real parts have opposite signs; i.e $R1 * R2 < 0$ and $I1$, $I2 = 0$.
2. *Repelling node*, imaginary parts are zero and the real parts are both positive; i.e $R1$, $R2 > 0$ and $I1$, $I2 = 0$.
3. *Attracting node*, imaginary parts are zero and the real parts are both negative; i.e $R1$, $R2 < 0$ and $I1$, $I2 = 0$.
4. *Repelling focus*, imaginary parts are non zero and the real parts are positive; i.e $R1$, $R2 > 0$ and $I1$, $I2 \neq 0$.
5. *Attracting focus*, imaginary parts are non zero and the real parts are negative; i.e $R1$, $R2 > 0$ and $I1$, $I2 \neq 0$.

If the real part of the eigen values is zero the type of the flow is determined by higher order terms of the approximation of the flow in the neighborhood of the critical point.

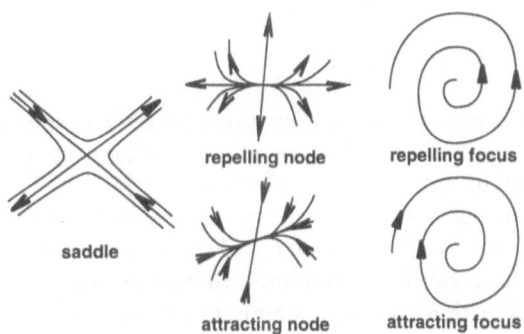

Fig. 2. Five different types of critical points.

Streamlines traced in the direction of the eigenvectors of the velocity gradient tensor will divide the flow field in distinct regions.

Implementation aspects of this technique can be found in [4].

3   Multi-level flow topology

Phenomena in turbulent flow fields are characterized by flow patterns of widely varying spatial scales. In terms of topological information, this means that a large set of critical points result from flow patterns at small spatial scales. However, the global structure of the flow can be described by a limited subset of all critical points. The governing idea of the multi-level flow topology method is that the displayed number of critical points should be limited to characterize only those flow patterns of a certain level of scale, while the other critical points are omitted.

In order to realize this idea, two distinct type of methods can be employed: implicit and explicit methods (see figure 3). Implicit methods are those that filter

48

the input data set to obtain a derived data set to which the original topology algorithm applied. Explicit methods are those which first compute the set of critical points from the input data set and then use a filter to prune this set.

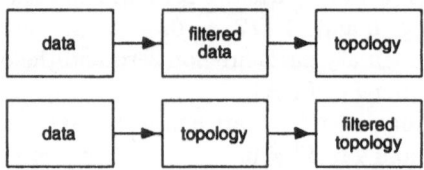

Fig. 3. Implicit (top) vs. explicit (bottom) methods of multi-level topology.

3.1 Implicit methods

In general, filters are used to enhance/suppress patterns in the data. For example, a low pass filter can be used to suppress high-frequency patterns; i.e. those flow patterns at small spatial scales. The idea is that by filtering the data and then doing the critical point analysis on the critical points caused by small disturbences in the flow will be filtered out. Nielson et al. obtained similar results with a method where wavelets where used to approximate the data [5].

For the images in this section, a simple box filter has been used: each data-point is replaced by an average over a small region in the orignal data. The motivation is that the box filter is a simple low-pass filter which will average out small scale patterns, while keeping large scale patterns intact.

Figure 4 illustrates the implicit method using a box filter. The left image shows the set of all 322 critical points of the original data set. The middle image shows the set of 179 critical points of the data set after being filtered with a 2x2 box filter. The right image shows the set of 40 critical points of the data set after being filtered with a 8x8 box filter. Although the implicit method is conceptually easy to understand and may achieve the desired effect in certain cases, a number of drawbacks can be mentioned. Due to filtering, there is no direct relation between critical points in the original and derived set. The position and type of a critical point can change after the filter is applied.

3.2 Explicit methods

Explicit methods use filters which prune the set of all critical points defining the topological structure. This is achieved by using specific knowledge about the characteristics of the flow topology.

For this purpose, a *pair distance filter* has been designed. The motivation of this filter is that an often occurring small disturbance of the flow is caused by

Fig. 4. Three views of the implicit method using a box filter. Left: original data set (322 critical points). Middle: 2x2 box filter (179 critical points). Right: 8x8 box filter (40 critical points).

pairs of critical points. For example, the topological structure of a two dimensional vortex consists of a focus (repelling or attracting) or a center combined with a saddle point (see figure 5). The size of the vortex is determined by the distance between the pair of critical points. Removing the pair from the topology does not influence the global structure of the flow.

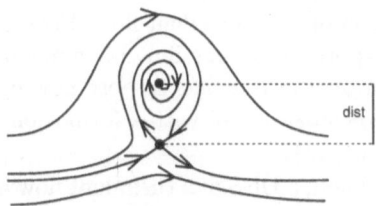

Fig. 5. An example of a critical point pair; a focus and saddle point forming a vortex. The distance defines the spatial scale of the point pair.

The pair distance filter can be implemented as follows: The set of all critical points are located and classified. Using the distance between saddles and attracting/repelling nodes, a pairwise distance matrix is constructed. The pair with the smallest distance is removed from the matrix. This is iteratively continued until a distance threshold is reached.

Figure 6 shows the pair distance filter in practice. The left image shows the set of all 322 critical points of the original data set. The middle image shows the critical points of the data set after being filtered with a distance threshold of $0.0001 * H$, in which H is the height of the set, resulting in a set of 114 critical points. The right image shows the critical points of the data set after being filtered with a distance threshold of $0.001 * H$, resulting in a set of 34 critical points.

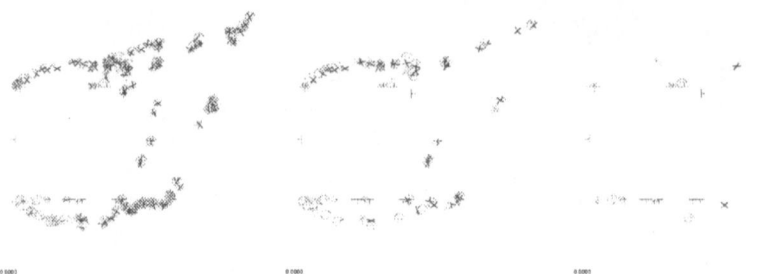

Fig. 6. Three views of the explicit method using the pair filter. Left: original data set (322 critical points). Middle: 0.0001 * H distance (114 critical points). Right: 0.001 * H filter (34 critical points).

4 Direct Numerical Simulation of Turbulent Flow

4.1 Problem and Data Set

We applied our method to the output direct numerical simulation (DNS) of a turbulent flow generated by A. Veldman and R. Verstappen [6]. DNS is an accurate technique for computing turbulent flow. Flow experts use the resulting visualizations to test hypotheses about flow phenomena and – after a detailed inspection of the animation – as a means to pose new hypotheses. Of particular interest is the detailed visualization of vortex formation and the transition from laminar to turbulent flow.

In this particular problem, a DNS of a turbulent flow around a square cylinder at Re = 22,000 (at zero angle of attack) has been performed. The size of the resulting data set is impressive: the resolution of the rectilinear grid is 314x538x64; the grid was finest near the cylinder. The number of time steps saved on disk are 7500.

4.2 Results

Figure 7 shows a view of the flow. This is the same 2D slice as in figure 1. The pair distance filter is used with a distance threshold of 0.001 * H, in which H is the height of the data set. Now, streamlines can be drawn without excessive cluttering of the image while, simultaneously, maintaining the the global structure of the flow. Note, for example, the large vortex (consisting of a saddle and attracting node) behind the square cylinder.

Figure 8 shows a zoomed in view of the previous image. The distance threshold for the pair distance filter is adjusted to reflect structures at a smaller scale.

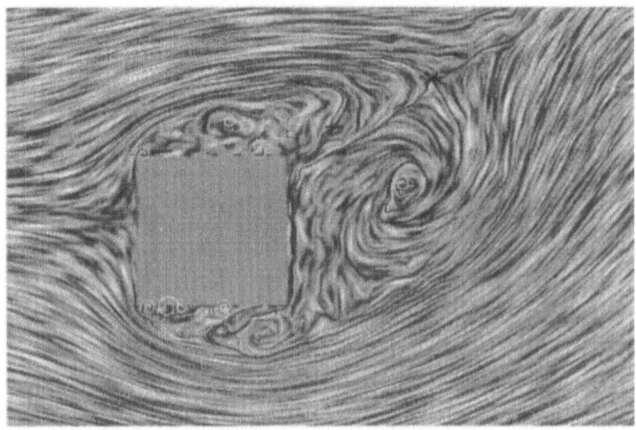

Fig. 7. A view of global flow structure around a square cylinder. The pair distance filter is used with a distance of 0.001 $*$ H.

4.3 Evaluation

This application clearly benefits from the added value of the multi-level flow topology technique, [7]. The data set contains an abundance of detailed information. Using traditional topology visualization methods on these data sets, excessive cluttering can not be avoided. With the multi-level approach, simplified views of the topology can be obtained without cluttering.

The multi-level topology method has been parallelized, so that interactive rates can be obtained. The interactive multi-level topology method can be used combination with interactive spot noise [8] to realize real-time animation of the time-dependent field and interactive zooming into details one time step.

In the near future DNS can be applied to flows with a Reynolds number in the order of 10^5. The increased size and detailed information in the resulting data sets will require multi-level flow visualization techniques.

5 Conclusion

In this paper, a technique for the visualization of multi-level topology in flow data sets was presented. It provides the user with a mechanism to visualize the topology without excessive cluttering while maintaining the global structure of the flow. Two methods have been introduced. Implicit methods can be used to filter data in order to suppress flow patterns at small spatial scales. Explicit methods can be used to prune a set of critical points, making use of specific knowledge about characteristics of the flow topology. The technique has been succesfully applied to a complex data set resulting from a direct numerical simulation.

In the future, two enhancements to the explicit method are envisioned. First, filtering will take not only distance information between critical points into ac-

52

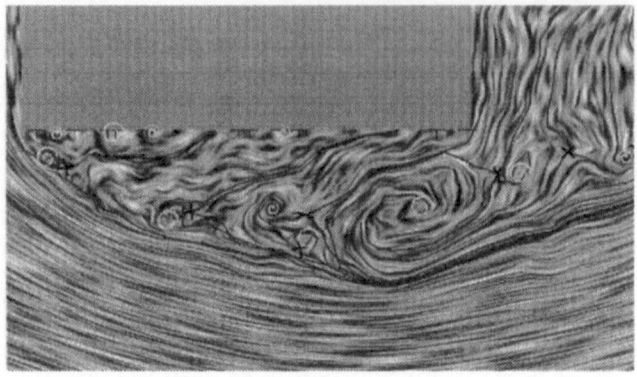

Fig. 8. A zoomed in view of flow topology around a square cylinder. The pair distance filter is used with a distance of $0.0002 * H$.

count, but also topological information about the computed links (i.e. stream-lines) between the critical point pairs. Second, temporal information about critical points can be used. Due to flow patterns at small temporal scales, the existance of a critical point may vary over time. By tracking the critical points over time, the lifetime of a point can be determined. The threshold metric can be the minimum lifetime of a critical point.

References

1. V.I. Arnold. *Ordinary Differential Equations*. MIT Press, 1973.
2. J.L. Helman and L. Hesselink. Visualizing vector field topology in fluid flows. *IEEE Computer Graphics and Applications*, 11(3):36–46, May 1991.
3. D. Asimov. Notes on th topology of vector fields and flows. Technical Report Technical Report RNR-93-003, NASA Ames Research Center, 1993.
4. A. Globus, C. Levit, and T. Lasinski. A tool for visualizing the topology of three-dimensional vector fields. In G.M. Nielson and L.J. Rosenblum, editors, *Proceedings Visualization '91*, pages 33–40. IEEE Computer Society Press, Los Alamitos (CA), 1991.
5. Gregory M. Nielson, Il-Hong Jung, and Junwon Sung. Wavelets over curvilinear grids. In David Ebert, Hans Hagen, and Holly Rushmeier, editors, *Proceedings Visualization '98*, pages 313–315. IEEE Computer Society Press, 1998.
6. R.W.C.P. Verstappen and A.E.P. Veldman. Spectro-consistent discretization of navier-stokes: a challenge to RANS and LES. *Journal of Engineering Mathematics*, 34:163–179, 1997.
7. A.E.P. Veldman. Private communication.
8. W.C. de Leeuw and R. van Liere. Divide and conquer spot noise. In *Proceedings Super Computing '97 (http:// sczy.tc.cornell.edu / sc97 / program / TECH / DELEEUW / INDEX.HTM)*, 1997.

Editors' Note: see Appendix, p. 315 for colored figures of this paper

Geometric Methods for Vortex Extraction

I. Ari Sadarjoen* and Frits H. Post

Faculty of Information Technology and Systems, Delft University of Technology,
Zuidplantsoen 4, 2628 BZ Delft, The Netherlands
e-mail: `Ari.Sadarjoen@mcc.ac.uk` / `Frits.Post@cs.tudelft.nl`

Abstract. This paper presents two vortex detection methods which are based on
the geometric properties of streamlines. Unlike traditional vortex detection methods, which are based on point-samples of physical quantities, one of our methods
is also effective in detecting weak vortices. In addition, it allows for quantitative
feature extraction by calculating numerical attributes of vortices. Results are presented of applying these methods to CFD simulation data sets.

Keywords: flow visualization, vortex detection, feature extraction

1 Introduction

Vortices are among the most important features of fluid flows. In aerodynamics, vortices
directly affect the flying characteristics of airplanes [5]. In turbomachinery design, vortices are to be avoided or minimized during design [9]. In oceanography, the evolution
of vortices in space and time is important for scientists' understanding of ocean circula-
tions [14]. Therefore, detecting and visualizing vortices is an important topic. Unfortunately, there is no formal definition of a vortex, which makes it difficult to detect them.
Therefore, vortex detection methods are based on heuristic criteria.

Vortex detection methods fall into two classes. The first class is based on physical
quantities evaluated at isolated points in the field, e.g. grid nodes, or in an infinitesimal
neighbourhoods of points. Examples of these quantities are pressure, vorticity magnitude, and helicity [1, 9]. The problem is that none of these methods works in all cases.
Their main deficiency is that they often do not find weak vortices, characterized by slow
rotation and low velocity magnitudes.

The second and relatively new class of vortex detection methods are the *geometric
methods*, which are not based on physical quantities, but on geometric properties of
curves, typically streamlines or pathlines.

The purpose of this paper is to present two geometric methods for vortex detection,
and a method for calculating numeric attributes of vortices. The first method tries to find
vortex cores by determining the curvature centre of the osculating circle of streamlines
through many sample points in the grid. The second method tries to find entire vortices
using the winding-angle concept, to select streamlines that are sufficiently rotational to
be part of a vortex.

---

* Present address: Manchester Visualization Centre, Manchester Computing, University of
Manchester, Oxford Road, Manchester M13 9PL, United Kingdom

We have extended the work in [12] to a technique for automatic feature extraction, characterizing the features by calculating a set of quantitative attributes, such as position, size, and rotation speed and direction. This is done by clustering the selected streamlines and determining numerical attributes of the vortices. This has the advantage that vortices can be described by a small set of attributes, which naturally causes a dramatic data reduction.

The structure of this paper is as follows. In Section 2, we give an overview of related work. Then, we describe our geometric methods for vortex detection: the curvature centre method in Section 3, and the winding-angle method in Section 4. In Section 5, we show some results of applying our methods to CFD simulations.

2 Related Work

The first class of vortex detection methods is typically based on point samples of physical quantities. The quantities involved are usually pressure, velocity, quantities derived from the velocity vector, or quantities derived from the velocity gradient tensor. All of these quantities are based on the assumption either that vortices are regions with a high amount of rotation, or that there exists a pressure minimum at vortex cores. Banks & Singer [1] and Roth & Peikert [9] have surveyed a number of quantities, and concluded that they often fail to capture all vortices. An important cause is that vortices are *regional* features, but these criteria are strictly based on point samples, or first-order approximations in infinitesimal regions. Recently, Roth & Peikert [10], recognizing the deficiencies of first-order approximations, proposed a higher-order method which is also able to detect bent vortices.

The second class of vortex detection methods is *geometric*, i.e. based on geometric properties of streamlines. De Leeuw [6] described an interactive way to detect vortices using a box-shaped probe in which sample points are taken. For all the sample points in the box, a number of properties were calculated, including the centre of curvature of the streamline through the sample point. When the box contained a vortex, the centres of curvature would accumulate near a point, otherwise they would be scattered.

Portela [7] has developed a formal mathematical framework for defining vortices in 2D, which corresponds to the intuitive notion of swirling motion around a central set of points. To define a central set of points, he proposed so-called Jordan structures; to define swirling motion, he used the winding-angle concept known from differential geometry.

In [12], we applied two geometric techniques to several hydrodynamic cases. One used curvature centres to find vortex cores, the second used a simplified winding-angle.

The present paper reviews the first technique, giving more details on its problems and pitfalls, and extends the second technique to a more quantitative one, by clustering the streamlines and calculating numeric attributes.

3 The Curvature Centre Method

The curvature centre method tries to detect vortices in 2D by sampling the field at many points, typically at all grid nodes. For each sample point, the *centre of curvature* is determined, which is the centre of the osculating circle of the streamline through that point [2]. In vortical regions of the field, the centres of curvature should accumulate at a point, as in Figure 1a. The samples taken on this perfectly circular streamline all project to the same centre of curvature. In non-vortical regions of the field, the centres of curvature will be scattered, as in Figure 1b.

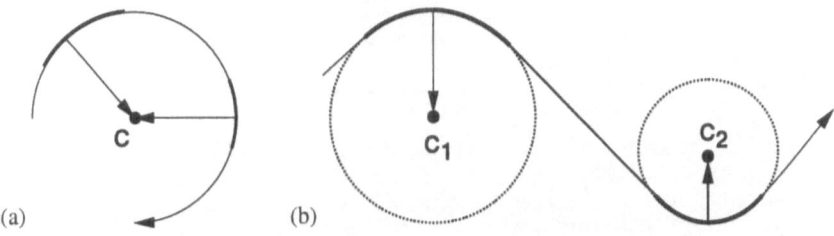

Fig. 1. (a) Circular streamline with coinciding curvature centres C and (b) Non-circular streamline with scattered curvature centres.

In this way, a set of curvature centre points is obtained, which are accumulated into a new grid, as illustrated in Figure 2. The number of curvature centres in each cell constitutes a new scalar field which we call the *Curvature Centre Density* (CCD) field.

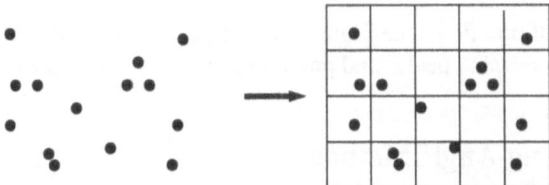

Fig. 2. Curvature centre points are accumulated into a new grid, resulting in the Curvature Centre Density (CCD) scalar field.

Figure 3 shows an example of a CCD field. The 2D data set originates from a numerical flow simulation of the Pacific Ocean, which models the west coast of North America [14]. The grid used is a rectilinear 2D grid of 117×84 nodes, at each of which the velocity has been calculated. The figure shows streamlines released from every grid node. The CCD field has been rendered as a white height field. Thresholding has been applied to select only the highest peaks of the field: $CCD > 0.8 \cdot CCD_{max}$.

This method works, but has the same limitations as traditional point-based detection methods. There are some false and some missing peaks. Some of the false peaks may be filtered out by thresholding or filtering. Also, supersampling may be applied to get more samples per grid cell [11]. An important cause of these problems are the streamlines which are not perfectly circular, but elliptical or elongated; this is often due to interaction between adjacent vortices. The effect is shown in Figure 4: in perfectly circular flow (see Figure 4a), there is a clear peak in the CCD field (see Figure 4b). How-

Fig. 3. Pacific Ocean with global streamlines and a white height field of the curvature centre density.

ever, in slightly elliptic flow (see Figure 4c), the peak is 'spread out' (see Figure 4d). This causes many missing peaks, and possibly also some false peaks.

4   The Winding-Angle Method

Another geometric method for detecting vortices in 2D, inspired by [7], builds upon the intuitive idea of a swirling pattern around a central set of points. The method tries to detect vortices by selecting looping streamlines and then clustering them. Selection is performed using a simplified winding-angle criterion and a distance criterion.

Let S_i be a 2D streamline, consisting of points $P_{i,j}$ and line segments $(P_{i,j}, P_{i,j+1})$, and let $\angle(A, B, C)$ denote the angle between line segments AB and BC. Then, the *winding-angle* $\alpha_{w,i}$ of streamline S_i is defined as the cumulative change of direction of the streamline segments:

$$\alpha_{w,i} = \sum_{j=1}^{N-1} \angle(P_{i,j-1}, P_{i,j}, P_{i,j+1}) \tag{1}$$

See Figure 5. We use signed angles, with positive rotation for a counterclockwise-rotating curve, and negative rotation for a clockwise-rotating curve. Obviously, $\alpha_{w,i} = \pm 2\pi$ for a fully closed curve; lower values may be used to find winding streamlines which do not make a full revolution.

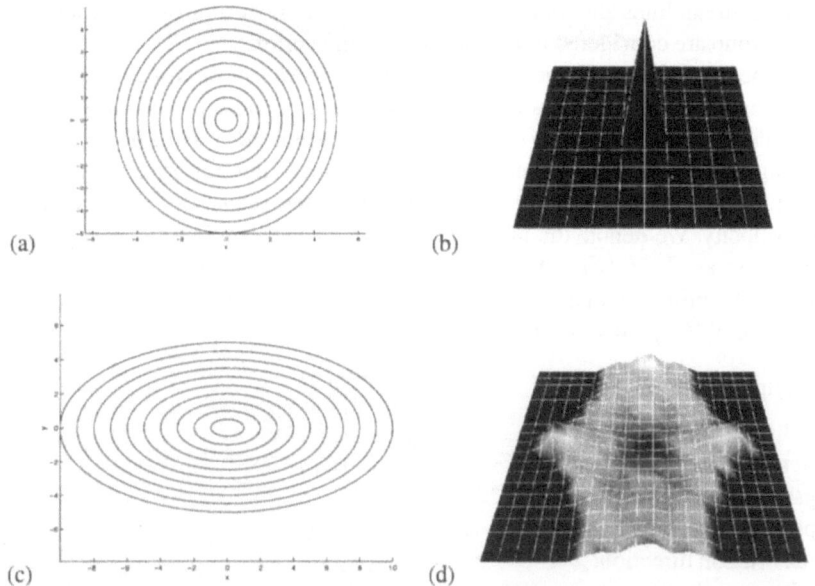

(a)

(b)

(c)

(d)

Fig. 4. In circular flow (a), there is a peak in the CCD field (b). In elliptic flow (c), the peak in the CCD field is spread out (d).

The selection process tries to find the streamlines that belong to a vortex by using two criteria: (1) the winding-angle of a streamline should be $k \cdot 2\pi$, with $k \geq 1$, and (2) the distance between the starting and final point of the streamline should be relatively close.

We have extended the work described in [12] from a visual, qualitative selection technique, to a more quantitative feature extraction technique. We now use the selected streamlines for automated vortex extraction and for determining numerical vortex attributes. This is done in two stages: *clustering* and *quantification*.

The purpose of *clustering* is to group those streamlines together which belong to the same vortex. Rather than clustering streamlines, it is easier to cluster points.

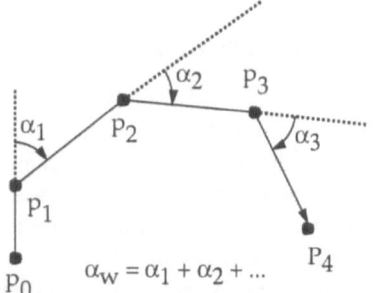

Fig. 5. The winding-angle $\alpha_{w,i}$ is the sum of the angles between the edges.

To this end, each streamline is mapped to a point by determining the centre point, or geometric mean, of all sample points on the streamline. These centre points are then clustered as follows. The first cluster is formed by the first point. For each subsequent point, it is determined which previous cluster lies closest. If the point is not within a predetermined radius of all the existing clusters, it constitutes a new cluster. In this way, the selected streamlines are combined into a distinct number of groups. Streamlines of the same group are considered to be part of the same vortex.

Once the streamlines have been clustered, *quantification* of the vortices is performed by calculating numeric attributes of the corresponding streamline clusters. We approximate the shape of the vortices by *ellipses*. Fitting an ellipse to a set of points is done by calculating statistical attributes, such as mean, variance, and covariance, of the points [8, 13]. In addition, we calculate specific vortex attributes, such as rotation direction and angular velocity. We denote the number of points on a streamline S_i as $|S_i|$, a cluster of streamlines as $C_k = \{S_{k,1}, S_{k,2}, \ldots\}$, where $S_{k,l}$ is streamline #l in cluster #k, the number of streamlines in that cluster as $|C_k|$, and all the points on all the streamlines in that cluster as $\Psi(C_k)$. Now, we can calculate the following attributes for each vortex:

- streamline centre: $\quad\quad\quad\quad \bar{S}_i = \frac{1}{|S_i|}\sum_{j=1}^{|S_i|} P_{i,j}$
- cluster centre: $\quad\quad\quad\quad\quad \bar{C}_k = \frac{1}{|C_k|}\sum_{l=1}^{|C_k|}(\bar{S}_{k,l})$
- cluster covariance: $\quad\quad\quad M_k = cov(\Psi(C_k))$
- ellipse axis lengths: $\quad\quad \lambda_k = eig(M_k)$
- ellipse axis directions: $\quad \mathbf{d}_k = eigvec(M_k)$
- vortex rotation direction: $\; d_k = sign(\alpha_{w,k})$
- vortex angular velocity: $\quad \omega_k = \frac{1}{|C_k|\Delta t}\sum_{l=1}^{|C_k|}\alpha_{w,l}$

The vortices can be visualized by mapping their attributes to icons: the first three statistical attributes are used to calculate the axis lengths and directions of an ellipse which approximates the size and orientation of a vortex. The rotation direction of a vortex is visualized by small arrows. Finally, the angular velocity of a vortex is visualized by adding wheel spokes to the ellipse, the number of which is made proportional to the angular velocity: fast rotation is suggested by many spokes, slow rotation by few.

5 Results

5.1 The Bay of Gdańsk

The first example uses a data set of a simulation performed at WL | Delft Hydraulics of the Bay of Gdańsk, a coastal area in Poland. The goal of the simulation was to investigate the flow patterns induced by wind, the inflow of the Wisła (Vistula) river, and turbulence. The model is defined on a curvilinear grid of $43 \times 28 \times 20$ nodes, indexed by (i, j, k). Each node contains a 3D velocity vector $\mathbf{v}$, an eddy-diffusivity scalar E, and its gradient ∇E.

Figure 6 shows the result of applying the winding-angle method. The global flow patterns in the data set are visualized by the grey streamlines released from every grid node in a horizontal grid slice at the centre of the grid ($k = 9$). Selected streamlines are drawn in black. The ellipse icons visualize the approximate size and shape of the vortices, with the ellipse axes drawn in dashed lines. Arrows indicate the rotation direction of the vortices. The numbers of spokes indicate the strength of the vortices: the higher the number of spokes, the faster the rotation.

It can be seen that this method captures all visible vortices, including elongated and weak ones. An impression of the vortex strength (rotation speed) is immediately visible

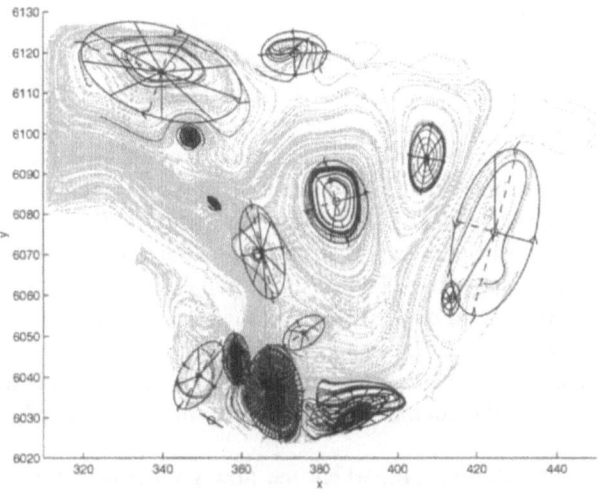

Fig. 6. Vortices in the Bay of Gdańsk. The number of spokes is proportional to ω.

from the number of spokes. It is also interesting to see that adjacent vortices rotate in opposite directions.

Figure 8 (see Appendix) shows a slightly different visualization, which uses rendered ellipses. Here, the angular velocity is not visualized, but the rotation direction of the vortices is indicated by the colour of the ellipses: red indicates clockwise rotation, and green counterclockwise rotation.

Once the vortices have been found, numerical attributes may be determined for each of them. Table 1 shows some of the results. Notice the differences between the largest and the smallest vortex (approx. factor 20), and between the fastest and the slowest one (approx. factor 15). There does not seem to be any correlation between the size and the rotation speed of the vortices.

number of clusters	15
number of CW vortices	5
number of CCW vortices	10
min. radius [km]	0.991
max. radius [km]	21.2
min. ω [s^{-1}]	5.3810^{-5}
max. ω [s^{-1}]	8.9310^{-4}

Table 1. Some numerical attributes of the vortices in the Bay of Gdańsk.

z-plane	#vortices	max λ_{CW}	max λ_{CCW}
16	1	-	0.5305
17	1	-	0.5832
18	1	0.6216	0.2076
19	2	0.8446	0.4968
20	2	1.1138	0.6117
21	1	0.9370	-

Table 2. Statistics of the vortices in the flow past a tapered cylinder.

5.2 Flow Past a Tapered Cylinder

The second example uses a data set of a simulation performed at NASA-Ames Research Center which concerns a laminar flow past a tapered cylinder [4]. This tapered cylinder has a variable radius depending on the z-coordinate, which influences the vortex shedding frequency at that height. The grid used is a structured, cylindrical grid with $64 \times 64 \times 32$ nodes, each of which contains density, x, y, z-momentum, and stagnation. The simulation is time-dependent, but we used only one time step.

We have applied the winding-angle method to extract vortices from six different z-slices, to show different patterns. Table 2 shows numerical statistics of these vortices, where max λ_{CW} is the maximum axis length of the clockwise vortex, and max λ_{CCW} of the counterclockwise vortex. It can be seen that for $z < 18$, there is only one counterclockwise vortex, for $18 \leq z \leq 20$, there are two vortices, rotating in opposite directions. For $z > 20$, the counterclockwise vortex has disappeared, leaving only a large clockwise vortex.

Figure 7 visualizes these slices, where the flow goes from left to right, past the cylinder which is drawn as a semi-circle on the left. Again, the global flow pattern is shown by grey streamlines, the selected streamlines are drawn in black, and the vortices are approximated by ellipse icons. Spokes indicate the rotation speed, and arrows the rotation direction

Figure 9 (see Appendix) shows a colour visualization of the slice at $z = 20$. As in the previous colour figure, the rotation direction of the vortices is indicated by the colour of the ellipses: green and red indicate opposite rotation. The grid slice has been coloured with the λ_2 scalar quantity. This quantity is defined as the second-largest eigenvalue of the tensor $S^2 + \Omega^2$, where $S = \frac{1}{2}(\nabla \mathbf{v} + (\nabla \mathbf{v})^T)$ and $\Omega = \frac{1}{2}(\nabla \mathbf{v} - (\nabla \mathbf{v})^T)$ are the symmetric and anti-symmetric parts of the velocity-gradient tensor $\nabla \mathbf{v}$. Highly negative values are supposed to indicate the presence of vortices [3]. However, in this example, the lowest (blue) values for λ_2 are observed at the front rather than behind cylinder, without any obvious vortices in the streamline pattern. Therefore, in this example, our winding-angle criterion turns out to be better than λ_2.

6 Conclusions and Future Work

We have described two geometric vortex detection methods. The curvature centre method has limitations similar to traditional methods. The winding-angle method is useful for finding both strong and weak vortices. Another important advantage is that it also allows for quantification, which leads to data reduction.

Future work includes incorporating critical points in the winding-angle method, to trace streamlines only in the neighbourhood of critical points, rather than globally in the entire field. Another useful application of the numerical attributes of vortices would be to perform spatial matching and temporal tracking of vortices. Matching and connecting ellipses found in adjacent (x/y/z) slices allows us to find 3D vortices, as long as they project reasonably to the slices. Temporal tracking allows us to study the evolution of vortices in time. Finally, we intend to extend the winding-angle technique to 3D. This is not trivial, because the winding-angle can be only determined in a 2D plane, and in vortices with a strong forward velocity component, it is not easy to choose such a plane.

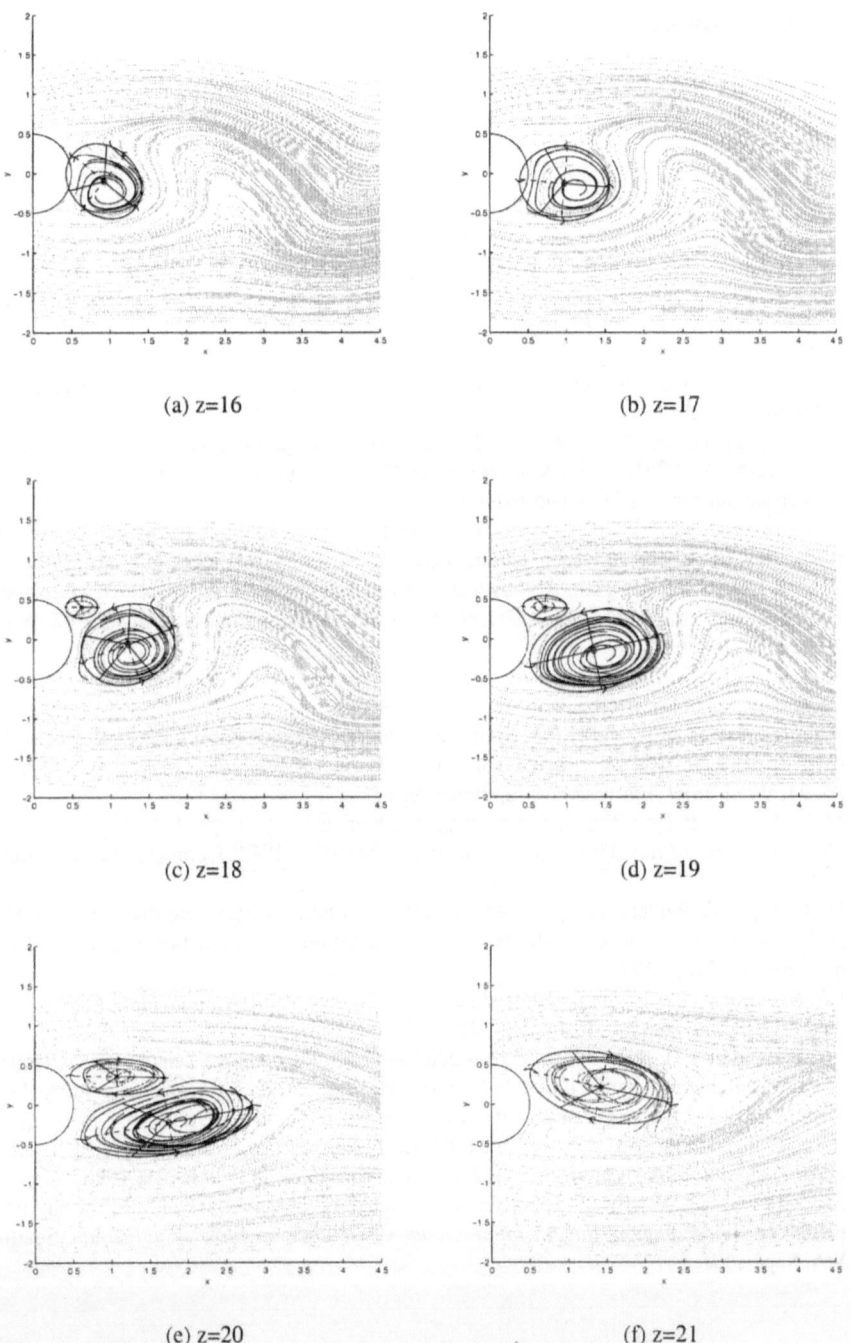

(a) z=16

(b) z=17

(c) z=18

(d) z=19

(e) z=20

(f) z=21

Fig. 7. Flow past a tapered cylinder; different z-planes show different vortices.

62

Acknowledgment

We thank Freek Reinders for his efforts in preparing the visualization of the tapered cylinder.

References

1. D.C. Banks and B. A. Singer. A predictor-corrector technique for visualizing unsteady flow. *IEEE Transactions on Visualization and Computer Graphics*, 1(2):151–163, June 1995.
2. G. Farin. *Curves and Surfaces for Computer Aided Geometric Design*. Academic Press, 1990.
3. J. Jeong and F. Hussain. On the identification of a vortex. *Journal of Fluid Mechanics*, 285:69–94, 1995.
4. D.C. Jespersen and C. Levit. Numerical simulation of flow past a tapered cylinder. Technical Report AIAA 91-0751, NASA Ames R esearch Center, Reno, NV, January 1991. 29th AIAA Aerospace Sciences Meeting and Exhibit.
5. D. Kenwright and R. Haimes. Vortex identification - applications in aerodynamics: A case study. In R. Yagel and H. Hagen, editors, *Proc. Visualization '97*, pages 413–416, 1997.
6. W.C. de Leeuw and F.H. Post. A statistical view on vector fields. In M. Göbel, H. Müller, and B. Urban, editors, *Visualization in Scientific Computing*, Eurographics, pages 53–62, Wien, 1995. Springer-Verlag.
7. L.M. Portela. *On the Identification and Classification of Vortices*. PhD thesis, Stanford University, School of Mechanical Engineering, 1997.
8. F. Reinders, F.H. Post, and H.J.W. Spoelder. Feature extraction from Pioneer Venus OCPP data. In W. Lefer and M. Grave, editors, *Visualization in Scientific Computing '97*, pages 85–94, Wien, 1997. Eurographics, Springer Verlag.
9. M. Roth and R. Peikert. Flow visualization for turbomachinery design. In R. Yagel and G.M. Nielson, editors, *Proc. Visualization '96*, pages 381–384. IEEE Computer Society Press, 1996.
10. M. Roth and R. Peikert. A higher-order method for finding vortex core lines. In D. Ebert, H. Hagen, and H. Rushmeier, editors, *Proc. Visualization '98*, pages 143–150. IEEE Computer Society Press, 1998.
11. I.A. Sadarjoen. *Extraction and Visualization of Geometries from Fluid Flow Fields*. PhD thesis, Delft University of Technology, 1999.
12. I.A. Sadarjoen, F.H. Post, B. Ma, D.C. Banks, and H.G. Pagendarm. Selective visualization of vortices in hydrodynamic flows. In D. Ebert, H. Hagen, and H. Rushmeier, editors, *Proc. Visualization '98*, pages 419–423. IEEE Computer Society Press, 1998.
13. T. van Walsum, F.H. Post, D. Silver, and F.J. Post. Feature extraction and iconic visualization. *IEEE Transactions on Visualization and Computer Graphics*, 2(2):111–119, 1996.
14. Z.F. Zhu and R.J. Moorhead. Extracting and visualizing ocean eddies in time-varying flow fields. In *Proceedings of the 7th International Conference on Flow Visualization*, Seattle, WA, Sept. 1995.

Editors' Note: see Appendix, p. 316 for colored figures of this paper

Attribute-Based Feature Tracking

Freek Reinders[1], Frits H. Post[1], and Hans J.W. Spoelder[2]

[1] Dept. of Computer Science, Delft University of Technology
email: {k.f.j.reinders, f.h.post}@cs.tudelft.nl
[2] Fac. of Sciences, Div. of Physics and Astronomy, Vrije Universiteit
email: hs@nat.vu.nl

Abstract. Visualization of time-dependent data is an enormous task because of the vast amount of data involved. However, most of the time the scientist is mainly interested in the evolution of certain features. Therefore, it suffices to show the evolution of these features. The task of the visualization system is to extract the features from all frames, to track the features, i.e. to determine the correspondences between features in successive frames, and finally to visualize the tracking results.
This paper describes a tracking system that uses feature data to track the features and to determine their evolution in time. The feature data consists of basic attributes such as position, size, and mass. For each set of attributes a number of correspondence functions can be tested which results in a correspondence factor. This factor makes it possible to quantify the goodness of the match between two features in successive time frames. Since the algorithm uses only the feature data instead of the grid data, it is feasible to perform an extensive multi-pass search for continuing paths.

Keywords: Time-Dependent Data, Feature Extraction, Attribute-Based Feature Tracking.

1 Introduction

Numerical simulations are increasingly focusing on the investigation of time-dependent phenomena. In general this results in a vast amount of data that is hard to process, interpret, and visualize. Off-line generation of images (using a global visualization technique) and creation of a playback animation, can give an overview of the evolution of the data [4]. However, this has a number of deficiencies: it cannot be performed interactively since the amount of data is too large to handle, it leaves the extraction of apparent features to the visual inspection by the scientist, and it does not give a quantitative description of the evolving phenomena. In order to overcome these flaws one should use an automatic way of tracking coherent structures (features) in the data. This requires a matching criterion between features in successive time steps (frames). The problem of matching two features is also cited as *the correspondence problem* [3].

Once the correspondence problem is solved, the sequence as a whole can be described. For each feature a description can be made of its lifespan, describing it's motion, growth, interaction with other features, etc. This allows us to select one feature and visualize the evolution of it, or to detect certain events like unusual changes or a specific interaction. The tracking process results in entirely new ways of visualizing evolving phenomena.

The correspondence problem is an issue in several scientific disciplines such as image processing, computer vision, and scientific visualization. Although each discipline has its own perspective to the problem, many approaches are similar to each other. Roughly, the methods can be classified in two categories: image-based methods and feature-based methods.

In image-based methods, the displacement of coherent structures is determined on a pixel-to-pixel basis. Examples of image-based techniques are (from image processing) the maximization of the cross-correlation between two images [8], and methods (from computer vision) using optical flow [1]. Since these methods involve an optimization process using the original data, they are suitable for 2D-images, but are inefficient for 3D fields.

In feature-based methods, first feature extraction takes place for each frame. The resulting features are then matched to the features in subsequent frames. Features may include points, lines, surfaces, or volumes, and can be extracted in numerous ways depending on the application domain and the type of feature. After the extraction, the correspondence problem can be solved using two mechanisms: region-matching, or attribute-matching. Region-matching is achieved by matching the regions of interest, for instance by overlap [2, 11, 12], or by a linear affine model [5]. These methods process the (binary) grid-data and consequently are memory consuming. Contrary to this, attribute-matching [9, 10] only works with certain attributes describing the features, which is a small set of numbers. Hence, attribute-matching costs less memory, one can even afford multi-pass searching for the optimal correspondence between features. However, since the attributes are a reduced model of a feature, the problem is finding a suitable set of correspondence criteria.

This paper describes an attribute-based feature tracking system that solves the correspondence problem based on primary attributes of features, such as position, size, and mass. The key of the algorithm is the use of a prediction scheme and the use of a multi-pass search for continuing paths. The process is highly interactive; the scientist can guide the tracking process by changing criteria and parameters, resulting in different tracking solutions.

The paper is organized as follows, Section 2 summarizes our feature extraction procedure, Section 3 defines a number of components of the tracking algorithm, and Section 4 describes the algorithm itself. Then, two applications are described in Section 5, and finally, some conclusions and topics for future research are given.

2 Feature Extraction

The first step in feature tracking is the extraction of features from each frame. Feature extraction is a set of techniques in scientific visualization aiming at algorithmic, automated extraction of relevant features from data sets [13]. The tracking algorithm described here works on features that are 3D amorphous 'objects' of a size (much) smaller than the data set domain, but larger than a grid cell. The two essential attributes of such features are position and size. In principle, these can be obtained with any feature extraction method, but in this work we have used a method based on selective visualization, which is described in more detail in [13] and [7].

The result of the feature extraction is a number of features, each quantitatively described by a set of attributes, which can be visualized by an iconic representation. The collection of the attribute sets of all features is called *feature data*. The feature data lifts the original grid data field to a higher level of abstraction; data describing the features in contrast to data containing interesting features. We developed an object-oriented data structure in C++ which stores the feature data and which allows operations on the features. The possibility of manipulating the features as individual entities has important consequences: the features can be quantitatively compared, and for our tracking purpose the grid data is not needed anymore.

The transformation from grid data to feature data signifies a data reduction in the order of 1000. The feature data is a drastically reduced model of the original data. Therefore, one has to verify the accuracy of the calculated attributes before they can be used for tracking. In [7] we showed that the determination of position and size is very accurate and robust even for noisy data. Thus, the feature data is reliable and can be used for time tracking.

3   Tracking Components

3.1   Prediction

Our tracking algorithm is based on a simple assumption: **features evolve consistently**, i.e. their behavior is predictable. This implies that once a path of an object is found, we can make a prediction to the next frame and search for features in that frame that correspond to the prediction. A prediction can be made for the next frame at the end of the path, but also for the preceding frame at the beginning of the path. This means we can search forward and backward in time. Figure 1 shows six matched features that form the path of an object (dark objects), it shows the prediction at the end of the path (transparent object), and it shows a candidate feature (light object). Clearly, the candidate feature corresponds very well to the prediction and should be added to the path.

The concept of making a prediction and searching for candidate features corresponding to this prediction has two major advantages. Firstly, it allows tracking of fast moving small objects that do not overlap. Secondly, it allows tracking in two directions of time: forward and backward. Still, a correspondence must be detected between the prediction and the candidates in the next frame.

3.2   Correspondence functions

In order to detect a correspondence between the prediction and a candidate feature, we evaluate a number of *correspondence functions*. These correspondence

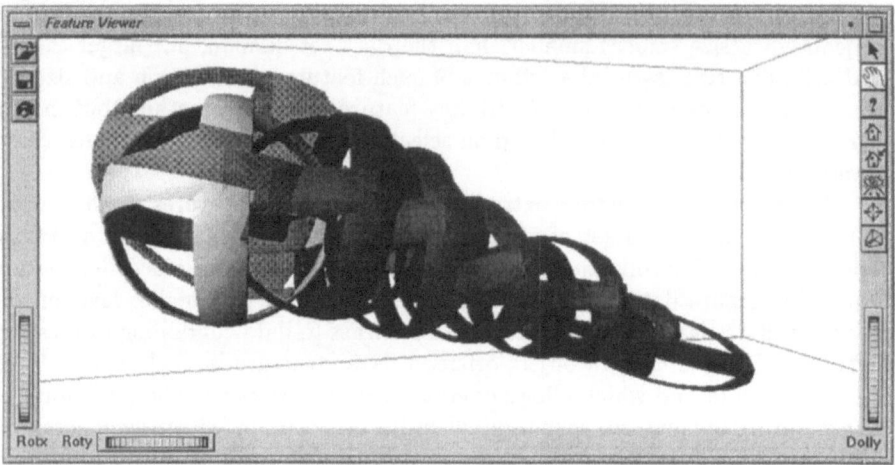

Fig. 1. A visualization of a path (dark objects), its prediction (transparent object) and one of the candidates (light object).

functions C_{func} are based on certain consistency rules like consistent growth, speed, rotation, etcetera. Each function is accompanied by a tolerance T_i and a weight W_i. The tolerance allows a deviation from the prediction and the weight indicates the importance of the function ($W_i \leq 0.0$ means no evaluation of the function at all). The correspondence factor returned by a function is $0 \leq C_{func} \leq 1.0$ if the candidate deviates from the prediction within the tolerance, and it is $C_{func} < 0.0$ if the deviation is larger then the tolerance (equation 1). The total correspondence between a prediction and a candidate is expressed as the weighted sum over all contributing functions (equation 2). The result has similar behavior as equation 1, hence a positive match between a prediction and a candidate is found when $Corr \geq 0.0$.

$$C_{func} = \begin{cases} 1 & \text{Exact match} \\ 0 & \text{Limit tolerance} \\ < 0 & \text{No match} \end{cases} \tag{1}$$

$$Corr = \frac{\sum_{i=1}^{N_{func}} C_i * W_i}{\sum_{i=1}^{N_{func}} W_i} \tag{2}$$

The number of possible correspondence functions depends on the attribute sets in a feature. Each attribute set yields a number of correspondence functions. For instance an ellipsoid fit contains information about size, position, and orientation. Therefore, the accompanying correspondence functions (table 1) are based on the consistency of growth, speed, and rotational speed. Consistency of growth is tested by the two volumes V, consistency of speed is tested by the distance between the two centroids $dist(p, c)$, and the rotational speed is tested by the dot scalar product of the main-axes of the two ellipsoids $= cos(\angle(\bar{p}, \bar{c}))$. It should be noted that the semantics of the correspondence functions depend on the physical phenomena underlying the dynamics of the features.

Correspondence function	Correspondence factor	consistency rule
$\dfrac{\|V_p - V_c\|}{\max(V_p, V_c)} <= T_{vol}$	$C_{vol} = 1 - \dfrac{\frac{\|V_p - V_c\|}{\max(V_p, V_c)}}{T_{vol}}$	consistent growth
$dist(p,c) <= T_{pos}$	$C_{pos} = 1 - \dfrac{dist(p,c)}{T_{pos}}$	consistent speed
$1 - \|cos(\angle(\bar{p}, \bar{c}))\| <= T_{angle}$	$C_{angle} = 1 - \dfrac{1 - \|cos(\angle(\bar{p}, \bar{c}))\|}{T_{angle}}$	consistent rotational speed

Table 1. Correspondence functions linked to the ellipsoid fit attribute set, p = prediction, and c = candidate.

The same correspondence factors are used to determine a *confidence index* for a complete path. Each connection (edge) between two features in this path gives two correspondence factors (forward and backward), except for the connection to end-nodes and from starting-nodes of the path. These only have one factor since only a prediction can be made in one direction. The confidence index of a complete path is calculated as follows:

$$Conf(\text{path}) = 1 - e^{-\frac{t}{\tau}} \tag{3}$$

$$\text{with } t = \sum_{i=1}^{edges} C_i$$

where τ is a growth factor (it can be taken equal to the minimal path length discussed in section 4.1). The confidence factor increases as the length of the path increases, which is convenient since our confidence in a path increases for longer paths. The confidence index is used when a choice has to be made between two paths sharing the same feature.

4 Tracking Algorithm

With the prediction, correspondence factor and confidence index we have all the components needed for a successful tracking algorithm. In order to obtain a prediction, we still have to initialize a path before we are able to continue it with corresponding candidates in consecutive frames.

4.1 Initialization

The initialization of a path is achieved by assuming a correspondence between two features in two successive frames. This assumption leads to a prediction that can be compared to candidates in the third frame. If there is a candidate in the third frame that corresponds to the prediction, a new path is created and the path is continued into subsequent frames. Since a match in the third frame may be found coincidentally, an additional test on the resulting path is performed to ensure genuineness: the path length must be greater than a minimal path length (normally taken 4 or 5 frames).

Algorithm 1 Initialization, starting a new path

```
StartPaths()
{
    for_all (frame[i])
        for_all (unmatched features in frame[i])
            for_all (candidates in Frame[i+1]) {
                assume connection between feature_i and candidate_{i+1}
                calculate prediction_{i+1}
                for_all (candidates in Frame[i+2]) {
                    if (Correspondence(prediction_{i+1}, candidate_{i+2}))
                        create new path
                        ContinuePath(path, frame[i+3])
                    if (path->length >= MinPathLength)
                        add path to graph
                }
            }
}
```

Furthermore, two options are possible for the candidates: 1) all features in the next frame are candidates, and 2) only unmatched features in the next frame are candidates. The first option allows multiple solutions for one feature, but also requires an exhaustive search. The second choice limits the search because the number of candidates decreases when more paths are found, but it also removes a feature as possibility once a feature has been added to a path, i.e. the tracking solution depends on the order in which the features are tested. The pseudo code of the initialization is shown in Algorithm 1 (forward tracking). The worst case complexity of the initialization is of $O(nm^2)$, with n the number of frames and m the number of features per frame which is usually much less than 100. It should be noted that the number of unmatched features from which a path is started, decreases rapidly once a number of paths are found. So in practice the complexity will be much lower.

Once a path has been initialized, it is continued recursively until the path ends or until the last frame is reached. Obviously, this scheme may lead to multiple paths sharing the same feature: e.g. multiple candidates satisfy the prediction, or a candidate was already added to another path. This happens especially if the tolerances are relaxed. In case of multiple correspondences the path with the best confidence index gets the advantage.

4.2 Multiple Passes

The initialization can be performed forward and backward in time. This will yield different tracking results because the prediction is different in each direction. Correspondences may be found in one direction, but not in the other direction. Therefore, it is useful to go back and forth through the frames in several passes. If a pass is started with an existing tracking solution, the existing paths are first tested for extension before new paths are initialized. There are two ways to extend a path: first paths may be joined; if two paths start and end in successive frames, the two paths may be each other's continuation. Second, a path may be extended with unmatched features in the next frame.

Thus, multiple passes can be conducted on an existing correspondence solution. This also allows multiple passes with changed tracking parameters, for instance with different tolerances. Experience shows that good tracking results are obtained when the tracking is started with strict tolerances and for each successive pass relax the tolerances a bit (with only unmatched features as candidates). First, the obvious paths are found, and then more indistinct paths are found, thus the order of the tested features has less influences.

4.3 Feature Graph

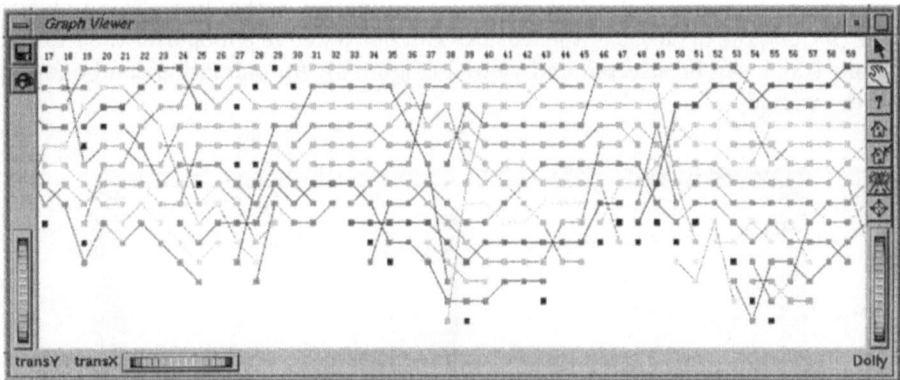

Fig. 2. The feature graph: a node represents a feature in a certain frame and an edge between two nodes represents a positive correspondence between the two features.

The result of the tracking algorithm can be visualized in a directed graph (Figure 2). The frames are plotted horizontally and the features are plotted vertically. Each frame is represented as a level in the graph and all features in that frame are depicted as a node at that level. An edge between two nodes visualizes an established correspondence between two features. Once two features are matched they belong to the same object, at two different instances of time. Therefore, all connecting nodes identify the path of the same object and the features in this path can be given the same object-id (and color). The path describing the motion of an object is stored in the graph as a collection of nodes and edges;

also, information is stored about the paths passing through each frame. The graph data structure holds both path-information and frame-information, i.e. it provides possibilities to walk through the graph in both longitudinal and transverse direction.

5 Applications

5.1 Flow Past a Tapered Cylinder

The first application is a flow past a tapered cylinder[1] consisting of 400 frames of 2.6 Mb each (Plot3D data, grid size 64x64x32, total data size > 1Gb). Features were extracted from the enthalpy (enthalpy $\geq$ -0.6997), and for each feature an ellipsoid fit was calculated and stored in a feature data file (total size 476Kb). The resulting features are highly interacting regions in the wake of the cylinder. After executing six passes with increasing tolerances for the position and size functions, 280 paths were found and only 329 of the original 4121 features remained unmatched (more than 92% matched). These unmatched features can be explained by the fact that the continuation rules do not apply for feature interaction events such as split/merge. Figure 2 shows a small part of the complete graph, the features are vertically sorted by their size. Black nodes indicate unmatched features, and colors indicate the different paths. In a second viewer (the player), the 3D icons of the features can be viewed in a loop over the frames, or one frame can be selected and viewed (Figure 3). An animation can be found on the web [6].

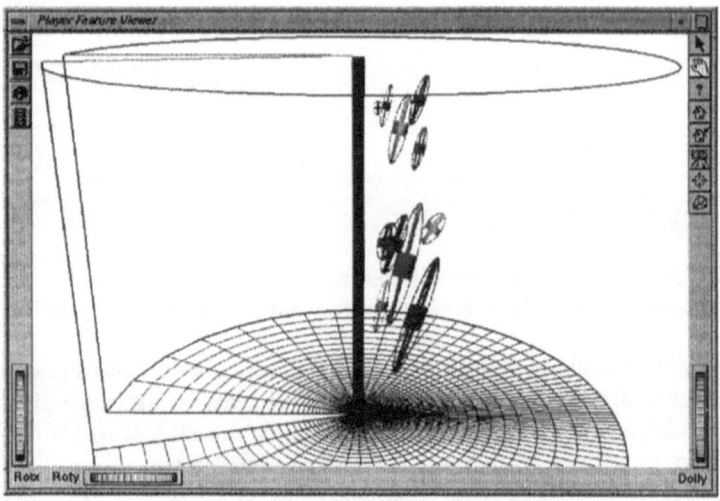

Fig. 3. Flow past a tapered cylinder.

[1] Data courtesy NASA Ames Research Center, http://science.nas.nasa.gov/Software/DataSets/

5.2 Turbulent Vortex Structures

The second application is a CFD simulation with turbulent vortex structures. We obtained the feature data as described in [11]. The feature data consisted of 100 frames with for each vortex structure its position, volume, and mass attributes. The 100 frames contain a total of 4903 features (an average of almost 50 per frame). After the tracking 277 paths were found and only 314 features remained unmatched. This example was a good test case for our tracking algorithm, because of the high average and strongly varying number of features per frame. We found that the algorithm had no problems in processing the data (each pass took less than a minute on an

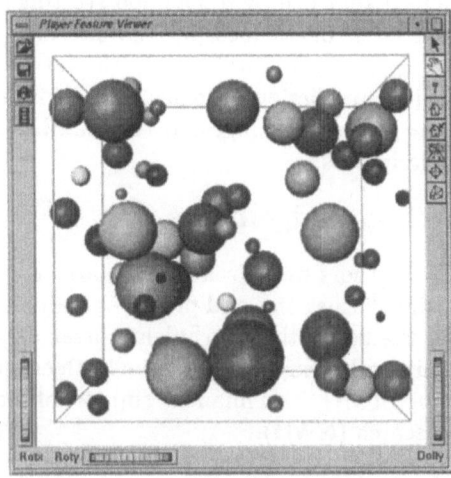

Fig. 4. Turbulent vortex structures.

SGI Onyx with one 75 MHz R8000 processor and 256 Mb of memory). An animation of this example can also be found on the web [6].

6 Conclusions and Future Research

Time-dependent data can be analyzed and visualized by the feature tracking system discussed in this paper. The system first extracts interesting features from the data, then calculates basic attributes such as position and size for each feature, and solves the correspondence problem using these attributes. The solving-algorithm is based on the assumption that features evolve consistently and therefore their behavior is predictable. Using simple behavioral rules, we predict the position and other attributes of each feature in the next frame. This prediction is used to find matching features in the next frame; all candidates in that frame are tested for correspondence with the prediction. Correspondence is found when a number of correspondence functions are satisfied. Each attribute set is associated with a number of correspondence functions that is tested with a certain tolerance and weight.

The initialization of a path is achieved by testing all possible connections until a match is found in the third frame. The resulting path is extended until no more connections can be found, or the path reaches the last frame. When the path satisfies a minimal path length it is added to the feature graph and all the features in the path are given an object-id. This process of finding continuing paths can be conducted forward and backward in time, thus allowing multiple-pass searching. A convenient scheme is to start with strict tolerances (finding obvious paths) and relax the tolerances each pass (finding less obvious paths). The applications show that this tracking system works fine, even with chaotic data with many interacting features.

Although the algorithm is currently limited to finding continuing paths, we believe that it provides a good starting point for finding events like split/merge, and birth/death (see [9]). Events may be detected by applying a new set of correspondence functions. In the future, also effort will be put in more sophisticated prediction schemes, other visualization and interaction techniques for user guided tracking, and tracking based on other attributes, such as skeleton descriptions.

Acknowledgements

The authors wish to thank dr. Jarke J. van Wijk of TU Eindhoven for the stimulating discussions, and prof. Deborah Silver and Xin Wang of Rutgers University for the use of their turbulent vortex data.
This work is supported by the Neth erlands Computer Science Research Foundation (SION), with financial support of the Netherlands Organization for Scientific Research (NWO).

References

1. G. Adiv. Determining 3D Motion and Structure from Optical Flows Generated by several Moving Objects. *IEEE Trans. on PAMI*, 7:384–401, 1985.
2. Y. Arnaud, M. Debois, and J. Maizi. Automatic Tracking and Characterization of African Convective Systems on Meteosat Pictures. *J. of Appl. Meteorology*, 31:443–453, May 1992.
3. D.H. Ballard and C.M. Brown. *Computer Vision*. Prentice-Hall, 1982. Chapter 7.
4. J. Becker and M. Rumpf. Visualization of Time-Dependent Velocity Fields by Texture Transport. In D. Bartz, editor, *Visualization in Scientific Computing '98*, pages 91–101. Springer Verlag, 1998.
5. D.S. Kalivas and A.A. Sawchuk. A Region Matching Motion Estimation Algorithm. *CVGIP: Image Understanding*, 54(2):275–288, Sep 1991.
6. F. Reinders. http://wwwcg.twi.tudelft.nl/~freek/Tracking. Web page with feature tracking examples.
7. F. Reinders, H.J.W. Spoelder, and F.H. Post. Experiments on the Accuracy of Feature Extraction. In D. Bartz, editor, *Visualization in Scientific Computing '98*, pages 49–58. Springer Verlag, 1998.
8. W.B. Rossow, A.D. Del Genio, and T. Eichler. Cloud-Tracked Winds from Pioneer Venus OCPP Images. *J. of Atmos. Sci.*, 47(17):2053–2082, Sep 1990.
9. R. Samtaney, D. Silver, N. Zabusky, and J. Cao. Visualizing Features and Tracking Their Evolution. *IEEE Computer*, 27(7):20–27, July 1994.
10. I.K. Sethi, N.V. Patel, and J.H. Yoo. A General Approach for Token Correspondence. *Pattern Recognition*, 27(12):1775–1786, Dec 1994.
11. D. Silver and X. Wang. Volume Tracking. In R. Yagel and G.M. Nielson, editors, *IEEE Proc. Visualization '96*, pages 157–164. Computer Society Press, 1996.
12. D. Silver and X. Wang. Tracking Scalar Features in Unstructured DataSets. In D. Ebert, H. Hagen, and H. Rushmeier, editors, *IEEE Proc. Visualization '98*, pages 79–86. Computer Society Press, 1998.
13. T. van Walsum, F.H. Post, D. Silver, and F.J. Post. Feature Extraction and Iconic Visualization. *IEEE Trans. on Visualization and Computer Graphics*, 2(2):111–119, 1996.

Editors' Note: see Appendix, p. 317 for colored figures of this paper

New Approaches for Particle Tracing on Sparse Grids

Christian Teitzel and Thomas Ertl

Computer Graphics Group, University of Erlangen
Am Weichselgarten 9, 91058 Erlangen, Germany
{teitzel,ertl}@informatik.uni-erlangen.de
http://www9.informatik.uni-erlangen.de

Abstract. Flow visualization tools based on particle methods continue to be an important utility of flow simulation. Additionally, sparse grids are of increasing interest in numerical simulations. In [14] we presented the advantages of particle tracing on uniform sparse grids. Here we present and compare two different approaches to accelerate particle tracing on sparse grids. Furthermore, a new approach is presented in order to perform particle tracing on curvilinear sparse grids. The method for curvilinear sparse grids consists of a modified Stencil Walk algorithm and especially adapted routines to compute, store, and handle the required Jacobians. The accelerating approaches are on the on hand an adaptive method, where an error criterion is used to skip basis functions with minor contribution coefficients, and on the other hand the so-called combination technique, which uses a specific selection of small full grids to emulate sparse grids.

1 Introduction

The idea of the sparse grid technique was developed in the 1960s by Babenko [1] and Smolyak [12]. In 1990 sparse grids were introduced to the field of numerical computation by Zenger [15]. By means of these grids, it is possible to reduce the total amount of data points or the number of unknowns in discrete partial differential equations. Due to these benefits, sparse grids are more and more used in numerical simulations nowadays [2, 3, 5–7].

On the other hand, it is rather difficult to visualize the results of the simulation process directly on sparse grids, since evaluation and interpolation of function values is quite complicated. Because of this, the results of numerical simulations on sparse grids are usually interpolated to the associated full grid. Then all known visualization algorithms on full grids can be performed, e.g. particle tracing, iso-surface extraction, and volume rendering. However, a major drawback of this procedure is the fact that the advantage of low memory consumption of sparse grids comes to nothing using the associated full grid for the visualization step.

Therefore, visualization tools working directly on sparse grids are an important topic of research. Recently, Heußer and Rumpf presented an algorithm for

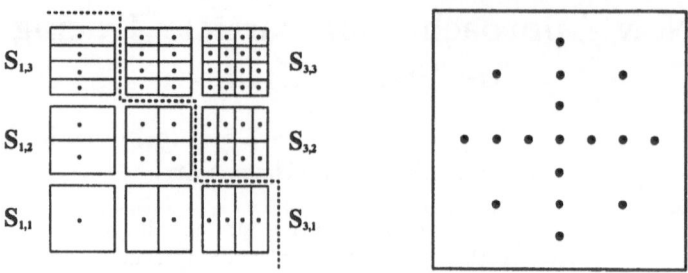

Fig. 1. On the left hand side a two-dimensional hierarchical subspace decomposition is shown and on the right hand side you can see the respective sparse grid.

iso-surface extraction on sparse grids [9]. In a previous work [14], we introduced particle tracing on uniform sparse grids and showed that sparse grids can be used for data compression in order to visualize huge data sets even on workstations with a limited amount of main memory. Now we present two major improvements of particle tracing on sparse grids. In order to accelerate sparse grid particle tracing, an adaptive approach for the interpolation process (Subsect. 2.1) and the so-called combination technique (Subsect. 2.2) are explained. Afterwards, we are going to introduce particle tracing on curvilinear sparse grids (Subsect. 3.3).

2   Sparse Grids

In this section a very brief introduction to the basic ideas of sparse grids is given. For a detailed survey of sparse grids we refer to [2, 15], for a brief summary to [14]. We describe only three-dimensional grids, whereas the sketches always reveal the two-dimensional situation.

If a smooth function f is used in a numerical computation it has to be discretized, which means that only function values at certain positions of a spatial grid are stored. On a uniform mesh this can be done by a hierarchical basis decomposition where piecewise tri-linear finite elements are used as basis functions. In the two-dimensional situation these basis functions are bi-linear hat-functions reaching their maxima at the dots on the left hand side of Fig. 1 and with disjunct supports denoted by the rectangles. Then, the interpolated function $\hat{f}_n$ is given by

$$\hat{f}_n = \sum_{i_1=1}^{n} \sum_{i_2=1}^{n} \sum_{i_3=1}^{n} f_{i_1,i_2,i_3} \tag{1}$$

$$\text{where} \quad f_{i_1,i_2,i_3} = \sum_{k_1=1}^{2^{i_1}-1} \sum_{k_2=1}^{2^{i_2}-1} \sum_{k_3=1}^{2^{i_3}-1} c_{k_1,k_2,k_3}^{(i_1,i_2,i_3)} \cdot b_{k_1,k_2,k_3}^{(i_1,i_2,i_3)} \quad . \tag{2}$$

The values $c_{k_1,k_2,k_3}^{(i_1,i_2,i_3)}$ are called contribution coefficients and $b_{k_1,k_2,k_3}^{(i_1,i_2,i_3)}$ denotes the mentioned basis functions.

The concept of sparse grids is to calculate the interpolated function $\tilde{f}_n$ by using only certain basis functions:

$$\tilde{f}_n = \sum_{i_1+i_2+i_3 \leq n+2} f_{i_1,i_2,i_3} \quad . \tag{3}$$

It can be shown that by interpolating this way a very small loss of accuracy is rewarded with a huge amount of saved storage. The number of nodes of the underlying grids are given by $O\left(2^{3n}\right)$ in case of full and by $O\left(2^n \cdot n^2\right)$ in case of sparse grids (compare Section 4).

2.1 Adaptive Evaluation of Sparse Grids

In order to improve on the rather time consuming standard sparse grid interpolation as described above, an adaptive approach for the function evaluation is presented in this subsection. Since our goal is to decrease the computing time of the sparse grid interpolation, we introduce an adaptive traversal of the standard sparse grid in order to compute function values. The idea is to omit contribution coefficients with a norm below a given error criterion during the interpolation process. Going into the details, we have to distinguish between adaptivity with regard to the L^2 and the L^∞ norm. Although they generate the same sparse grid, these norms lead to slightly different adaptive approaches.

Analyzing the situation with respect to the L^∞ norm, we find that the contribution of one basis element of subspace S_{i_1,i_2,i_3} to the function value is given by (compare Eq. (2))

$$\left\| c_{k_1,k_2,k_3}^{(i_1,i_2,i_3)} \cdot b_{k_1,k_2,k_3}^{(i_1,i_2,i_3)} \right\|_\infty = \left| c_{k_1,k_2,k_3}^{(i_1,i_2,i_3)} \right| \cdot \left\| b_{k_1,k_2,k_3}^{(i_1,i_2,i_3)} \right\|_\infty = \left| c_{k_1,k_2,k_3}^{(i_1,i_2,i_3)} \right| \quad . \tag{4}$$

Then, the greatest possible contribution of subspace S_{i_1,i_2,i_3} is

$$\max_{k_1,k_2,k_3} \left| c_{k_1,k_2,k_3}^{(i_1,i_2,i_3)} \right| \quad \text{with} \quad 1 \leq k_j \leq 2^{i_j-1} \tag{5}$$

and the maximum contribution of level n is

$$\sum_{i_1+i_2+i_3=n+2} \left(\max_{k_1,k_2,k_3} \left| c_{k_1,k_2,k_3}^{(i_1,i_2,i_3)} \right| \right) \quad . \tag{6}$$

For vector fields the absolute value $|\cdot|$ has to be replaced by an appropriate norm of the Euclidean space $\mathbf{R}^m$ and we apply the maximum norm of $\mathbf{R}^m$ in order to ensure maximum accuracy for all components of the vector field.

The actual concept of the adaptive grid traversal is that basis functions that have contribution coefficients with an absolute value below a given error bound are left out during the interpolation process. This results in a function evaluation that considers the local structure of the data set. That is, regions with a high variation in data values and, therefore, large contribution coefficients primarily contribute to the result, whereas coefficients of smooth regions are likely to be omitted.

As a second modification of the plain sparse grid algorithm, we have integrated a preprocessing step, which computes and stores the maximum contribution of each subspace (see Eq. (5)). This preprocessing step is performed during the conversion of a full grid to the appropriate sparse grid. This kind of adaptive grid traversal leads to a function evaluation with direction dependent accuracy, because different subspaces of the same level exhibit different resolutions in the three coordinate directions (reconsider the hierarchical subspace decomposition on the left hand side of Fig. 1). Omitting an entire level of the sparse grid (compare Eq. (6)) is totally independent of the spatial structure of the data set and, therefore, a contradiction to adaptive approaches in general.

Now let us discuss the adaptive approach based on the L^2 norm. A straightforward calculation shows that the contribution of one basis element of subspace S_{i_1,i_2,i_3} to the function value is given by

$$\left\| c_{k_1,k_2,k_3}^{(i_1,i_2,i_3)} \cdot b_{k_1,k_2,k_3}^{(i_1,i_2,i_3)} \right\|_2 = \left| c_{k_1,k_2,k_3}^{(i_1,i_2,i_3)} \right| \sqrt{\left(3 \cdot 2^{i_1+i_2+i_3-1}\right)^{-3}} \ . \tag{7}$$

Since $i_1 + i_2 + i_3 - 1 = n + 1$ with n denoting the current level, the square-root term only depends on the level and is, therefore, constant for all subspaces of the same level. Hence, this term is also a factor of the maximum contribution of the corresponding subspaces.

In contrast to the L^∞ norm, the L^2 norm generates an adaption strategy that considers not only the absolute value but also the level of a contribution coefficient. Contribution coefficients of higher levels are more likely to be omitted than coefficients of lower levels. Later on in Subsect. 4, we are going to see that the different properties of these adaptive grid traversals yield different results.

2.2 The Combination Technique

Since both the standard and the adaptive sparse grid interpolation of function values are quite complicated and rather time consuming, we have also implemented the so-called combination technique, which was introduced by Griebel, Schneider, and Zenger in 1992 [6]. Actually, the combination method has been used in numerical simulations in order to combine partial solutions computed on smaller, suitable full grids to the desired sparse grid solution. However, we start with a data set given on a sparse grid and decompose the grid such that the data set is represented on certain uniform full grids of low resolution. Now the fast and simple tri-linear interpolation can be performed on each of these full grids. The resulting value is computed by linear combination of the tri-linear interpolated full grid results. Specifically, it can be proven that the three-dimensional interpolated function $\tilde{f}_n \in \tilde{L}_n$ is given by

$$\tilde{f}_n = \sum_{i_1+i_2+i_3=n+2} f_{i_1,i_2,i_3}^c - 2 \cdot \sum_{i_1+i_2+i_3=n+1} f_{i_1,i_2,i_3}^c + \sum_{i_1+i_2+i_3=n} f_{i_1,i_2,i_3}^c \tag{8}$$

where f_{i_1,i_2,i_3}^c denotes the tri-linear interpolation of function values on the respective full grid. Figure 2 reveals the two-dimensional situation, which also

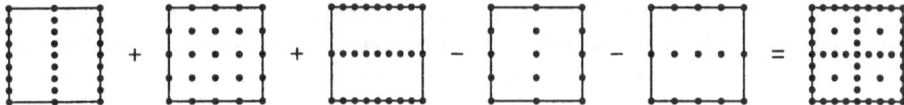

Fig. 2. A two-dimensional sparse grid of level 3 can be reconstructed by linear combination of five full grids of low resolution.

shows that the used full grids consist of the same nodes as the corresponding sparse grid.

Investigating the benefits of the combination technique, we find that the total number of summands of the standard sparse grid interpolation on a three-dimensional sparse grid of level n is given by $\frac{1}{6}n(n+1)(n+2)$ (compare Eq. (3)), whereas the total number of tri-linear interpolations of the combination method adds up to $\frac{3}{2}n(n-1)+1$ (see Eq. (8)). That is, the number of tri-linear interpolations of the combination method is one order of magnitude lower than the number of summands of the standard interpolation. However, the main advantage of the combination technique is the fact that uniform full grids are used. Thus, it is possible to implement the interpolation routine in terms of tight for-loops (see Sect. 3.2), which makes the combination technique an order of magnitude faster than the standard approach even for the lower levels.

3 Particle Tracing

Lagrange visualization techniques of vector fields are based on the numerical solution of an initial value problem for the ordinary differential equation: $dx/dt = v(x,t)$. Usually, a numerical integration method is used to obtain a solution. All such methods have in common that they must evaluate the vector field v at certain positions, which are in general not at grid points. Therefore, the value of v at such a position has to be interpolated. As mentioned in Subsect. 2, this interpolation on sparse grids is different from that one on full grids, whereas most other parts of the particle tracing algorithm can remain unchanged. Further exceptions are the routines required for handling curvilinear grids (see Subsect. 3.3).

Our sparse grid particle tracing modules are implemented as IRIS Explorer modules (compare also [14]). In order to visualize the particles, we have chosen lines, bands, tubes, balls, and tetrahedra as geometrical primitives. Of course, all kinds of traces can visualize an additional scalar value by means of color coding. Moreover, balls and tetrahedra can reveal another scalar value by their size. Besides that, bands and tetrahedra display the local vorticity of the flow via rotating around the actual streak line (see colored figures in the Appendix). As integration methods for the particle tracing algorithm, we use the integration schemes that we have already implemented in our full grid particle tracer. A comparison of these schemes can be found in [13]. An adaptive Runge-Kutta method of order 3 (RK3(2)) is used for the tests described in Sect. 4.

In contrast to tri-linear full grid interpolation, sparse grid interpolation does not operate locally, because one basis function in every subspace contributes to the function value. Since the tri-linear interpolation is one of the most time consuming operations during the particle tracing process on full grids [10], the complicated sparse grid interpolation is all the more time consuming. Therefore, it is important to execute the interpolation as fast as possible.

The contribution coefficients of the sparse grid are usually stored in a binary tree [2,3,9]. Then, a recursive tree traversal has to be performed in order to interpolate the function value. This tree traversal is very slow. Although caching strategies can increase the efficiency of the traversal [9], the computation of the values remains rather time consuming. In order to avoid the tree traversal and to accelerate the access to the contribution coefficients, we have developed a very efficient data structure based on arrays (see [14]). In the next two subsections our new approaches for further cutting the interpolation time are described.

3.1 Adaptive Grid Traversal

In order to perform an adaptive grid traversal as described in Subsect. 2.1, our former class hierarchy [14] has been slightly modified. The interpolation process has been enhanced in such a way that contribution values smaller than a given error bound are omitted. In addition, the preprocessing step for creating the actual sparse grid has been modified. During this process the contribution coefficients are computed from an analytic function or a full grid data set, or they are directly read from a sparse grid data set. Because we often deal with vector fields, each basis function does not contain a single contribution coefficient but an array of coefficients. For the purpose of adaptive function evaluation, the mentioned array has been extended by one component in order to store the maximum absolute value of the contribution coefficients. Moreover, a variable has been added for storing the maximum contribution coefficient of the entire subspace. Of course, all these additional variables storing maximum contribution coefficients are initialized during the creation of the sparse grid.

3.2 Combination Technique

Equation (8) determines the interpolation process for the combination technique, which combines full grids of low resolution to a resulting sparse grid. Since these full grids are uniform grids, the function values can be stored in three-dimensional arrays and derived by tri-linear interpolation. Thus, the interpolation method of the appropriately derived class can address the necessary function values in a tight loop. This fact makes the combination technique an order of magnitude faster than the previously described sparse grid interpolation even for low levels.

3.3 Curvilinear Sparse Grids

The underlying concept of curvilinear sparse grids is the same as for curvilinear full grids. In the case of uniform full grids, only the function values are stored in an array, whereas in case of curvilinear grids, the function values and the coordinates of the grid points as well are saved. If a curvilinear sparse grid is considered, the contribution coefficients of the coordinates of the grid points are stored as additional components of the basis functions. For the combination technique, the coordinates of the grid points are stored in the arrays of the small full grids accordingly.

Particle tracing in arbitrary non-uniform grids requires the so-called *point location* to be performed for each integration step, in order to find the cell containing the actual particle position. For the case of curvilinear grids, particle tracing algorithms can be divided into P-space and C-space methods. Sadarjoen et al. [11] showed that P-space algorithms are in general preferable to C-space methods. Hence, we have implemented a P-space algorithm appropriately adapted for sparse grids. The *stencil walk* algorithm introduced by Buning [4] has been modified in the following way. First of all, we initialize the desired C-space position r_c by starting in the center of our volume in C-space. In order to improve this guess, the C-space position is transformed into P-space. This is done by a sparse grid interpolation using either plain sparse grids, adaptive sparse grids, or the combination technique. If the difference of the transformed guess and the current position in P-space is small enough, we accept the C-space position. Otherwise the difference is transformed back into C-space via the inverse Jacobian and then added to the previous guess. Thereafter, the procedure is iterated until the appropriate position in C-space is located. On full grids, the stencil walk algorithm usually needs less than five iterations to find the correct C-space position.

As yet, the modifications of the stencil walk algorithm seem to be very moderate. But the main question is how to calculate the inverse Jacobian. On full grids, this is done on the fly by tri-linear interpolation of the eight Jacobians at the vertices of the current cell and subsequent matrix inversion. The Jacobians at the vertices are computed by finite differences. However, tri-linear interpolation is not possible on sparse grids. Thus, we have to use sparse grid interpolation and we have to store the inverse Jacobian, i.e. the respective contribution coefficients, at each sparse grid point. This memory overhead can only be justified with the fact that sparse grids themselves are very storage efficient.

Now we describe how the contribution coefficients of the inverse Jacobians are computed and stored. In case of uniform sparse grids, we compute the contribution coefficients in a preprocessing step from the input data and store the coefficients in the sparse grid structure, which is done by the setCoeff(...) methods. The data sets usually do not contain the Jacobians explicitly, thus, the Jacobians and their inverse matrices have to be calculated as well as their contribution coefficients. Since a contribution coefficient can not be computed from a single function value but from a specific collection of function values, the inverse Jacobians have to be stored somewhere. We have modified the setCoeff(...) methods in such a way that in a first pass the contribution coefficients of the function values and the inverse Jacobians are computed and stored in the sparse grid structure. In a second pass, the contribution coefficients of the inverse Ja-

Table 1. Computing times in CPU-seconds for integrating nine stream ribbons over 55 time steps in an analytic vortex flow using an adaptive RK3(2) scheme.

level	3	4	5	6	7	8
# of full grid points	9^3	17^3	33^3	65^3	129^3	257^3
uniform full grid	0.67 s	1.18 s	1.89 s	2.28 s	2.66 s	
uniform sparse grid	0.24 s	0.33 s	0.68 s	0.93 s	4.51 s	5.91 s
uniform combination	0.07 s	0.12 s	0.20 s	0.30 s	1.15 s	1.61 s
curvilinear full grid	0.70 s	1.30 s	2.58 s	5.28 s	10.59 s	
curvilinear sparse grid	1.56 s	3.28 s	6.82 s	9.31 s	22.72 s	31.16 s
curvilinear combination	0.64 s	1.19 s	2.02 s	3.02 s	6.05 s	8.49 s

cobians are computed and stored over the original components of the inverse Jacobians. The second pass traverses the levels beginning with the highest level and ending with level 1. It is done this way because the contribution coefficients only depend on the current function value and on the function values of lower levels. Hence, it is possible to overwrite the original components of the inverse Jacobians successively. Notice that level 0 is not part of the mentioned second pass because in level 0 contribution values and function values coincide.

4 Results

In order to compare our sparse grid particle tracing modules with a full grid particle tracer, several data sets were used. For uniform grids we used the data set of a cavity flow, which was provided by S. H. Enger from the Department of Fluid Mechanics and is given on a full grid with 129^3 nodes, which corresponds to level 7 in sparse grid terminology. The data set contains the velocity, pressure, and temperature at each vertex requiring more than 40 MB. The same data set with a resolution of 257^3 (level 8) would need more than 320 MB, which is probably too much for most workstations. On the other hand, this data set stored on a sparse grid consumes only 175 kB for level 7 and 415 kB for level 8 (compare Table 2). For curvilinear grids the NASA blunt fin data set was used (see Fig. 4). In addition, we used several analytic test data sets on uniform and different curvilinear grids. Therefore, we were able to create sparse and full grids in any resolution only limited by the main memory of the used machine. These vector fields were chosen for our quantitative performance tests, with the results being comparable to the ones obtained from the discrete given data sets. All tests were performed on an SGI with a 250 MHz R10000 processor. For each experiment nine stream ribbons consisting of about 500 particles were integrated. The CPU-times were measured in seconds and are listed in Table 1. The measured times show that interactive particle tracing is possible even on sparse grids of level 8 by using the combination technique.

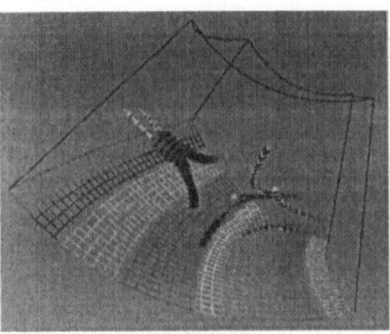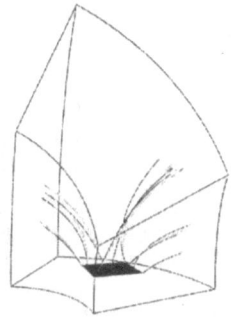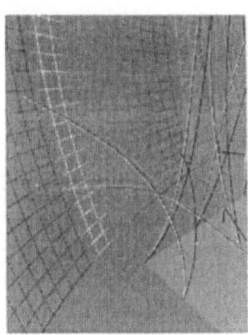

Fig. 3. In the image on the left hand side, streak tetrahedra in an analytically given flow on a curvilinear sparse grid of level 5 are shown. In the next two pictures streak lines depict a vortex flow field; yellow lines display the flow on a full grid, blue, green, and red lines on a curvilinear sparse grid of level 2, 3, and 4 respectively. In the closeup on the right hand side, it can be seen that the traces computed on the full grid and the sparse grid of level 4 are almost identical.

Investigating the accuracy of sparse grid particle tracing, the traces computed on sparse grids are compared with their counterparts resulting from full grids. Theory tells us that the difference should be rather small. In fact, the results of particle tracing on the analytic data set confirm this estimation, since the lines computed on uniform full and sparse grids coincide on screen for levels greater than 3 (compare [14]) and for levels greater than 4 on curvilinear grids (see Fig. 3). However, for the derivation of the upper bounds for the interpolation errors, a certain smoothness of the data was a prerequisite. Since discrete data sets are not smooth at all, these estimations do not hold in this case. Indeed, for discrete data we found out [14] that the particle traces computed on sparse grids converge rather slowly to the full grid solution. Nevertheless, due to the great advantage of low memory consumption, it is possible to use a sparse grid of a sufficiently high level to overcome this problem.

For the adaptive grid traversal, several experiments have shown that it is important whether the L^2 norm is used for the adaptive traversal or the L^∞ norm. Employing the L^∞ norm leads to a marginal decrease in computing time but to a significant loss in accuracy. However, by using the L^2 norm, it is possible to decrease computing time by about 20 per cent and to achieve nearly the quality of the corresponding plain sparse grid.

The next approach for accelerating particle tracing on sparse grids is the combination technique. The first advantage of this technique compared to adaptive grid traversal is the fact that there is no loss in accuracy at all. Combination technique and plain sparse grid interpolation create exactly the same particle path. The second and more important benefit is that the combination technique is almost by a factor of four faster than plain sparse grid interpolation (compare Table 1). That is, particle tracing based on the combination method is a lot faster and also more accurate than particle tracing based on adaptive sparse grid

Table 2. Memory consumption of a typical data set.

level	5	6	7	8	9	10
uniform full grid	640 kB	5 MB	40 MB	320 MB	2.5 GB	20 GB
uniform sparse grid	29 kB	73 kB	175 kB	415 kB	970 kB	2.2 MB
uniform combination	110 kB	295 kB	760 kB	1.8 MB	4.5 MB	10.5 MB
curvilinear full grid[1]	1 MB	8 MB	64 MB	512 MB	4 GB	32 GB
curvilinear sparse grid[2]	99 kB	248 kB	595 kB	1.4 MB	3.2 MB	7.5 MB
curvilinear combination[2]	375 kB	1 MB	2.5 MB	6.1 MB	15.3 MB	35.7 MB

[1] including coordinates but excluding inverse Jacobians
[2] including coordinates and inverse Jacobians

interpolation. Hence, the combination technique should be used for interpolation on sparse grids.

Now let us turn to curvilinear sparse grids. We have implemented curvilinear grids with all three sparse grid interpolation methods, but in consideration of the last paragraph we usually employ the combination technique in connection with curvilinear sparse grids. For a first test of particle tracing in curvilinear sparse grids we have used the NASA blunt fin data set (see Fig. 4). Additionally, our module has been verified with several analytic data sets (compare Fig. 3). On the one hand side these tests have confirmed that smooth data sets are more appropriate for using sparse grid methods than discontinuous data. On the other hand these tests have revealed that particle traces calculated on curvilinear sparse grids converge slower to the corresponding full grid trace than particle traces computed on uniform sparse grids. The reason for this decline in accuracy might be due to a less accurate point location caused by an intensive use of sparse grid interpolation in the stencil walk algorithm. Finally, time measurements have shown that particle tracing on curvilinear sparse grids is about five times slower than tracing on uniform grids. This is roughly the same deceleration as on full grids.

The great advantage of the sparse grid technique is the low number of required grid points. Table 2 demonstrates this benefit listing the memory consumption for various grid levels on the assumption that a typical data set resulting from a numerical flow simulation is given. Such data usually contain five floating point values, namely three velocity components, pressure, and temperature, at each grid node. Then, these floating point values add up to 20 bytes per node. In case of curvilinear grids three more floating point numbers for the coordinates and nine additional values for the inverse Jacobians are stored at each grid node. Nevertheless, sparse grids are very suitable for compressing huge data sets. This opens up the potential to visualize them even on workstations with a limited amount of main memory.

5 Conclusion

We have introduced particle tracing on curvilinear sparse grids and presented competing approaches for accelerating the time consuming sparse grid interpolation process. Technically, we have implemented adaptive sparse grids with error monitoring and the combination technique. This allows to carry out flow visualization directly on sparse grids without prior transformation to the associated full grids. This is an important step for the broader application of the sparse grid method, since in real applications it is often impossible to load full grids of more than 128^3 nodes into the main memory of a workstation for visualization purposes. Furthermore, the sparse grid approach can be used as a compression method in order to realize particle tracing in huge data sets on workstations with a small amount of main memory.

Feasible directions of future work are texture based algorithms and iconic methods combined with feature extraction. In addition, our sparse grid particle tracer could be extended to multi-block data sets in the same way as it was done in our full grid particle tracing module [8]. In a parallel work we are investigating further visualization techniques on sparse grids, namely hardware assisted volume rendering and fast iso-surface extraction.

References

1. K. I. Babenko. Approximation by trigonometric polynomials in a certain class of periodic functions of several variables. *Soviet Mathematics*, 1:672–675, 1960. Translation of Doklady Akademii Nauk SSSR.

2. H.-J. Bungartz. *Dünne Gitter und deren Anwendung bei der adaptiven Lösung der dreidimensionalen Poisson-Gleichung*. PhD thesis, TU Munich, 1992.

3. H.-J. Bungartz and T. Dornseifer. Sparse Grids: Recent Developments for Elliptic Partial Differential Equations. In *Multigrid Methods V*, Lecture Notes in Computational Science and Engineering. Springer-Verlag, 1998.

4. P. Buning. Numerical Algorithms in CFD Post-Processing. In *Computer Graphics and Flow Visualization in Computational Fluid Dynamics*, number 1989-07 in Lecture Series, Brussels, Belgium, 1989. Von Karman Institute for Fluid Dynamics.

5. M. Griebel, W. Huber, U. Rüde, and T. Störtkuhl. The combination technique for parallel sparse-grid-preconditioning or -solution of PDE's on multiprocessor machines and workstation networks. In L. Bougé, M. Cosnard, Y. Robert, and D. Trystram, editors, *Second Joint International Conference on Vector and Parallel Processing*, pages 217–228, Berlin, 1992. CONPAR/VAPP, Springer-Verlag.

6. M. Griebel, M. Schneider, and C. Zenger. A combination technique for the solution of sparse grid problems. In P. de Groen and R. Beauwens, editors, *International Symposium on Iterative Methods in Linear Algebra*, pages 263–281, Amsterdam, 1992. IMACS, Elsevier.

7. M. Griebel and V. Thurner. The efficient solution of fluid dynamics problems by the combination technique. *Int. J. Numer. Methods Heat Fluid Flow*, 5:251–269, 1995.

8. R. Grosso, M. Schulz, J. Kraheberger, and T. Ertl. Flow Visualization for Multiblock Multigrid Simulations. In P. Slavick and J. J. van Wijk, editors, *Virtual*

84

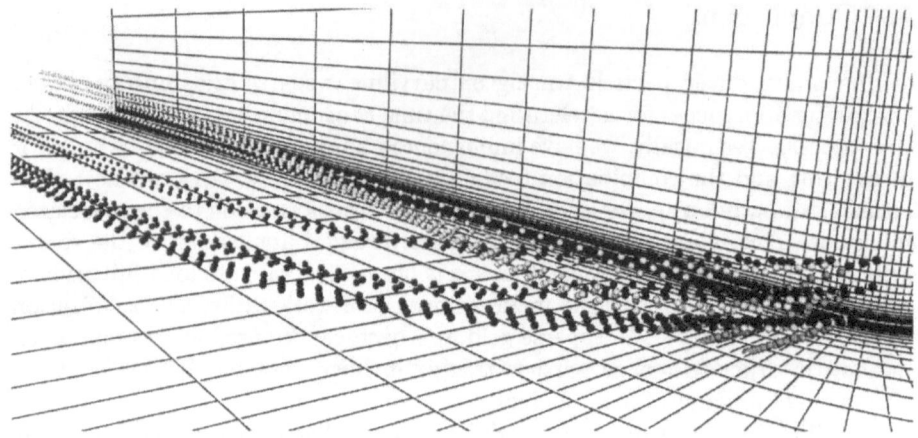

Fig. 4. Streak balls display the flow in the blunt fin data set; the red balls are computed on a curvilinear sparse grid of level 4, the yellow ones on a grid of level 3, and the green ones on a grid of level 2.

  Environments and Scientific Visualization '96, Heidelberg, 1996. Springer-Verlag. Proceedings of the Eurographics Workshop in Prague, Czech Republic.

9. N. Heußer and M. Rumpf. Efficient Visualization of Data on Sparse Grids. In H.-C. Hege and K. Polthier, editors, *Mathematical Visualization*, pages 31–44, Heidelberg, 1998. Springer-Verlag.

10. D. N. Kenwright and D. A. Lane. Optimization of Time-Dependent Particle Tracing Using Tetrahedral decomposition. In G. M. Nielson and Silver D., editors, *Visualization '95*, pages 321–328, Los Alamitos, CA, 1995. IEEE Computer Society, IEEE Computer Society Press.

11. A. Sadarjoen, T. van Walsum, A. J. S. Hin, and F. H. Post. Particle Tracing Algorithms for 3D Curvilinear Grids. In *Fifth Eurographics Workshop on Visualization in Scientific Computing*, 1994.

12. S. A. Smolyak. Quadrature and interpolation formulas for tensor products of certain classes of functions. *Soviet Mathematics*, 4:240–243, 1963. Translation of Doklady Akademii Nauk SSSR.

13. C. Teitzel, R. Grosso, and T. Ertl. Efficient and Reliable Integration Methods for Particle Tracing in Unsteady Flows on Discrete Meshes. In W. Lefer and M. Grave, editors, *Visualization in Scientific Computing '97*, pages 31–41, Wien, April 1997. Springer-Verlag. Proceedings of the Eurographics Workshop in Boulogne-sur-Mer, France.

14. C. Teitzel, R. Grosso, and T. Ertl. Particle Tracing on Sparse Grids. In D. Bartz, editor, *Visualization in Scientific Computing '98*, pages 81–90, Wien, April 1998. Springer-Verlag. Proceedings of the Eurographics Workshop in Blaubeuren, Germany.

15. C. Zenger. Sparse grids. In *Parallel Algorithms for Partial Differential Equations: Proceedings of the Sixth GAMM-Seminar*, Kiel, 1990.

Editors' Note: see Appendix, p. 318 for colored figure of this paper

Volume Visualization

(Research Papers)

A Methodology for
Comparing Direct Volume Rendering Algorithms
Using a Projection-Based Data Level Approach

Kwansik Kim and Alex Pang

Computer Science Department
University of California, Santa Cruz, CA 95064
ksk@cse.ucsc.edu, pang@cse.ucsc.edu

Abstract. Identifying and visualizing uncertainty together with the data is a well recognized problem. One of the culprits that introduce uncertainty in the visualization pipeline is the visualization algorithm itself. Uncertainties introduced in this way usually arise from approximations and manifest themselves as artifacts in the resulting images. In this paper, we focus on comparing different direct volume rendering (DVR) algorithms and their artifacts as a result of DVR algorithm selections and their associated parameter settings. We present a new data level comparison methodology that uses differences in intermediate rendering information. In particular, we extend the traditional image level comparison techniques to include data level comparison techniques. In image level comparisons, quantized pixel values are the starting point for comparison measurements. In contrast, data level comparison techniques have the advantage of accessing and evaluating the intermediate 3D information during the rendering process. Our data level approach overcomes limitations of image level approaches and provide capabilities to compare application dependent details as well as general rendering qualities. One of the key challenges with our data level comparison approach is finding a common base for comparing the rich variety of DVR algorithms. In this paper, we present how a projection algorithm can be used as a base for comparing other DVR algorithms. In addition, a set of projection-based metrics are derived to quantify the comparison measurements among DVR algorithms. The results presented in this paper complement our earlier findings where a ray-based approach was used as the base for comparing other DVR algorithms.

Key Words and Phrases: Scientific visualization, direct volume rendering, uncertainty, error, difference, similarity, metrics.

1 Introduction

Although a large number of visualization methods have been developed, few provide the foundations and methodologies to compare and evaluate them against each other. This shortcoming has been raised as a critical issue for the objective interpretations of scientific data [2, 9, 21, 22]. Direct volume rendering (DVR) is one of the most popular methods for visualizing 3D scalar data sets. These methods have been extensively investigated resulting in a rich variety of algorithms. Some of these can be found in [5, 6, 8, 10, 13, 15–17]. Unfortunately, this plethora of DVR methods produces images that are different from each other. In critical applications, it can be very disconcerting to have even slight differences in images rendered by various volume rendering algorithms. It is therefore necessary for both users and developers to be able to do in-depth study on the differences. Important differences stem from varying degrees of approximation in reconstruction, material classification (e.g. transfer functions), and accuracy in physical simulation of light and material interactions. Because it is a perceptual issue and DVR algorithm variations are often arbitrarily non-linear, it is very difficult to quantify errors (or uncertainties) that are introduced in the final rendered image. In addition, unlike quality issues in the image synthesis and image processing communities, criteria for measuring the quality often depend on the purpose of the particular visualization. They are not necessarily about how realistic or aesthetically satisfactory the final images are.

Fortunately, more and more DVR papers address the important issue of volume rendering qualities and comparisons [9, 11, 20, 21]. In those that address the issue, the norm is to use image level comparisons, and sometimes at the image summary level at best. Combinations of image analysis methods and summary statistics have been used to compare rendered images. For example, wavelet based image metrics [1] have been proposed to help determine perceptual similarities between volume rendered images. However, there are limitations to image metrics such as summary statistics and perceptual metrics as well. Williams and Uselton [21] first described some of the difficulties and limitations of image comparison. They presented the need for providing rigorous specifications of the volume rendering parameters, a set of image difference classifications and corresponding metrics for each category. However, there are inherent limitations to image level comparisons. While image level comparisons can provide information as to the location and degree by which two images differ, they often do not provide enough information as to why the two resulting images differ in general.

This paper addresses this shortcoming by proposing the use of data level comparison techniques. The goal is that if two images differ, then we want to provide an explanation for the causes for such difference. The name *data level comparison* was inspired by the work of Pagendarm and Post [12]. Data level methods incorporate intermediate and auxiliary information in the rendering process and use this information to generate a data level comparison image. Another distinguishing factor is that image level comparisons quantize data values then compare, while data level comparison methods compare data values then quantize thereby resulting in a greater dynamic range of comparison values. As in the Figure 1(a), we can compare volume data, intermediate rendering data, or final rendered images. In this paper, we use data level comparison to take advantage of any intermediate information in the volume rendering pipeline before the final image is generated.

Because of the wide variety of DVR algorithms, some of them have drastically different approaches and therefore difficult to obtain registered, intermediate rendering information for comparisons. Hence, we compare algorithms by first mapping them to a base or reference algorithm and then deriving metrics natural to the base algorithm (Figure 1(b)). We consider this mapping as invertible and thus we can experiment with multiple base algorithms and develop their corresponding metrics. In this paper, we use the projection algorithm as our base algorithm, and focus our attention on DVR algorithms applied to regularly gridded data.

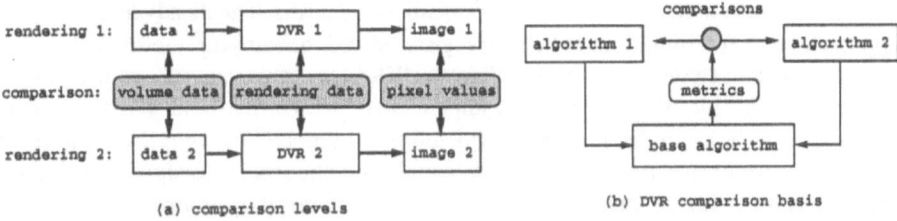

Fig. 1. Types of comparisons (a) and basis for comparing algorithms (b). (a) highlights three different areas where one can perform comparison: data, rendering information, and pixel values. Data level comparison includes comparison of data and rendering information, while image level comparison works with pixel values. (b) shows the architecture for comparing two different DVR algorithms via a common base algorithm.

2   Previous Work

In our earlier work [3,4], we described how DVR algorithms can be simulated using ray casting as a base or reference algorithm (Figure 1(b)). For example, projection-based algorithms were simulated by intersecting the ray with the set of projected polygons of each cell. We also developed several ray-based metrics such as the distance the ray traveled, the number of samples along each ray, and similarity measures for sample colors and data along each ray. Here, we complement our earlier work by using a projection-based algorithm as our base or reference algorithm. This involves mapping or simulating other DVR algorithms, such as ray casting, into projection-based DVR algorithms.

3   Data Level Approach

Figure 1(a) delineates image level comparison (using pixel values) from data level comparison (using volume and rendering data). In DVR, the intermediate information may include items related to the rendering process. Examples include: distribution of opacities, values that contribute to a pixel, and minimal or maximal sample values along the viewing direction. Intermediate information may also include items related to the volume rendering algorithm. Transfer functions, sampling locations and frequency, interpolation functions, opacity threshold, and projection filters are examples of information related to the volume rendering algorithm. It should be noted that in some cases this distinction is blurred. In either case, these information and others can be used to generate metrics for data level comparison.

There are two different approaches for collecting these intermediate information. One approach is to collect them directly from the different DVR algorithms. The other approach is to first map other DVR algorithms to a base or reference algorithm, then collect the metrics from this base algorithm. Both approaches have their advantages and disadvantages. The main advantage of the first approach is the accuracy of the intermediate information since we do not need to simulate or map the algorithms to a base algorithm and thereby possibly introducing some errors. The disadvantages of this approach are: (a) difficulty in collecting registered and meaningful intermediate information that can be compared across different DVR algorithms, and (b) the need to put data collection code in different DVR implementations. On the other hand, the main advantages of the second approach are: (a) ability to compare different DVR algorithm on the same set of intermediate data as collected from the reference algorithm, and (b) multiple choices for the reference algorithm since the mapping is invertible. That is, as demonstrated by this paper and our earlier work [3], a ray-based DVR algorithm can be mapped to a projection-based DVR algorithm, and vice versa. The second approach also has a number of disadvantages: (a) difficulty of completely specifying all the rendering parameters to precisely control the mapping of some DVR algorithms to the base algorithm, and (b) the need to map to the base algorithm or simulate other DVR algorithms. The critical thing to address the first difficulty of the latter approach is that rigorous specifications of all rendering parameters are necessary in order to faithfully simulate a given DVR algorithm. In the absence of this rigor, emphasis should be placed on specifying the more important rendering parameters. Our basic assumption is that the major differences from different DVR images result from those important rendering parameters.

4 Projection as a Base Algorithm

A popular model for computing the color intensities as the light passes through a sequence of translucent material is to assume that the light is attenuated by the opacities of the material. This is often described by the compositing equation below:

$$C_{out,i} = \alpha_i C_i + (1 - \alpha_i) C_{in,i} \tag{1}$$

α_i is the opacity and C_i is the color of the ith object to be composited. $C_{in,i}$ is the color intensities before the ith object is rendered (or composited) and $C_{out,i}$ is the result of rendering (or compositing) the ith object. Integrating this equation yields the so called *volume rendering integral*:

$$C(a,b) = \int_a^b E(s) e^{-\int_a^s \delta(x)dx} ds \tag{2}$$

$C(a,b)$ is the color intensity contributions through a line from position a to b. E is the color emission function and δ is the differential opacity function. One may view different DVR algorithms as variations on how to approximate the emission function E and the solution of the volume rendering integral. For example, Eqn. 2 has a closed form solution if we assume constant color emission and opacity between the interval of integration. Image order algorithms, like ray casting, discretize the volume rendering integration process and compute multiple compositions (Eqn. 1) of the integrated colors

($C(a, b)$ in Eqn. 2) of small adjacent sampling intervals. The simplest approximation is simply $C(a, b) = E(a)$ or $E(b)$ at each sampling point. Our challenge then is to make a projection-based algorithm general enough to facilitate simulations of the different variants of DVR algorithms.

The structure of our base algorithm is the same as that of any object order algorithm like *coherent projection* [17] or *splatting* [16, 5]. The basic idea of these object order algorithms is to compute the contributions of the sub-volumes and composite them to the images in the proper order. We conveniently collect any desired metrics as the algorithm calculates contributions to the final rendered image. The structure of our comparison algorithm is:

Procedure **project**:

1. Determine the order of volume cells or voxels to be projected.
2. For each cell or voxel, in the appropriate order
   (a) Compute the contribution of the cell or voxel and composite.
   (b) Collect *metrics*.

The term *cell* refers to the cube formed by the 8 neighboring data points in a regular grid volume. The term *voxel* refers to the region around a data point. The shapes of the voxel depend on the data model. In *splatting* algorithms, the shape (and thus its 2 dimensional contribution to the final rendered image) is defined as a function of the distance from the data point.

Projection-based algorithms [17, 14] usually take advantage of hardware polygon shading found in modern workstations. However, the projection algorithms are general enough to be extended and simulate effects of other DVR algorithms if they are implemented in software using *volume scan conversion*. Therefore, in step (2a) of procedure **project**, we implemented volume scan conversion to process a set of projected polygons in software. This allows us to:

- simulate other variations of DVR algorithms (e.g. raycasting) by varying the volume integration method at each pixel in the volume scan conversion procedure.
- derive projection-based *metrics* to compare a wide range of algorithms on a common basis.

As pointed out in [19], scan conversion of surface polygons has been studied extensively, but the process in volume rendering is a different problem. In scan converting surfaces, the surface (or polygon) has the color properties to be scanned. However, in scan converting a volume cell (or voxel), we need to process the material (or data) between volume cell faces that has the color properties. In the following description of the procedure **composite_sub-volume**, the details of our volume scan converting is explained. The structure of our base algorithm is the same as other volume scan converting procedures but it carries additional information such as relative locations of front and back points for simulations of other algorithms (at lines in 2(c)(2) of procedure **composite_sub-volume** below).

Procedure **composite_sub-volume**:

1. Decompose the given sub-volume into a set of projected polygons
2. For each polygon **P**,
   (a) Get the range **R** of the scanlines that **P** occupies in screen space and set up the edge table.
   (b) Empty the active edge table.
   (c) For each scanline of the range **R** in bottom-to-top order,
       (1) Update active edge table.
       (2) For each pixel from the left edge to the right edge,
           * Update front and back pixel information.
           * Compute the contribution between the front and back sub-volume at the pixel.

For projection based algorithms such as *Coherent Projection* [17] or *splatting* [5, 16], the footprint of a cell is often approximated as a set of polygons. The typical usage of these projection algorithms can be easily simulated with our base algorithm by using surface scan conversion only. In this case, the contribution from a cell (or voxel) is simply a color intensity interpolated within the footprint polygon. Sampling and reconstruction of volume data and color mappings are done only at the vertices of projected polygons as in [5, 16, 17].

Image order algorithms, such as variations of ray casting, can be simulated by changing the sampling patterns and the approximations to the volume rendering integral (i.e. last line of Procedure **composite_sub-volume**). Figure 2 (color plate 1) illustrates the simulation of variants of ray casting while projecting a volume cell. The figure shows a cross sectional view perpendicular to the projection plane. For each pixel, the base algorithm updates the information needed to project a volume cell, such as data values and relative locations of the front and back points within the cell. Using the specifications of rendering parameters, such as sampling patterns, we can calculate the color contributions of the cell along the ray and composite them to the final image. Pixel (a) of figure 2 shows values used by a ray casting algorithm that samples at cell face intersections. The pink colored squares represent the sample points. The color contribution for pixel (a) is calculated using Eqn. 2 assuming a constant color value between the sampling interval (often the average of front and back sample colors is used) and composited to the screen (using Eqn. 1). Pixel (b) shows the values used by another ray casting algorithm that uses regular sampling. Pixels (c) and (d) together show yet another variation as used by volume slicing. Here, it is simulated as a regular sampling pattern with the first sample point starting from the plane closest to the screen. For all the sample points in the figure 2, either their data can be reconstructed and colors mapped from the transfer function, or their sample colors are interpolated using the pre-mapped colors of the eight cell corners.

In the context of scan converting polygons for projection-based DVR algorithms, we use the following to classify several popular DVR algorithms:

1. **Data model** – this distinguishes whether data is defined at vertices or at voxel centers. An associated interpolation or distance function is also specified for each.
2. **Interpolation value** – this distinguishes whether data values are reconstructed or color values are being interpolated at the polygon vertices or sample locations. That is, the color intensities (E) in Eqn. 2 are calculated either by

        – interpolating the data values and then evaluating the transfer function, or

        – evaluating the transfer function first, then interpolating the colors.

3. **Scan conversion procedure** – this specifies whether polygon or volume scan conversion is used to render the polygonal decompositions of volume cells. In polygon scan conversion, a software Gouraud shading is used to render polygons.

4. **Sampling strategy** – this distinguishes whether samples are taken regularly or only at cell face intersections, and how they should be distributed throughout the entire volume data.

Based on this four level classification scheme, we can identify several DVR algorithms that can be simulated as projection-based algorithms. This classification is not meant to be exhaustive but simply illustrates how different DVR algorithms can be viewed in terms of their variants. For complete specifications of DVR algorithms, more detailed rendering parameters within each criterion should be specified. Used in this manner, Table 1 shows how some algorithms are distinguished by their data model, interpolation value, scan conversion procedure, and sampling pattern.

Data Model	Interpolation Values	Scan conversion	Sampling Pattern	DVR Algorithm
Cell model with Tri-linear Interpolation	data or color	volume	regular	ray casting
		volume	cell face	ray casting
	color	volume	regular	volume texture
		volume	cell face	shear-warp
	color	polygon	cell face	coherent projection
Voxel model with distance function	color	polygon	irregular (depends on distance function)	splatting

Table 1. Illustration of how different DVR algorithms can be expressed in terms of projection-based algorithm using different combinations of the data model, value being interpolated, scan conversion type, and sampling pattern.

The data model in column (1) comes with either an interpolation function or distance function. When the data model assumes values are defined at vertices (*cell model*), an interpolation function is often used. In such situations, tri-linear interpolation seems to be the interpolation method of choice in many DVR implementations. However, other possibilities include higher order interpolations or adaptive reconstruction [7]. When the data model assumes values are defined at voxel centers (*voxel model*), a distance function is often used to model how data vary within the confines of the voxel. Other distance functions, as well as simpler voxel modeling using nearest neighbors, can also be incorporated.

The simulations of various DVR algorithms that we just described have varying degrees of accuracy. More precise specifications need to be made if an exact simulation is desired. For example, it is important to note that different methods of polygonalization may lead to different looking images (see Figure 3, color plate 1). Thus, the complete polygonalization policy must also be specified if a faithful simulation is desired. Another thing to note is that projection-based algorithms often rely on hardware scan conversion of the polygons. Therefore, it is also possible to notice some differences, especially along the boundaries of projected polygons.

We verified our approach and implementation with a renderer called *mdh* [17, 18] which has multiple choices of algorithms. We made the rendering parameters of our simulation as identical as possible to those for the algorithms in *mdh* and made sure that the differences in the intermediate rendering information are within a given tolerance. That is, we took differences in colors, locations and data values for each front and back vertices of the projected polygons of all volume cells between our simulations and the projection algorithm of *mdh* and made sure that the differences are negligible (less than 10^{-7} in scale of 0 to 1.0 for each color channel).

5 Metrics for Projection-based DVR Comparisons

In this section, we present a set of data level metrics derived from our projection algorithm basis and proposed for comparing DVR algorithm simulations. These metrics should reveal information about the volume data (or color intensities) as well as the behavior of DVR algorithms. The idea is to identify differences in algorithms that may not be revealed from image level metrics alone. Note that there are numerous other useful intermediate information (e.g. gradient and normal calculations, and surface classification) that can also be collected and used as metrics.

– **Threshold-based Metrics**
  The following metrics are obtained to examine the behavior of a DVR algorithm for a given threshold condition. Each metric is measured at each pixel when the accumulated (or sample) color components reaches the threshold condition. While opacities are often used for specifying threshold levels, other color components can also be used. In our current implementation, a user can give a threshold condition that combines color and opacity threshold values.
  1. **Number of cells**
    Different algorithms use different rendering parameters and thus each algorithm may require a different *number of cells* to satisfy the given threshold condition.
  2. **Distances**
    It is useful to measure the distance traveled into the volume before reaching a specified threshold condition at each pixel. Distance can be measured from the user's eye position (*eye distance*) or from the bounding box of the data volume in the viewing direction (*volume distance*).
– **Contribution Metrics**
  The following metrics measure the contribution of each cell to the final rendered image. The user specifies a pixel in the image to probe, then metrics are measured and visualized to show which cells contributed to the selected pixel and by how much. These metrics can be used with or without specifying an opacity or color

threshold condition. Contribution metrics for each cell may either be *absolute* or *additive*. Absolute contribution is the amount contributed by a data cell to a pixel as if there is nothing between the cell and the image plane. Additive contribution is the actual amount of contribution by a data cell to a pixel because its absolute contribution is attenuated by the accumulated opacity so far. Looking at the front-to-back composition equation,

$$C_{acc,new} = C_{acc,old} + (1 - \alpha_{acc,old})C_{cell,i} \qquad (3)$$

where $C_{acc,new}$ is the new color intensity after composition, $C_{acc,old}$ is the accumulated color before composition with the ith cell, $C_{cell,i}$ is the color contribution by the ith cell, and $\alpha_{acc,old}$ is the opacity component of $C_{acc,old}$. The *absolute* contribution metric uses the term $C_{cell,i}$. On the other hand, the *additive* contribution metric uses the term $(1 - \alpha_{acc,old})C_{cell,i}$.

1. **Pixel Probe**
   This measures the amount of color intensity contributed by each cell to the pixel being probed.
2. **Cell Probe**
   This is similar to the pixel probe but shows other information (e.g. averages, minimum, maximum, and standard deviation) by the contributing cells to the target pixel. These statistics are collected based on how each data cell contributed to the pixels in the final image. For example, when the contribution from each cell is distributed unevenly across several pixels, a measure of spread can be calculated for that cell.

– **Data Probe**
  Similar to cell probe, except the user selects a particular data cell and is shown the contribution made by that cell on the different pixels of the DVR image. Note that this is different from the projection filter.
– **Difference Metrics**
  While *threshold metrics* and *contribution metrics* probe how each algorithm individually behaves, differences of these metrics can show where and how two or more algorithms differ. For example, in addition to the difference and statistical measures (e.g. average, minimum, maximum, and standard deviations) between two algorithms, differences of the *cell probe* metric includes the correlation of color intensities generated by the two algorithms for the pixels that are covered by the selected cell.

6 Examples

In this section, we show some examples of applying our comparative visualization methods and discuss the applicability of our metrics. Figure 4 (color plate 1) shows volume rendered images generated by two different algorithms. Both algorithms render a 64^3 Hipip (High Potential Iron Protein) volume data. The image (a) is generated by ray casting (simulation with our base algorithm) with sampling and reconstruction of data at the cell faces. The image (b) shows an image generated using a polygon projection algorithm. The image in (c) shows the absolute differences between image (a) and image (b). Color intensities in the difference image (c) are magnified to show the difference clearly (each color intensity is multiplied by 10). This image based comparison method can show location and magnitudes of differences but not much more.

Figure 5 (color plate 2) demonstrates how our data level metrics can provide more insight. Image (a) shows the colormapped visualization of the *number of cells* needed to reach an opacity level of 0.11 using the ray casting algorithm that generated image (a) of Figure 4. It shows that the regions around the red and blue molecules require more number of cells to be examined before reaching the given opacity. Image (b) shows the number of cells needed to reach an opacity level of 0.2 using the polygon projection algorithm that produced image (b) of Figure 4. Image (c) shows differences in the number of cells to reach opacity 0.15 between the two algorithms. It shows a higher difference near the boundary regions of the red and blue molecules. Aside from the opacity threshold, users can also try threshold conditions composed of other color channels. Users can confirm that one source for the differences in Figure 4 is due to the different number of cells used by each algorithm. The users can further compare algorithms using metrics such as *pixel probe* for specific pixels of interests. Images (d) and (e) of Figure 5 show absolute and additive color contributions from all the cells contributing to a specific pixel. In particular, the data cells that contributed to the red component of the selected pixel are highlighted. The absolute contributions in (d) show that the red intensities composited were prominently higher in a small region in the volume data, but the additive contributions in (e) are more widely spread. The yellow arrows show the viewing directions and the front-to-back traversal of the algorithm.

Figure 6 (color plate 2) illustrates a hypothetical case study using our metrics. The two DVR images are generated using our scanline simulation of a ray-based DVR method, but with different transfer functions. The volume data is from a CAT scan of a human head. In column (a), the location of a hypothetical tumor in the brain is identified to be within the box region. The upper image shows the tumor but the lower image does not show it. The visualizations in column (b) show the *volume distance* metric associated with each image. The opacity threshold is set to 0.31 in both cases. Black color is assigned to pixels where the opacity does not reach the opacity threshold. In the lower image of column (b), near the region of interest, an almost flat, blue wall indicates that the pixels accumulated enough opacity without traveling through many layers of data cells. On the other hand, the upper image of column (b) shows an almost uniformly black region where the blue wall used to be. With the exception of the tumor, the region around it did not produced sufficient opacity. This tool is especially useful if the person understands how the DVR algorithms work and how they are affected by the different rendering parameters. However, other metrics can also help users to understand and verify the rendering methods. Columns (c) and (d) show data level analyses using contribution-based metrics. A pixel of the tumor is first selected to be examined. In column (c), the absolute amount of opacity contributions are visualized for all sub-volume cells that contribute to the given pixel. In column (d), the additive opacity values are shown. The visualizations show that the transfer function for the lower row produced opacity values that are too high to reach the tumor. On the other hand, on the top row, both absolute and additive contributions are highest where the hypothetical tumor is located. From the comparisons of the final rendered images alone, it may not be obvious whether the opacity for brain tissue is set too high or the data range for the tumor tissues is not set properly to be detected by the transfer function. This case demonstrates that our comparative visualization techniques can provide more insight into why two rendered images are different and how rendering parameters (such as transfer function) affect the resulting images.

7 Conclusion

We presented a new data level framework that solves difficulties of comparing different DVR algorithms and demonstrated how different DVR algorithms can be simulated using the projection algorithm basis. From this base or reference algorithm, several new data level comparison metrics were then presented that highlight different aspects of the volume data and the DVR algorithms. These metrics, used individually and in combination, provide additional information beyond how two different DVR images are different – they seek to provide clues as to *why* and *how* the two images may be different. We also gave examples of using our metrics and a hypothetical case study that demonstrates the applicabilities of the metrics. Our new methodology overcomes the limitations of conventional image level comparisons and helps us to perform more in-depth comparisons which are closely related to the purpose of a particular visualization application as well as general rendering quality comparisons. These results are important to scientists to help them interpret different visualization results objectively.

Acknowledgements

We would like to thank Craig Wittenbrink and Suresh Lodha for collaborative work on uncertainty visualization and Sam Uselton for comments and feedback on image level comparisons. We would also like to thank Chang Sung Jeong and Carol Mullane for help with proofreading this paper. This projected is partially supported by NSF grant IRI-9423881, DARPA grant N66001-97-8900, NASA NCC2-5281, ONR grant N00014-96-1-0949, and LLNL/DOE grant B347879.

References

1. Ajeetkumar Gaddipatti, Raghu Machiraju, and Roni Yagel. Steering image generation with wavelet based perceptual metric. In *Computer Graphics Forum*. Eurographics, September 1997.
2. A. Globus and S. Uselton. Evaluation of visualization software. *Computer Graphics*, pages 41–44, May 1995.
3. Kwansik Kim and Alex Pang. Ray-based data level comparison of direct volume rendering algorithms. Technical Report UCSC-CRL-97-15, UCSC Computer Science Department, 1997.
4. Kwansik Kim and Alex Pang. Ray-based data level comparative visualization of direct volume rendering algorithms. In *to appear in Scientific Visualization, Dagstuhl Workshop Proceedings*. Springer, 1998.

5. David Laur and Pat Hanrahan. Hierarchical splatting: A progressive refinement algorithm for volume rendering. In *Computer Graphics (ACM Siggraph Proceedings)*, volume 25, pages 285–288, July 1991.

6. Marc Levoy. Efficient ray tracing of volume data. *ACM Transactions on Graphics*, 9(3):245–261, July 1990.

7. R. Machiraju and R. Yagel. Reconstruction error characterization and control: A sampling theory approach. *IEEE Transactions on Visualization and Computer Graphics*, pages 364–378, December 1996.

8. Tom Malzbender. Fourier volume rendering. *ACM Transactions on Graphics*, 12(3):233–250, July 1993.

9. S. R. Marschner and R. J. Lobb. An evaluation of reconstruction filters for volume rendering. In *Proceedings of 1994 Symposium on Volume Visualization*, pages 100–107. ACM, October 1994.

10. Nelson Max. Optical models for direct volume rendering. *IEEE Transactions on Visualization and Computer Graphics*, 1(2):99–108, June 1995.

11. Torsten Moller, Raghu Machiraju, Klaus Mueller, and Roni Yagel. Classification and local error estimation of interpolation and derivative filters for volume rendering. In *Symposium on Volume Visualization*, pages 71–78, San Francisco, CA, October 1996. ACM/IEEE.

12. Hans-Georg Pagendarm and Frits H. Post. Comparative visualization – approaches and examples. In M. Gobel, H. Muller, and B. Urban, editors, *Visualization in Scientific Computing*, pages 95–108. Springer-Verlag, 1995.

13. P. Sabella. A rendering algorithm for visualizing 3D scalar fields. In *Computer Graphics*, pages 51–58, August 1988.

14. P. Shirley and A. Tuchman. A polygonal approximation to direct scalar volume rendering. In *1990 Workshop on Volume Visualization*, pages 63–70, San Diego, CA, December 1990.

15. Allen Van Gelder and Kwansik Kim. Direct volume rendering with shading via 3D textures. In *ACM/IEEE Symposium on Volume Visualization*, pages 22–30, San Francisco, CA, October 1996.

16. L. Westover. Footprint evaluation for volume rendering. In *Computer Graphics*, pages 367–376, August 1990.

17. Jane Wilhelms and Allen Van Gelder. A coherent projection approach for direct volume rendering. In *Proceedings of SIGGRAPH 91*, pages 275–284, 1991.

18. Jane Wilhelms and Allen Van Gelder. Multi-dimensional trees for controlled volume rendering and compression. In *Proceedings of the Symposium on Volume Visualization*, pages 27–34, color plate 125, Washington, D.C., 1994.

19. Jane Wilhelms, Allen Van Gelder, Paul Tarantino, and Jonathan Gibbs. Hierarchical and parallelizable direct volume rendering for irregular and multiple grids. In *Proceedings of IEEE Visualization '96*, pages 57–64, 1996.

20. Peter L. Williams, Nelson L. Max, and Clifford M. Stein. A high accuracy volume renderer for unstructured data. Technical Report UCRL-JC-126942, Lawrence Livermore National Laboratories, September 1997.

21. Peter L. Williams and Samuel P. Uselton. Foundations for measuring volume rendering quality. Technical Report NAS-96-021, NASA Numerical Aerospace Simulation, 1996.

22. Craig M. Wittenbrink, Alex T. Pang, and Suresh Lodha. Verity visualization: Visual mappings. Technical Report UCSC-CRL-95-48, Univ. of Cal. Santa Cruz, 1995.

Editors' Note: see Appendix, p. 319 for colored figure of this paper

Parallel Multipipe Rendering for Very Large Isosurface Visualization

Tushar Udeshi and Charles D. Hansen*

Department of Computer Science
University of Utah
50 S. Campus Center Drive Rm. 3190
Salt Lake City, Utah
84112-9205
{tudeshi, hansen}@cs.utah.edu

Abstract. In exploratory scientific visualization, isosurfaces are typically created with an explicit polygonal representation for the surface using a technique such as *Marching Cubes*. For even moderate data sets, Marching Cubes can generate an extraordinary number of polygons, which take time to construct and to render. To address the rendering bottleneck, we have developed a multipipe strategy for parallel rendering using a combination of CPUs and parallel graphics adaptors. The multipipe system uses multiple graphics adapters in parallel, the so called *SGI Onyx2 Reality Monster*. In this paper, we discuss the issues of using the multiple pipes in a Sort-Last fashion which out performs a single graphics adaptor for a surprisingly low number of polygons.

1 Introduction

Many applications generate scalar fields $\rho(x, y, z)$ which can be viewed by displaying *isosurfaces* where $\rho(x, y, z) = \rho_{iso}$. Ideally, the value for ρ_{iso} is interactively controlled by the user. When the scalar field is stored as a structured set of point samples, the most common technique for generating a given isosurface is to create an explicit polygonal representation for the surface using a technique such as *Marching Cubes*[8]. This surface is subsequently rendered with an attached graphics hardware accelerator, such as the SGI Infinite Reality. For even moderate data sets, Marching Cubes can generate an extraordinary number of polygons, which take time to construct and to render. For very large (i.e., greater than several million polygons) surfaces the isosurface extraction and rendering times limit the interactivity. One approach to address this issue is to exploit parallelism.

The use of parallelism in computer graphics hardware is widely known. Most current generation graphics adaptors utilize parallelism in their design and implementation[1, 2, 9]. While these systems are extremely proficient at rendering geometry with superb rendering capabilities, such as texture mapping, the bottle neck for rendering large polygon sets is the speed at which polygons can be

---

* Contact Person for this paper

sent through the graphics pipeline[1]. Since there is a single thread which can send polygons to the graphics adaptor, the majority of rendering applications are serial and implicitly exploit the parallelism inherent in the graphics adaptor.

For scientific visualization of very large data sets, 512^3 and higher, parallel isosurface extraction and rendering techniques have been studied[3, 6, 7, 11]. These techniques exploit the large memory and parallelism of massively parallel computers to deal with the data explosion caused by scientific visualization of the large simulations running on the same machines. These techniques mimic, in software, the parallelism in the graphics hardware and achieve speedup although not at interactive rates. One of the problems with these approachs is the lack of an attached framebuffer or graphics adaptor for displaying and interacting with the image.

Clearly what is desired is a combination of these approachs which takes advantage of the large memory and parallelism provided by large-scale parallel computers and the interactive rendering capabilities provided by graphics hardware. Fortunately, we have recently seen the convergence of these two with the SGI ONYX2 which is an SGI Origin 2000 with attached InfiniteReality graphics adaptors[10]. SGI ONYX2 with multiple InfiniteReality graphics adaptors are called Reality Monsters. While these promise acceleration based on parallelism on both the macro scale (multiple graphic adaptors) and the micro scale (internal to each graphics adaptor), these systems are new and methods for exploiting them have not been studied. This paper addresses an approach to exploiting the multiple graphics adaptors for polygon rendering.

2   Parallel Graphics Hardware Approach

One approach to the classification of parallel rendering algorithms is to categorize based upon whether the parallelism is achieved in image-space or in object-space. However, many recent algorithms obtain performance by utilizing both image-space and object-space parallelism. A more useful taxonomy for parallel rendering which classifies rendering methods based on where data are sorted from object-space to image-space was presented by Molnar et al.[5]. A typical rendering process performs some *geometric* processing followed by some *rasterization* processing. Parallelization can take place during the geometric processing, during the rasterization processing, as well as pipelined parallelism between the two stages. At some point, primitives are sorted from object-space to image-space. Looking at where this sort takes place provides a useful method for classifying parallel rendering techniques. The sort to screen-space can take place before the geometric processing, after the geometric processing but before the rasterization, or after both the geometric processing and the rasterization. These methods are referred to as Sort-First, Sort-Middle, or Sort-Last.

The Sort-First approach exploits some *a priori* knowledge about which part of the screen the primitives will fall. This is utilized to send the primitives, pos-

---

[1] For 3D texture mapping based volume rendering, the bottleneck is typically pixel fill-rate. We limit our application to explicit polygon rendering.

sibly polygons, to the appropriate processor elements. Frame-to-frame coherence provides such knowledge. The screen is subdivided in some manner, typically interleaved, and each region is assigned a processing element which performs both geometry and rasterization without any need for communication. Sort-First suffers from load inbalance for both the geometry processing stage as well as the rasterization stage if primitives are not evenly distributed across the screen partitions.

The Sort-Middle approach needs no *a priori* knowledge and primitives, typically polygons, are distributed in some fair scheme between all the processing elements. Each processing element performs the geometric operations on its portion of the data. Following this stage, the transformed primitives are sent to the processor element responsible for the portion of the screen into which these primitives fall. Sort-Middle is typically well balanced during the geometry processing stage but suffers from load inbalance during the rasterization stage if primitives are not evenly distributed across the screen partitions. Additionally, there can be a communication bottleneck if all geometry processors are sending data to a single rasterization processor. The SGI Infinite Reality graphics adapter uses Sort-Middle parallelism internally in its graphics pipeline. Using current technology, Sort-Middle parallelism cannot be exploited with multiple pipes since each pipeline is totally independent from the rest.

Sort-Last rendering is sometimes referred to as an image compositing system. Primitives are distributed in some fair scheme among the processing elements. Each processing element performs both geometric processing as well as rasterization independent of all other processor elements. A local image is rendered on each processor element and the images are composited together to form a final image. In some systems, only active pixels from each subimage participate in the compositing phase. Sort-Last behaves particularly well with respect to load balancing since all primitives are fairly distributed at the beginning. However, the communication load for the image compositing phase can be quite severe and requires very high speed networks. In addition, transparency is non-trivial for Sort-Last systems. Still, the scalability for Sort-Last rendering makes it an attractive alternative for very large polygonal data sets such are those generated from large scientific data.

We have implemented a technique for exploiting the parallel (up to eight) Infinite Reality graphics adapters. One can use a Sort-First style approach by dividing the image and assigning each graphics adapter a subimage for rendering. This approach uses either preculling of visible polygons (software based frustum culling) or uses the hardware based visibility culling in the graphics adapters themselves. The problem with software based visibility culling is the dependency on spatial hierarchies for the underlying surface. For exploratory scientific visualization where different isosurfaces are generated often and at interactive rates, such pre-processing for each resulting isosurface would be time prohibitive. The difficulty with using the hardware based visibility culling is that each graphics adapter needs to process the entire set of polygons for its subimage. Using a Sort-Last rendering scheme provides a different approach. In multipipe Sort-Last rendering, each graphics adapter renders only a portion of the polygons and the resulting partial images are combined, using depth comparison, to produce the final result. This is the approach we have chosen to use since it provides better interactivity without the burden of preprocessing the isosurface once it is created.

The basic idea behind our algorithm is to divide the renderable data among the available graphics adapters, render each subset separately and *locally*, and combine the resulting partial images in an incremental fashion. This technique is strongly related to composition based volume renderers in that each graphics adapter renders a portion of the final image and these are combined[4]. It is similar to the composition network approach of Pixel Flow[9].

2.1 Binary Swap

Assume n is the total number of polygons representing an isosurface and p is the number of graphics adapters. Typically p is a power of two although this can be relaxed through simple extensions. We assume the isosurface, represented as n polygons, exists in shared memory. Each graphics adapter loads and locally renders n/p polygons. At this point, each graphics adapter has a partially complete image. These images are then composited onto the graphics adapter used for the final display. We use the *binary-swap*[4] method for image composition which composites the image in $log_2(p)$ steps. At each step, the graphics pipes send the top half of their active image to their partner and receive the bottom half from their partner. The partners are determined as being 2^i away where i is the composting step. Thus when utilizing 8 graphics adapters, for the first composite the partners are $[(0,1), (2,3), (4,5), (6,7)]$ and for the second composite the partners are $[(0,2), (1,3), (4,6), (5,7)]$.

Psudo-code of the algorithm follows. Assume p = number of pipes(numbered $0,1,...., p-1$), q = the current pipe, H = height of image. For simplicity p is assumed to be a power of 2.

```
procedure binaryswap
begin
      let h = 0
      let levels = log₂(p)
      let height = H
      for l = 0,1... (levels-1)
      begin
            let factor = 2ˡ
            let A be a band of the image from h to (h + height)
            if (lᵗʰ bit of q is 1) /* odd pipe */
                  Send lower half of A to pipe (q - factor)
                  Recieve and composite an image from pipe (q - factor) onto upper half of A
                  let h = h + height/2
            else /* even pipe */
                  Send upper half of A to pipe (q + factor)
                  Recieve and composite an image from pipe (q + factor) onto lower half of A
            endif
            let height = height/2
            synchronize all pipes /* barrier */
      end
      send A to display pipe
end
```

Composition of the incoming image with the current image is done in hardware making use of the stencil buffer. This is achieved by the following sequence of steps:

1. Clear the stencil buffer to all zeros
2. Set the stencil operation to replace value to 1 if depth test passes and keep current value otherwise.
3. Draw depth buffer values. This will set the stencil buffer
4. Set stencil test to pass if stecil value = 1 and enable stencilling.
5. Draw the color buffer values.

This has the effect of maximizing hardware usage and scales better with image size than performing the composite with the CPUs. The active image is reduced by half at each step. At the end of $log_2(p)$ levels, the active image on each graphics adapter is composited into the display graphics adapter. Figure 1 shows the compositing steps from bottom to top. The bottom row shows the results of rendering the initial polygon distribution on each graphics pipe. The gray areas in the images are back facing polygons. The three compositing levels are the next three rows up. The compositing partners are shown with lines between the compositing levels.

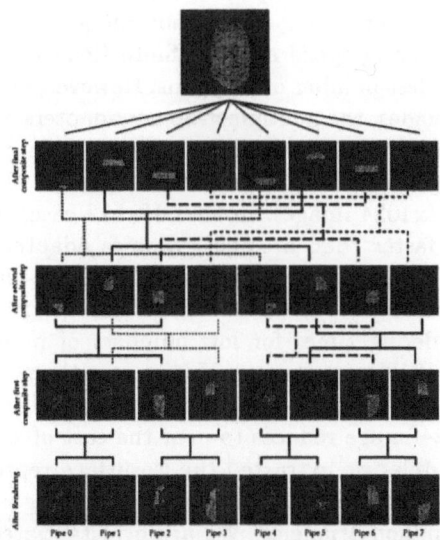

Fig. 1. The compositing levels are along the verticle axis and the graphics adapters are along the horizontal axis. After the final step (the top most level), the final image is composed from each of the partial images. (See color plate)

3   Experiments and Results

We have tested our technique with a variety of large polygon data sets. In this section, we present the results. In all tests, we compare our technique with the best (non-compositing) single pipe version. The rendering code is all written in OpenGL.

For the first example, we wanted to stress the multipipe rendering system to understand where the tradeoffs were for this technique. Recall, our goal is to render extremely large polygon data sets such as those one would see generated from 1GByte data sets. If we lower the number of polygons, we would expect the overhead from the multiple graphics adapters to limit the speed up. In fact, for sufficiently small polygon sets, we would expect the single graphics adapter to out-perform multiple graphics adapters. To test this, we take a 131,000 triangle isosurface generated from a CT scan. To test the scaling of polygon count, we instantiated multiple copies of this data, from 2 up to 8 copies. This provided a series of polygon data sets which varied from 131K to 1M polygons. We rendered these with 1 to 8 graphics adapters. To mitigate instantaneous timing anomalies, we rendered each frame 100 times. The results are shown in Figure 2 and Figure 3 for 512 x 512 and 1024 x 1024 images respectively. The X-axis is the size, in triangles, of the isosurface being rendered. The Y-axis is the time in seconds for rendering 100 images. As is shown in the plots, for the 1024x1024 image the single graphics pipe out performs the multple pipes for the isosurface containing 131,000 triangles. This is to be expected since the overhead of reading back the frame buffers and compositing in the multiple graphics adapters case adds overhead and the rendering speed of the Infinite Reality graphics adapters can easily handle that modest number of polygons. However, once the polygon count increases to 262K triangles, the multiple graphics adapters out perform the single graphics pipe for all cases. The improvement for multiple graphics pipes steadly increases as the polygon count increases. One can notice the difference between the 512x512 and 1024x1024 images. In the 512x512 case, the multiple graphics adapters are always faster than a single graphics adapter, even on this small number of polygons! The overhead of reading back and re-writing to the graphics adapters (the compositing step) for the larger images in the 1024x1024 case results in slower rendering times for low numbers of polygons. The overhead predominates for the 131K triangle case. However, as the polygon count increases to 1M triangles, the overhead becomes less of the overall time and the cost for rendering a 1024x1024 image reduces to near the cost of a 512x512 image.

For the next example, we extracted the isosurface representing the skin for the head portion of the visible woman data set (see Figure 1). The isosurface is composed of 1.4 million triangles. We again instantiated multiple copies of this data with 2, 3, and 4 copies resulting in 2.8M, 4.2M, and 5.6M polygon isosurface data sets. Figure 4 and Figure 5 show the results for both 512x512 and 1024x1024 images. Notice that the rendering times are very close with the 512x512 image being slightly faster for rendering with larger numbers of graphics pipes. This is due to the increased number of steps in the compositing operations with an increased number of graphics pipes. However, with even a modest 1.4M

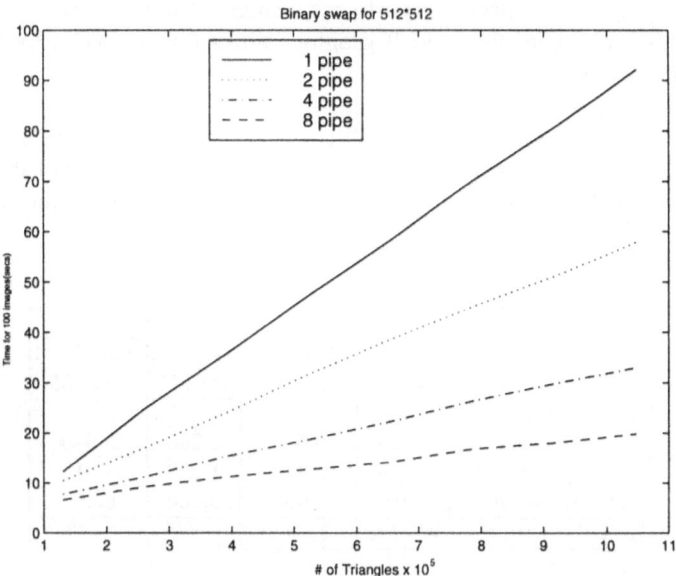

Fig. 2. Times for rendering a 512 x 512 image

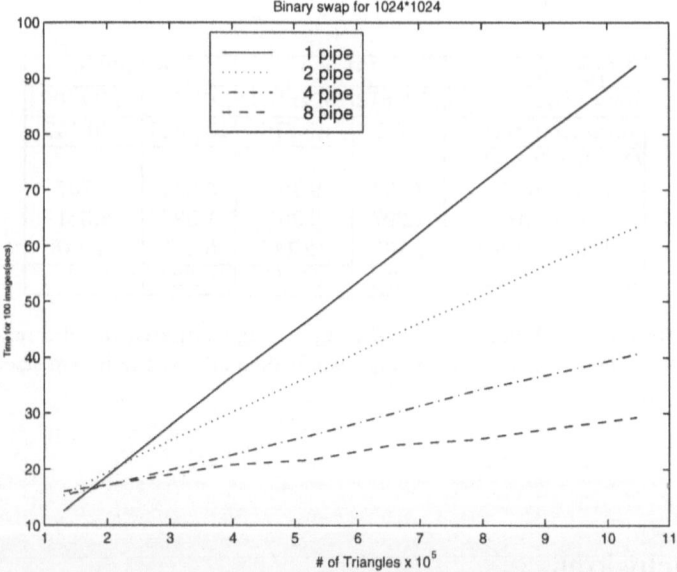

Fig. 3. Times for rendering a 1024 x 1024 image

polygons, 8 graphics adapters outperforms a single graphics adapter by a factor of 2. For 5.6M polygon data set, 8 graphics pipes outperforms, by 6 times, a single graphics pipe.

Tables 1 and 2 show the overhead caused due to the binary swap communication and compositing steps. For a small polygon count of 131,000, the overhead is fairly high. However for a polygon count of 5.6 M, these steps take less than 10% of the time thus resulting in huge improvements in time compared to the single pipe case.

Polygon count	131,000		5.6 Million	
Image size	512x512	1024x1024	512x512	1024x1024
Total time(secs)	10.4237	15.305	241.317	246.005
Overhead(secs)				
communication	1.248	5.640	1.248	5.640
composite	1.238	2.812	1.238	2.812
total overhead	2.486	8.452	2.486	8.452
% of total time	23.85%	55.2%	1.03%	3.04%

Table 1. Overhead for binary-swap and compositing compared to total rendering time for two pipes. The overhead is constant since it depends on the image size, not on the number of polygons.

Polygon count	131,000		5.6 Million	
Image size	512x512	1024x1024	512x512	1024x1024
Total time(secs)	7.709	15.276	136.846	140.756
Overhead(secs)				
communication	3.704	9.767	3.704	9.767
composite	1.397	2.381	1.397	2.381
total overhead	5.101	12.148	5.101	12.148
% of total time	66.2%	79.5%	3.72%	8.6%

Table 2. Overhead for binary-swap and compositing compared to total rendering time for four pipes. The overhead is constant since it depends on the image size, not on the number of polygons.

4   Conclusions

We have designed and implemented a multipipe parallel rendering system which takes advantage of multiple attached graphics adapters on high-end systems. For large polygon data sets, this method proves more interactive than utilizing

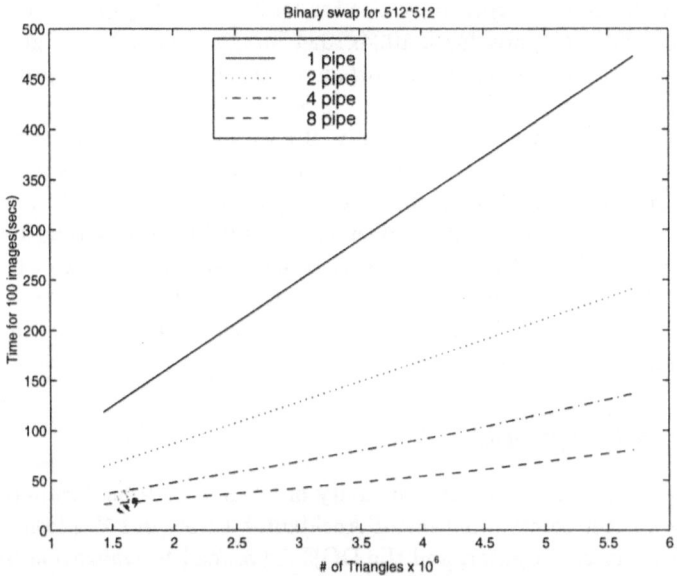

Fig. 4. Times for rendering a 512 x 512 image for 1.4 to 5.6 million polygons

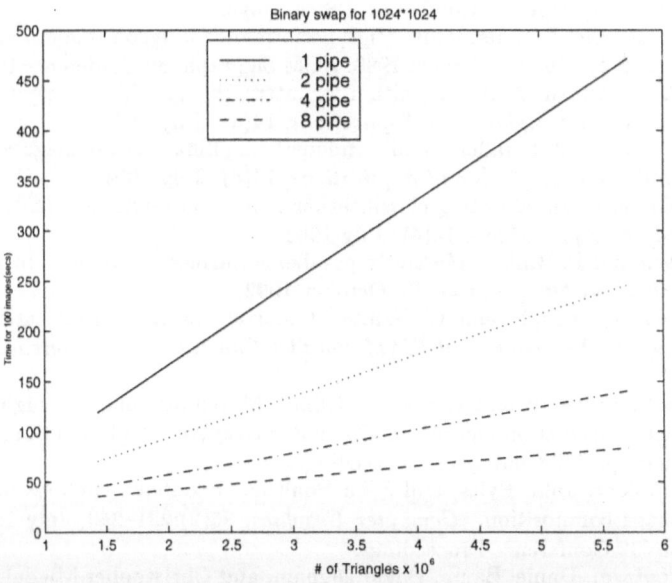

Fig. 5. Times for rendering a 1024 x 1024 image for 1.4 to 5.6 million polygons

a single graphics pipe. Surprisingly, the tradeoff for multi-pipe rendering occurs at less than 200k polygons for a 1024x1024 image. Given the large number of polygons extracted using typical isosurface methods, this clearly provides acceleration through the parallel rendering. We recognize that the cost is significantly higher yet many sites are taking delivery of such systems and we anticipate the usefulness of such parallel techniques particularly as data set sizes scale up. Since the cost of reading back the entire rendering window is fixed, based upon the Infinite Reality hardware, an enchancement would be to only read back the region of the image defined by the projection of the bounding box of the polygon set. Also, if it were possible to copy directly from graphics adapter to graphics adapter without caching the active images in shared memory, we would expect to see an increase in performance.

5   Acknowledgments

This work was supported by the University of Utah/SGI Visual Supercomputing Center, the Center for Simulation of Accidental Fires and Explosions, a DOE ASCI Level 1 Alliance Center, and the DOE Advanced Visualization Technology Center, a partnership between Univ of Utah, ANL, and LANL.

References

1. Kurt Akeley. RealityEngine graphics. *Computer Graphics*, 27:109–116, August 1993. ACM Siggraph '93 Conference Proceedings.
2. Kurt Akeley and Tom Jermoluk. High-performance polygon rendering. *Computer Graphics*, 22(4):239–246, August 1988. ACM Siggraph '88 Conference Proceedings.
3. David A. Ellsworth. A new algorithm for interactive graphics on multicomputers. *IEEE Computer Graphics and Applications*, 14(4), July 1994.
4. Kwan-Lui Ma et al. Parallel volume renderer using binary-swap image composition. *IEEE Computer Graphics and Applications*, 14(4), July 1994.
5. Steve Molnar et al. A sorting classification of parallel rendering. *IEEE Computer Graphics and Applications*, 14(4), July 1994.
6. C. Hansen and P. Hinker. Massively parallel isosurface extraction. In *Proceedings of Visualization '92*, pages 77–83, October 1992.
7. C. Hansen, M. Krogh, and W. White. Massively parallel visualization: Parallel rendering. In *Proceedings of SIAM Parallel Computation Conference*, February 1995.
8. William E. Lorensen and Harvey E. Cline. Marching cubes: A high resolution 3d surface construction algorithm. *Computer Graphics*, 21(4):163–169, July 1987. ACM Siggraph '87 Conference Proceedings.
9. Steven Molnar, John Eyles, and John Poulton. Pixelflow: High-speed rendering using image composition. *Computer Graphics*, 26(2):231–240, July 1992. ACM Siggraph '92 Conference Proceedings.
10. John Montrym, Daniel Baum, David Dignam, and Christopher Migdal. Infinitereality: A real-time graphics system. *Computer Graphics*, 26:293–302, August 1997. ACM Siggraph '97 Conference Proceedings.
11. F. Ortega, C. Hansen, and J. Ahrens. Fast data parallel polygon rendering. In *Proceedings of Supercomputing '93*, pages 709–718, November 1993.

Editors' Note: see Appendix, p. 321 for colored figure of this paper

Interactive Direct Volume Rendering of Time-Varying Data

John Clyne and John M. Dennis

Scientific Computing Division
National Center for Atmospheric Research

Abstract. Previous efforts aimed at improving direct volume rendering performance have focused largely on time-invariant, 3D data. Little work has been done in the area of interactive direct volume rendering of time-varying data, such as is commonly found in Computational Fluid Dynamics (CFD) simulations. Until recently, the additional costs imposed by time-varying data have made consideration of interactive direct volume rendering impractical. We present a volume rendering system based on a parallel implementation of the Shear-Warp Factorization algorithm that is capable of rendering time-varying 128^3 data at interactive speeds.

1 Introduction

In recent years Direct Volume Rendering (DVR) has proven to be a powerful tool for the analysis of time-varying, three-dimensional CFD datasets associated with the aerospace, atmospheric, oceanic and astrophysical sciences. The benefits of DVR, probabilistic data classification [6] and direct projection of volume samples, are key to producing insightful visualizations of simulation results, rich with amorphous and fluid-like features [5].

Visual data exploration is an inherently interactive process. In order to fully exploit the power of DVR as an analysis tool, interactive frame rates must be achieved. Static images or animations produced through a batch process, representing complex 3D phenomena, may be of limited value. Invariably, features of interest are obscured, improperly classified, or poorly lit. Much work has been done in the area of accelerating DVR methods for static data [8, 11, 18, 12, 15, 2, 13]. Interactive volume rendering of single time-steps is now within the reach of many researchers. However, temporal animations are typically produced through a batch process. Time-varying datasets impose additional costs that have made interactive rendering difficult to achieve. The principal culprit is I/O. Rendering even a moderately sized (256^3) dataset made up of 8-bit voxels at a frame rate of 10 Hz requires an input pipe capable of delivering a sustained bandwidth of over 150 MB/sec.

In this paper, we detail our efforts at making interactive DVR of time-varying data possible. We present a volume rendering system based on a parallel implementation of Lacroute's Shear-Warp Factorization algorithm [9]. The remainder of this paper is organized as follows: In Section 2 we discuss research efforts related to our own. In Section 3 we present an overview of the serial and parallel implementations of the Shear-Warp algorithms, and we discuss steps we have taken to support time-varying data. Our implementation is presented in Section 4. In Section 5 we address performance. In Section 6 we draw conclusions.

2 Related Work

There have been numerous successful efforts in the area of performing real-time DVR of moderately-sized (256^3) static data sets [3, 7, 2, 8, 13, 15]. The approaches taken can be classified into hardware and software based methods. Hardware methods include the development of custom hardware, dedicated to performing volume rendering [7, 13], and the clever exploitation of conventional polygon engines in order to accelerate the volume rendering task [3, 17]. Interactive volume rendering can be achieved via software methods by mapping serial algorithms onto general-purpose parallel computers. The increasingly wide-scale availability of commercial multiprocessors as general compute platforms has made this strategy popular and practical as evidenced by the abundance of research published on this subject [11, 18, 2, 8, 15].

More closely related to our efforts of visualizing time-varying data is the work of Shen and Johnson [16], and Chiueh and Ma [4]. Shen and Johnson address the I/O problem by applying data compression. A preprocessing step is performed on the volume data to create a *differential file*. The first time-step in a series serves as a basis function and is recorded in its entirety to the *differential file*. For each subsequent time-step, only the differential information between the time-step being processed and the preceding time-step is recorded. Shen and Johnson report on a second optimization as well: they adapt their volume renderer to only update those pixels that are affected by voxels that have changed between the preceding and current time-step.

More recently, Chiueh and Ma take another approach, pipelining the multi-timestep rendering process and exploiting a combination of *intra-volume* and *inter-volume* parallelism [4]. Conventional parallel volume renderers operating on static data dedicate multiple processors to render a single data volume. Chiueh and Ma term this *intra-volume* parallelism. Alternatively, *inter-volume* parallelism is employed by simultaneously rendering multiple data volumes on multiple processors. In the extreme case, the number of time-steps rendered in parallel equals the number of processors available.

Though both of these techniques are capable of greatly accelerating overall rendering rates, their utility as interactive tools may be restricted. Chiueh's and Ma's approach suffers from the pipeline effect: changes in viewing parameters invalidate the pipeline, requiring it to be drained and restarted. These restarts, which result in inter-frame delays, may be acceptable for short pipelines. For longer pipelines, this latency may be great enough to significantly impact interactivity. The differential volume renderer is also impacted by viewing parameter changes, such as changes in viewpoints or classification, which force the entire image to be updated. Additionally, differential volumes do not readily support random access of time-steps.

Our goal is to produce a portable, high-frame-rate, low-latency rendering system that is unencumbered by these restrictions and is thus better suited for interactivity. We exploit a number of advances in computing to accomplish this task. We implement a fast, parallel rendering algorithm, which includes several optimizations that address time-varying data. Our principal optimizations include a fast, table-based gradient estimator and the overlapping of I/O and computational tasks. We utilize a commercially-available, shared-memory multiprocessor. We take advantage of readily available, inexpensive, high-speed disk arrays to help address I/O bandwidth requirements. We utilize commodity, high-bandwidth (100Mb/sec) networking to deliver imagery from our rendering platform to a display host.

3   Shear-Warp Factorization Algorithm

The Shear-Warp Factorization algorithm operates by applying an affine viewing transformation to transform object space into an intermediate coordinate system that Lacroute calls *sheared object space*. Sheared object space is defined by construction such that all viewing rays are parallel to the third coordinate axis and perpendicular to the volume slices. Volumes are assumed to be sampled on a rectilinear grid. For a parallel projection, the viewing transformation is simply a series of translations. Note that the resampling weights are invariant within each slice because all voxels are translated by the same amount within each slice. After the slices have been translated and resampled, they may be efficiently composited in front-to-back order using the *over* operator to produce an intermediate, warped, *baseplane* image. The distorted, baseplane image must then be resampled into final image space.

Parallel implementations of the Shear-Warp algorithm, demonstrating good speedup, have been reported using both message-passing [2, 15] and shared-memory [8] systems. Our target platform is an HP Exemplar shared-memory multiprocessor. We therefore elected to extend an existing serial implementation of the algorithm, the VolPack library [10], following Lacroute's successful parallelization of this same package [8]. We discuss our parallel implementation, pointing out the deviations between Lacroute's efforts and our own. We next discuss steps taken aimed at accommodating time-varying data volumes.

The VolPack library implements two rendering algorithms that are of interest to us. The first operates on preprocessed, run-length encoded (RLE) data volumes. The pre-processing step computes the view-independent opacity and gradient information for the data samples. Viewing and shading parameters may subsequently be changed, but classification information is fixed. There are two advantages to the RLE algorithm: 1) Opacity and gradient information do not need to be computed during rendering. 2) The RLE data structure permits the algorithm to take advantage of data coherency by skipping over transparent voxels. For many datasets this optimization can yield a significant savings. The second algorithm simply operates on raw data. No preprocessing is performed. Consequently, the renderer does not utilize volume data coherency. For static data, the RLE algorithm generally has far superior performance than the raw data algorithm. However, as we discuss later, the raw data rendering algorithm has some attractions for time-varying data, namely, reduced I/O requirements.

Execution time for the RLE algorithm is dominated by two calculations: 1) projection of the volume into the baseplane image and 2) warping the baseplane image into the final image. The raw data algorithm must perform the additional step of computing gradient vectors and determining voxel opacity values. We parallelize each of these computational phases, synchronizing processors in between.

Projection of the 3D volume, which involves resampling and compositing the volume slices, is the most computationally expensive task in the RLE algorithm and also the most challenging to efficiently implement in parallel. To minimize processor synchronization, a task partitioning based on an image-space decomposition of the baseplane image is employed. Each processor is responsible for computing a number of scanlines in the baseplane image. The baseplane image is partitioned into small groups of contiguous scanlines. Processors are initially statically assigned groups of scanlines in cyclical fashion. As the calculation proceeds, dynamic task stealing is utilized to perform load balancing. Each processor maintains its own queue of groups of scanlines. As soon as a processor finishes all of its work, it tries to steal a group of scanlines from its neighbors. This process continues until projection is complete. Our task-stealing implementation differs from Lacroute's in that processors only look to a limited number of neighbors for additional work. While this may lead to load imbalance under rare pathological conditions, it simplifies termination logic and reduces synchronization requirements.

Resampling the baseplane image typically represents 5% of the total calculation for the RLE renderer. We partition the workload by dividing the post-warped image into P groups of adjacent scanlines (where P is the number of processors) and statically assign one group of scanlines to each processor. There are no pixel interdependencies in the post-warped image, thus no synchronization is required. Our implementation differs from Lacroute's which uses an interleaved assignment of tile-shaped tasks, in an effort to provide load balancing, instead of groups of contiguous scanlines.

The raw data algorithm requires the additional step of classifying the data and computing normal vectors. Classification is performed via a user-defined lookup table. Normal vectors are estimated using central differences. We easily parallelize these operations by partitioning the volume into contiguous blocks and assigning each processor a single block.

3.1 Time-Varying Data

Time-varying data incur additional rendering costs primarily due to I/O. The principal technique we employ to address these costs is to "hide" the I/O behind the rendering calculations by using a separate I/O thread to double-buffer *reads* of volume files. A parallel memory copy is then employed to move the input data from buffer space to user space. We note that double-buffering entails keeping two copies of data in memory, which may limit the dataset's size on machines with smaller memories.

There are tradeoffs to be considered between the RLE and raw data rendering algorithms. The RLE algorithm is typically faster because it exploits data coherency. However, the RLE data volumes, which contain three copies of the data and include gradient information, may be substantially larger than the original raw data volumes and have correspondingly greater I/O requirements. The exact size of the RLE volume is largely determined by the user-defined classification function: the more voxels whose opacity is mapped to zero, the smaller the RLE volume becomes. Finally, there is another issue that may make the raw data algorithm more attractive: RLE datasets prohibit changes in classification.

4 System Overview and Implementation

We have developed a comprehensive rendering system based on the extensions to the VolPack library discussed above. *Volsh* is an interactive application with an X11-based graphical user interface (GUI). Since large multiprocessor configurations typically do not have directly attached display devices, Volsh is implemented as two separate UNIX processes communicating via internet-domain sockets. One process runs on the parallel machine, performing rendering and managing the GUI. The second process runs on the display host and is responsible for ingesting post-warped imagery transmitted by the renderer, resampling the post-warped imagery to final imagery at the desired screen resolution, and posting the imagery to the frame buffer. The resampling and posting operations are performed via the OpenGL API. The rendering host and display host are networked via a 100Mb/sec FDDI ring. The network bandwidth requirements between the renderer and the display process are non-negligible. To minimize bandwidth requirements, the renderer transmits reduced-resolution post-warped imagery (typically 256x256 pixels) to the display system. As with the input stream, we dedicate a single thread to double-buffer the output image stream.

4.1 Hardware

The results reported in Section 5 were collected from an HP Exemplar SPP2000 X-Class technical server. The X-Class Exemplar SPP2000 is a cache coherent Non Uniform Memory Access (cc-NUMA) scalable shared memory computer consisting of as many as 512 180-MHz Hewlett-Packard PA-RISC 8000 processors, each with a 1 MB data and instruction cache. The Exemplar SPP2000 hardware architecture can be thought of as a tightly coupled cluster of Symmetric Multi-Processor (SMP) "hypernodes." Each hypernode consists of up to 16 PA-RISC 8000 processors, on 8 dual-processor boards, each connected to 8 memory boards via a crossbar interconnect.

We report results obtained from a single hypernode configured with 16 processors, 2 GB of memory, and a high-bandwidth disk array. The disk array is a level 0 RAID, implemented with software striping, consisting of 9 drives striped across 3 UltraSCSI controllers. The sustained bandwidth that we have measured for *read* operations on the disk array is 80MB/sec.

4.2 Software

Volsh is implemented in C and C++. Parallelization is accomplished using POSIX threads (pthreads), making the software portable to most shared-memory architectures.[1]

Voxels in raw datasets are 8-bit quantities. RLE voxels are 32-bits: the original sample value is represented by 8 bits, the remaining bits are used to represent gradient magnitude (8 bits) and an encoded representation of a gradient (13 bits). Shading is performed using a shade tree [1] representing the Phong lighting equations. The shade tree is implemented as a lookup table, indexed by the voxel's scalar value and encoded normal vector. The calculation of the shade tree itself is entirely based on lookup tables and does not contribute measurably to the overall computation time.

Fast Gradient Estimator Normal vectors are estimated prior to projection by applying the 6-neighborhood central-difference method. Before the normal vector can be applied in lighting equations and voxel classification, it must be normalized and its magnitude must be calculated. Computing these quantities directly requires a square root and three divides. We avoid these expensive operations by transforming the Cartesian coordinates into normalized, spherical coordinates and then mapping the longitude and latitude angles, λ and ϕ, back into normalized, rectangular form. Both of these transformations are performed via lookup tables. The mapping from Cartesian to spherical coordinate space and its inverse are given by equations (1) and (2) below, respectively.

$$\lambda = tan^{-1}\frac{y}{x}$$
$$\phi = tan^{-1}\frac{z}{\sqrt{x^2 + y^2}}, \tag{1}$$

$$x = cos\lambda \times cos\phi$$
$$y = sin\lambda \times cos\phi$$
$$z = sin\phi. \tag{2}$$

The angular distances returned by equation (1) are restricted to the range $-\pi \le \lambda < \pi$ and $-\pi/2 \le \phi \le \pi/2$. Because we are using 8-bit voxels, the

---

[1] Due to performance problems with the Exemplar pthread implementation, we were forced to use HP's native thread library when benchmarking on the Exemplar.

domain for each Cartesian coordinate a_i in equation (1) is restricted to the range $-255 \leq a_i \leq 255$. Consequently, lookup tables small enough to fit into cache may be readily constructed and used to evaluate the arc tangent and square root operations. Note that using a lookup table to normalize the Cartesian coordinates directly would require a table with 255^3 entries.

The arc tangent lookup tables return quantized, integer, angular measurements. We represent λ as a 7-bit quantity and ϕ as a 6-bit quantity, giving us a 13-bit encoded normal which can subsequently be mapped into normalized, floating-point, Cartesian space using a lookup table in place of equation (2). Empirical evidence suggests that the 13-bit normal representation is adequate and additional precision is not warranted (see Plate 1, top). The gradient's magnitude, used during classification to attenuate opacity, is approximated using a *Manhattan Distance* approximator, accurate to $\pm 8\%$ [14].

5 Performance

We use a number of different datasets to evaluate the performance of our rendering system. The Quasi-geostrophic (QG) data are computer simulation results depicting turbulence in the Earth's oceans. These data and the classification functions chosen for them are of particular interest to us in exploring performance characteristics because earlier time-steps are very dense, dominated by amorphous (semi-transparent) features, and exhibit little spatial coherence. Later time-steps are relatively sparse and opaque, exhibiting substantially more coherence (see Plate 2). Our second dataset was produced from a simulation of the wintertime stratospheric polar vortex. The polar vortex (PV) data are relatively sparse and opaque, exhibiting a great deal of coherence throughout the entire simulation (see Plate 1 (middle), (bottom)). Lastly, we include some results from the familiar UNC Chapel Hill Volume Rendering Test Dataset. Though these medical data are not time-varying, their performance characteristics are well-known to the volume rendering community (see Plate 1 (top)). Table 1 lists the datasets, their resolution, the size of each raw time-step, and the average size of each RLE time-step. We note from Table 1 that the relative size of raw and corresponding RLE datasets provide a measure of spatial coherency. The larger the RLE dataset relative to the raw dataset, the less coherency exists. The sizes of the RLE datasets also provides a measure of how much work is required to render them; larger RLE files require increased rendering time.

In the experiments reported below, all data are read directly from disk. The kernel buffer cache was flushed prior to running each experiment. Unless otherwise specified, all results are for Phong-illuminated, monoscopic, three-channel (RGB), 256x256 resolution imagery, rendered with a parallel viewing projection. Lighting is provided by a single light source with fixed lighting parameters.

5.1 Static Data

Figure 1 plots rendering rate in frames per second vs. number of processors for static data, using both the RLE and raw datasets. The frame rates shown are the averages for a 180-frame animation, with a 2-degree rotation about the Y axis between each frame. The time to display the image is not included. We see from Figure 1 that for static data, the RLE data algorithm performs significantly faster than the raw data algorithm. However, there are variations in the relative differences in performance. The highly coherent PV datasets exhibit the greatest increase in performance between the raw and RLE data (roughly an order of magnitude improvement, from 1.7Hz to 20Hz on 15 processors). Conversely, the QG dataset does not benefit as much from the RLE encoding (about a factor of 2 improvement from 5Hz to 10Hz on 15 processors). We also observe that the performance of our implementation on the Brain dataset is comparable to that reported by Lacroute [8], both in terms of speedup (approximately 11) and maximum frame rate (approximately 10Hz). [2]

Lastly, we note that all the performance curves dip after the 15-processor run. This event occurs when processors are forced to perform non-rendering tasks. For static data, these non-rendering tasks are comprised of normal UNIX system activities. In subsequent experiments involving time-varying data, we dedicate two processors for double-buffering (one for input and one for output), leaving 14 processors available for rendering.

5.2   Time-Varying Data

In this section we discuss our results with time-varying data. Unlike our static-data experiments, the experiments reported below include the time to display imagery. Unless otherwise stated, both the ingestion of data volumes and output of image streams are double buffered.

Figure 2 plots the frame rate in frames per second vs. the number of processors for the QG and PV datasets using both RLE and raw data. We observe that each of the RLE curves go flat at some point. This occurs when the task becomes I/O bound as evidenced in Figure 3 by the growth of the i_time parameter. The i_time parameter depicts the unmaskable, double-buffered I/O time for reading data. It is the amount of time the rendering processes must wait for input. When the task is computationally (render) bound, i_time is close to zero. As the rendering process is sped up through the addition of processors, the task can become I/O bound, and i_time will grow. The less voluminous raw data do

---

[2] Lacroute's benchmarks were performed using only a single color channel.

Dataset	Resolution	Raw Size (MB)	Average RLE Size (MB)
Brain128	128x128x84	1.38	3.34
Brain256	256x256x167	10.94	16.27
QG128	128x128x128	2.10	10.78
PV128	128x128x75	1.23	1.92
PV256	256x256x149	9.76	13.70

Table 1. Datasets, their respective voxel resolutions, individual time-step sizes for raw data, and average time-step sizes for RLE data.

not become I/O bound in this plot, and their corresponding performance curves display better speedup.

We also observe that as we would expect from our static data measurements, the highly coherent PV RLE data perform considerably better than the PV raw data. However, the QG RLE data initially perform much better than the QG raw data, but in higher processor runs the difference in performance is not as great. The rising QG raw curve is rapidly approaching the flat RLE curve. The QG RLE data very quickly become I/O bound because of their large size relative to the raw data. Witness the growth of the *i_time* parameter for QG RLE data in Figure 3 after 6 processors. Adding additional processors may speed up rendering, but it does not affect the dominating I/O time. On the other hand, though the raw data algorithm is more computationally expensive, the I/O requirements are much lower. Adding processors continues to reduce the rendering and gradient calculation time, which dominate the lower-processor-number runs for the raw data.

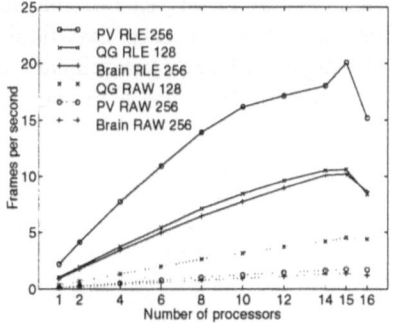

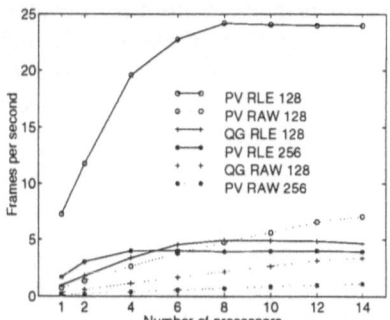

Fig. 1. Frame rates of static data runs using RLE (solid lines) and raw datasets (dashed lines)

Fig. 2. Frame rates of time-varying data runs using RLE (solid lines) and raw datasets (dashed lines)

5.3 Double Buffering

Figure 4 shows the effects on overall rendering performance of overlapping the I/O and computational tasks for the QG RLE dataset. The ideal speedup that may be achieved is given by

$$s = \frac{T_{input} + T_{output} + T_{compute}}{max(T_{input}, T_{output}, T_{compute})},$$

where T_{input}, T_{output}, and $T_{compute}$ are the times required for reading data, displaying imagery, and performing the computational tasks, respectively. We observe from this equation that the maximum possible speedup is three, and this maximum occurs when all T are equal. In our experiments, the input requirements are fixed and completely determined by the size of the volume dataset and the available bandwidth of the storage device. Similarly, the output bandwidth requirements, which are relatively low, are fixed and determined by the resolution of the post-warped image and the bandwidth of the display device (in our case, a FDDI-attached host).

Comparing the single-buffered and double-buffered runs in Figure 4, we observe that on fewer processors, double-buffering has little effect on overall performance which is dominated by the rendering calculation. For the single-processor run, the input and output time combined represent only about 25 percent of the total run-time. Hence the best speedup we can realize by double-buffering is only $1/0.75 \approx 1.3$. However, as the number of processors is increased, rendering time is reduced. By the six-processor run, rendering time and input time are nearly equal, and we see a speedup over the single-buffered performance of about two. Adding even more processors reduces rendering time, but does not affect I/O. Hence the overall performance indicated by the double-buffered curve goes flat after six processors. Contrarily, the performance indicated by the single-buffered curve continues to improve beyond six processors, but never reaches the performance of that of the double-buffered runs. Lastly, we observe that double buffering has completely masked the cost of output. Though insignificant on fewer processors, for greater processor runs the output time has a measurable impact on overall performance.

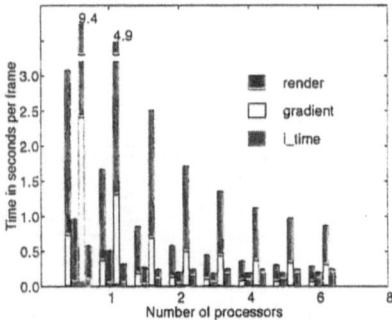

Fig. 3. Timing distribution of rendering process showing non-maskable read time, *i_time*; rendering time, *render*; and gradient calculation time for the raw data algorithm, *gradient*. Shown for each processor from left to right are: QG128 Raw, QG128 RLE, PV256 Raw, and PV256 RLE

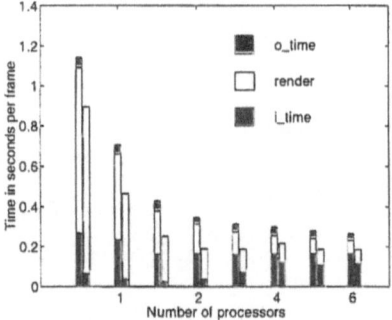

Fig. 4. Effects of double buffering on the QG RLE dataset. The first and second time distribution bar for each processor show single-buffered and double-buffered performance, respectively. The I/O times, *i_time* and *o_time*, include the overhead of double-buffering and non-maskable I/O

6   Conclusions

We have developed a Direct Volume Rendering system capable of interactively rendering moderately sized, time-vary datasets. The system is based entirely on commercially available components. The performance for a given volume resolution is largely determined by the amount of spatial coherency that exists within the data and the user-chosen classification function. For highly coherent, sparse data the pre-classified, RLE-encoded volumes are comparable in size to the raw data, and the RLE algorithm performs best. For data that are dense and exhibit little spatial coherency, the increased I/O costs associated with the more voluminous RLE-encoded files may make the raw data algorithm more attractive if sufficient processing power is available.

Double buffering was shown to be an essential component of our system. We have seen that the technique is maximally effective when the processing requirements of all the overlapped tasks are similar. In our experiments, the performance of the two I/O tasks is fixed. The third task we chose to overlap, the computational task, can be sped up by employing additional processors. For fewer processors, the computational task is the most expensive of the three overlapped tasks and can therefore mask the cost of I/O. As we increase parallelism, the computation task may be sped up until it no longer dominates. Once I/O begins to dominate, the limit of the benefit of adding processors has been reached. For the RLE data, this limit was generally reached on a relatively few processors. For raw data, which are computationally more expensive to render, the computational limit was never reached; the rendering task remained compute bound. We can view this result as somewhat encouraging. Given that improvements in microprocessor technology far outpace improvements in storage bandwidth, we speculate that the performance of the raw data algorithm may surpass the RLE data algorithm in the near future.

Although we predict that next-generation processors may permit the raw data algorithm to outperform the RLE algorithm, for the present the converse is true for the computing platform we tested. In the case that time-varying raw data cannot be rendered at sufficient frame rates, and the user is forced to work with RLE data, we do not view the inability to change classification of RLE data as a serious detriment. Our observations have been that researchers perform classification while working on a small number of time-steps, one static time-step at a time. Once a satisfactory classification has been arrived at, the researcher will then begin to explore the data temporally. At this point the classification function has been chosen, and the data may be RLE encoded for improved performance.

Lastly, we note that the RLE method in our current implementation stores three copies of the data for each time-step, one for each principal viewing axis. With some modification, significant savings in I/O requirements could be realized by loading only the RLE encoding required by the current viewing direction. Though this complicates the double-buffering scheme somewhat, we believe the potential performance improvements are well worth pursuing.

References

1. G. Abram and T. Whitted. Building block shaders. In *Computer Graphics*, pages 283–288, Dallas, TX, August 1990.
2. M. B. Amin, A. Grama, and V. Singh. Fast volume rendering using an efficient, scalable parallel formulation of the shear-warp algorithm. In *Parallel Rendering Symposium*, pages 7–14, Atlanta, GA, October 1995.
3. B. Cabral, N. Cam, and J. Foran. Accelerated volume rendering and tomographic reconstruction using texture mapping hardware. In *Symposium on Volume Visualization*, pages 91–97, Washington, D.C., October 1994.
4. T.-C. Chiueh and K.-L. Ma. A parallel pipelined renderer for the time-varying volume data. Technical Report 97-90, ICASE, Hampton, VA, December 1997.
5. J. Clyne, T. Scheitlin, and J. Weiss. Volume visualizing high-resolution turbulence computations. *Theoretical and Computational Fluid Dynamics*, 1998.
6. R. A. Drebin, L. Carpenter, and P. Hanrahan. Volume rendering. *Computer Graphics*, 22(4):51–58, August 1988.
7. K. Knittel and W. Straber. A compact volume rendering accelerator. In *Symposium on Volume Visualization*, pages 67–74, Washington, D.C., October 1994.
8. P. Lacroute. Analysis of a parallel volume rendering system based on the shear-warp factorization. *IEEE Transactions on Visualization and Computer Graphics*, 2(3):218–231, September 1996.
9. P. Lacroute and M. Levoy. Fast volume rendering using a shear-warp factorization of the viewing transformation. In *Computer Graphics Proceedings*, pages 451–458, Orlando, FL, July 1994.
10. P. G. Lacroute. *Fast Volume Rendering Using a Shear-Warp Factorization of the Viewing Transformation*. PhD thesis, Stanford University, Stanford, CA, September 1995.
11. K.-L. Ma, J.S. Painter, C.D. Hansen, and M.F. Krogh. A data distributed parallel algorithm for ray-traced volume rendering. In *Parallel Rendering Symposium*, pages 15–19, San Jose, CA, October 1993.
12. U. Neumann. Parallel volume-rendering algorithm performance on mesh-connected multicomputers. In *Parallel Rendering Symposium*, pages 97–104, San Jose, CA, October 1993.
13. H. Pfister and A. Kaufman. Cube-4 - a scalable architecture for real-time volume rendering. In *Symposium on Volume Visualization*, pages 47–54, San Francisco, CA, October 1996.
14. J. Ritter. A fast approximation to 3d euclidian distance. In A. Glassner, editor, *Graphics Gems*, pages 432,433. Academic Press, 1990.
15. K. Sano, H. Kitajima, H. Kobayashi, and T. Nakamura. Parallel processing of the shear-warp factorization with the binary-swap method on a distributed-memory multiprocessor system. In *Symposium on Parallel Rendering*, pages 87–94, Phoenix, AZ, October 1997.
16. H.-W. Shen and C. Johnson. Differential volume rendering: A fast volume visualization technique for flow animation. In *Visualization '94*, pages 180–187, Washington, D.C., October 1994.
17. R. Westermann and T. Ertl. Efficiently using graphics hardware in volume rendering applications. In *Computer Graphics Proceedings*, pages 169–177, Orlando, FL, 1998.
18. C. M. Wittenbrink and A.K. Somani. Permutation warping for data parallel volume rendering. In *Parallel Rendering Symposium*, pages 57–60, San Jose, CA, October 1993.

Editors' Note: see Appendix, p. 322 for colored figures of this paper

Efficiently Rendering Large Volume Data Using Texture Mapping Hardware

Xin Tong[1], Wenping Wang[2], Waiwan Tsang[2], Zesheng Tang[1]

[1] CAD Laboratory, Department of Computer Science, Tsinghua University,
Beijing, 100084, P. R. China
ztang@tsinghua.edu.cn
[2] Department of Computer Science, University of Hong Kong,
Pokfulam Road, Hong Kong
{wenping, tsang}@cs.hku.hk

Abstract. Volume rendering with texture mapping hardware is a fast volume rendering method available on high-end workstations. However, limited texture memory often prevents the method from being used to render large volume data efficiently. In this paper, we propose a new approach to fast rendering of large volume data with texture mapping hardware. Based on a new volume-loading pipeline, the volume data is preprocessed in such a way that only the volume data that contains object voxels are loaded into texture memory and resampled for rendering. Moreover, if classification threshold is changed, our algorithm classifies and processes the raw volume data accordingly nearly in real time. Our tests show that about 40% to 60% rendering time is saved in our method for large volume data.

1 Introduction

In scientific visualization, direct volume rendering is used to generate high quality semi-transparent images with details. As millions of voxels are usually involved in rendering, the amount of computation is enormous. Although much effort has been made to shorten the rendering time [3][5][7][10], the real-time direct rendering of the volume data still cannot be achieved by software.

The rapid progress of graphics hardware development offers new possibility for real time volume rendering [1]. Carbral [2] first proposed to use 3D texture mapping hardware to render volume data in 1994. Since then, several improved algorithms and implementations have been reported [4][11]. In these methods, the volume data is rendered via the texture mapping hardware in two steps. As shown in Fig. 1, in the first step for volume loading, volume data are read from a disk file (i.e., the first level) into volume buffer (i.e., the second level) in main memory. Then the data are loaded into texture cache (i.e., texture memory, the last level) and defined as 3D texture maps. In the second step for volume rendering, a set of polygons are used to resample the volume data by texture mapping and then these texture mapped polygons are blended from back to front to generate the final image. For volume data that can reside entirely into the texture cache, the volume loading step is done only once. In the following rendering step, the texture mapping and composition operations are performed by hardware very quickly, about ten frames per second or higher. However, due to limited texture memory supported by hardware, large volume data must be divided into several blocks and rendered one by one. For such large volume data, the increased time on resampling texture I/O will reduce the frame rate greatly.

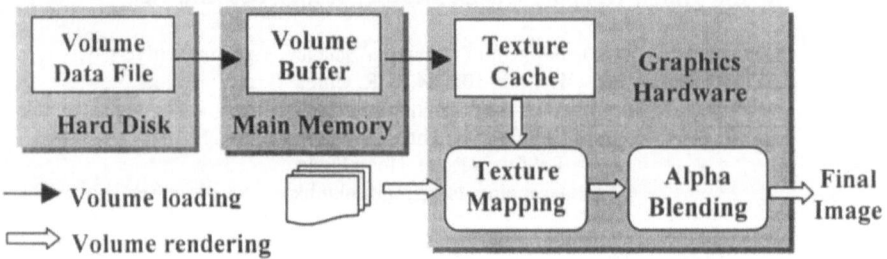

Fig. 1. The diagram of the conventional method

In fact, in many visualization applications, lots of voxels in volume data are so-called *empty voxels*, which belong to the background or parts of no interest. As the empty voxels are often treated as transparent (alpha=0) in rendering, skipping the empty regions may speed up volume rendering significantly, while not affecting image quality at all [12]. Unfortunately, in most existing methods [2][4][11], classification is performed by a hardware look-up table in the volume rendering step so that all the volume data must be loaded into the texture memory for rendering. In addition, since the texture mapping hardware requires that data in volume buffer be stored as a 3D array, volume data can not be processed efficiently using the three-level volume loading pipeline.

In this paper we propose an algorithm to speed up the volume rendering with 3D texture mapping hardware. Unlike conventional method, a four-level volume-loading pipeline is exploited to classify and preprocess volume data efficiently. After the volume data is classified into empty voxels and object voxels, only the parts that contain the object voxels are loaded into the texture memory for rendering, while the empty voxels will be kept in volume buffer but not loaded into texture memory. In this way, the volume rendering step is also accelerated by avoiding the unnecessary resampling operations.

The rest of the paper is organized as follows. In section 2, the related work is reviewed. Our algorithm is described in detail in section 3. Section 4 gives the experimental results. The paper is concluded in section 5.

2 Related Work

Skipping empty regions is a well-known acceleration technique that has been extensively exploited by ray casting type volume rendering algorithms. Marc Levoy [8] employed an octree of the binary volumes to enumerate the presence of objects in volume data. Thus for the ray casting method, the ray can skip the largest empty space containing the current sampling point by traversing the octree. However, the octree data structure of binary volumes not only occupies auxiliary memory space but also needs to be reconstructed whenever the classification threshold is altered. Moreover, since object voxels are represented by a set of octree nodes with different sizes, it is then difficult to render these nodes by texture mapping hardware.

In proximity-clouds methods [3][14], an auxiliary distance volume is constructed from original volume data. In the distance volume each voxel value is the distance from the voxel to the closest object voxel. If the current resample point on a ray is at a voxel, then based on the distance value of the voxel the next resample point can leap to object voxels quickly, so that many unnecessary resampling can be avoided. Although traversing the distance volume is more efficient than traversing the octree, the proximity-clouds methods need large additional memory to store the distance volume. In addition, it takes long time to compute the distance volume when the classification is modified.

The PARC approach [10] uses graphics hardware to identify the segments of rays that contribute to the final image. Given a classification threshold, the outer faces of object cells are presented by a set of rectangular polygons. Then for each view direction, the polygons are projected into the zbuffer twice to obtain the ray segments that possibly contribute to the final image and only these ray segments are traversed and resampled in ray casting. Unfortunately, because the depth test occurs after the texture mapping in the graphics pipeline, the PARC method cannot be used for volume rendering with texture mapping hardware. Although the envelope of object voxels can be used to clip the resampling polygons in volume rendering with texture mapping hardware, only resampling operations in the rendering step can be reduced and the time spent on volume loading step is not reduced at all.

The shear-warp method [5] takes advantage of spatial coherence of voxels to skip the empty regions in volume data. The method also cannot be applied in the volume rendering with texture mapping hardware, as the volume must be resampled by hardware slice by slice.

3 Efficient Rendering of Large Volume Data

3.1 System Overview

As shown in Fig. 2, our method also contains two steps: volume loading and rendering. However, a four-level volume-loading pipeline is adopted, where a new texture buffer is added between volume buffer and texture cache. In the volume loading step, the volume data is divided into several volume blocks and loaded into volume buffer first. Three min-max arrays are then built for each volume block. Given a classification threshold, empty voxels in each volume block are identified and trimmed off by the bounding boxes of object voxels. These trimmed volume blocks are transferred into a texture buffer and defined as texture blocks. Finally the texture blocks in the texture buffer are merged into bigger texture chunks to reduce the overload of texture memory I/O. In the volume rendering step, after the texture chunks are sorted according to a given view direction, they are loaded into texture cache from back to front. Then the texture blocks are resampled by a set of polygons that are clipped by the bounding boxes. After blending all the texture-mapped polygons to the frame buffer, a new image is generated for display.

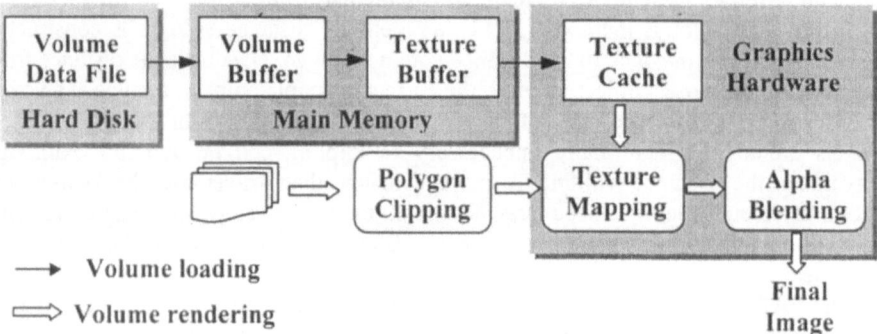

Fig. 2. The diagram of our new volume rendering method

In our method the parameters of volume-loading pipeline are determined by the configuration of texture cache, while texture cache that lies in the texture memory is configured by the operating system and special hardware implementation. In general, the following rules are compatible with texture cache, where the n, m, s and k are constant determined by the configuration of texture memory [9]:

- The texture data that will be loaded in the texture cache must be stored in an array.
- The texture maps in the texture cache is stored in a tiled page mode.
- The size of texture page is $2^n \times 2^m$ for 2D texture and $2^n \times 2^m \times 2^s$ for 3D texture.
- A texture page is the minimum unit for 3D texture I/O.
- At most 2^k texture pages can be resident in texture cache at same time.

We suppose that a volume data is axis-aligned and stored as an $l \times m \times n$ 3D array. The length of the volume (in voxels) in X, Y, Z direction is l, m, n respectively. Suppose $V_{i,j,k}$ denoted the voxel value of voxel (i, j, k) as in the 3D array.

3.2 Volume Loading

The volume loading step can be divided into three phases. The whole volume data is uniformly subdivided into several blocks and loaded into the volume buffer first (see section 3.2.1). Next the volume block is classified and trimmed by the bounding boxes of object voxels, and the trimmed blocks are then moved into texture buffer (see section 3.2.2). Finally, in order to optimize the I/O performance of texture cache, the texture blocks are merged (see section 3.2.3).

3.2.1 Volume Buffer loading and min-max array generation
In this stage, volume data is subdivided into several volume blocks. In order to avoid artifacts caused by seams between adjacent blocks in the final image, each block shares

its boundary voxel data with its neighboring blocks. Two factors are considered in selecting the block size. On one hand, the smaller is the block size, the more empty regions can be excluded. On the other hand, smaller blocks would result in more redundant data on the overlapping boundaries of the blocks and in fact the block size cannot be less than the size of the texture page. Moreover, the large number of texture fragments with irregular sizes after trimming would cause the overhead of graphics hardware, and it is also difficult to merge these small blocks into bigger blocks to optimize I/O of texture memory. On that account, the size of volume block is chosen in our method to be the size of the smallest block that satisfies the following two rules:

- The volume block must contain at least one texture page.
- The volume block must have equal length along each axis. This rule is used to simplify the following merging operation.

Given the size 2^d of volume blocks, volume data is loaded from a disk file into the volume buffer block by block. The number of blocks is

$$\left\lceil \frac{l-1}{2^d-1} \right\rceil \times \left\lceil \frac{m-1}{2^d-1} \right\rceil \times \left\lceil \frac{n-1}{2^d-1} \right\rceil \tag{1}$$

If the last block along an axis cannot be fully filled with the volume data, the remaining part of the block is supposed to be filled with empty voxels. An array is created for the 2^d data slices of the block along one axis; each element of the array records the minimum and maximum data value of the corresponding slice. Three arrays that correspond to three coordinate axes are set up for each volume block.

3.2.2 Data classification and volume block trimming

A block list is generated after the above preprocessing step. Each node of the list contains a pointer to a volume block, the min-max arrays of the block and the position of the volume block in world space. At this stage, most empty voxels in the volume block are excluded so that the block will be trimmed to a smaller size and moved into the texture buffer.

After a classification threshold is specified by the user, the three min-max arrays of the block are traversed to find three intervals $[L_x, R_x]$, $[L_y, R_y]$ and $[L_z, R_z]$ that are defined on X, Y, Z axes respectively, such that:

- $V_{i,j,k}$ are empty voxels for all $i < R_x$ and all $i > R_x$
- $V_{i,j,k}$ are empty voxels for all $j < R_y$ and all $j > R_y$
- $V_{i,j,k}$ are empty voxels for all $k < R_z$ and all $k > R_z$

The three intervals listed above bound the object voxels of the volume block in a box. According to the volume of this bounding box, there are three cases for one block. In the first case, the volume of the bounding box is zero. Obviously, all the voxels in the block are empty voxels in this case, so the block is marked as a white block. In the second case, the volume of the bounding box is the same as that of the block, which means that the block cannot be trimmed. The block is marked as black block in this case, and all the data in the volume block are moved into the texture buffer as a texture block. In the last case, part of the block can be trimmed off, and the block is marked as a grey block first. Then we choose the trimming axis to be the one along which the interval length is the shortest. After that, the block is trimmed along the trimming axis. The remaining part that contains object voxels are moved into the texture buffer as a texture block. A 2D example is shown in Fig. 3, where the trimming axis is the Y axis. Only the empty voxels $V_{i,j,k}$ with $i<L_y$ and $i>R_y$ are trimmed off. Although the empty voxels that lie out of the other four faces of the bounding box are retained, they will be skipped to avoid resampling in rendering step. Hence after being trimmed along the trimming axis s, the size of texture block is $2^d \times 2^d \times d_s$, where $d_s = R_s - L_s + 1$. Finally, the information about the bounding box is saved in the node of the block list.

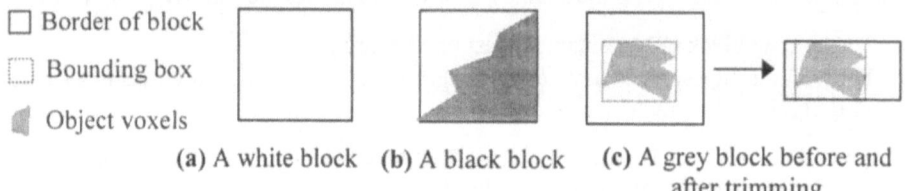

□ Border of block

□ Bounding box

◼ Object voxels

(a) A white block   (b) A black block   (c) A grey block before and after trimming

Fig. 3. The 2D illustration of the block trimming

After trimmed blocks are moved into the texture buffer, the volume data is divided into two parts. The texture blocks that contain object voxels are moved into the texture buffer, while the empty parts are trimmed off and stored in the volume buffer.

3.2.3 Texture block merging

Although the texture blocks could be directly loaded into texture memory for rendering now, the frequent texture I/O and binding operations would degrade the performance of the algorithm seriously. So we will try to merge the texture blocks into bigger texture chunks before rendering.

In this stage, the texture blocks are grouped into clusters first. A cluster contains $u \times v$ adjacent texture blocks with the same index along Z. If all of the texture blocks are black blocks, the size of the cluster is the same as the size of the texture cache. That is, the texture blocks in a cluster can reside in the texture cache at the same time. The constants u and v can be computed from the size of the volume block easily. If necessary, some white blocks are added into the last clusters along the axis. The position of the texture block in the texture cache (i.e. the texture coordinates) is also recorded in the block node.

After grouping, the texture blocks are merged in two steps: merging in cluster and merging between clusters. In each step, two basic merge operations are executed for a pair of texture blocks: horizontal merge and vertical merge. The merge operation and the procedure of merging blocks are described below.

3.2.3.1 Merge Operation

- Horizontal merge

As shown in Fig. 4(a), two blocks with the smallest difference of their thickness (block length along Z axis) are selected and merged into a $2^{d+1} \times 2^d \times d_m$ chunk, where $d_m = \max (d_{block1}, d_{block2})$. Thus for a thinner data block, the extended part will be filled with empty voxels. The original thickness of the block is saved in the block node. Although some extra empty voxels have to be added into the texture chunk, this would speed up texture memory I/O and the empty voxels can be skipped without resampling at the rendering step.

- Vertical merge

As shown in Fig. 4(b) and 4(c), If the thickness of the two blocks satisfy $d_{block1} + d_{block2} \leq 2^d$, the two blocks can be piled up together. The two original texture blocks will be directly merged into a chunk of size $2^d \times 2^d \times (d_{block1} + d_{block2})$. If one of source blocks has been horizontally merged into a chunk of size $2^{d+1} \times 2^d \times d_m$, the extended part in this chunk is excluded first. Then the other original block is piled up to it. The result is a chunk of size $2^{d+1} \times 2^d \times d_{new}$, where $d_{new} = \max(d_m, d_{block1} + d_{block2})$. If necessary, some empty voxels are used to fill in the extended part.

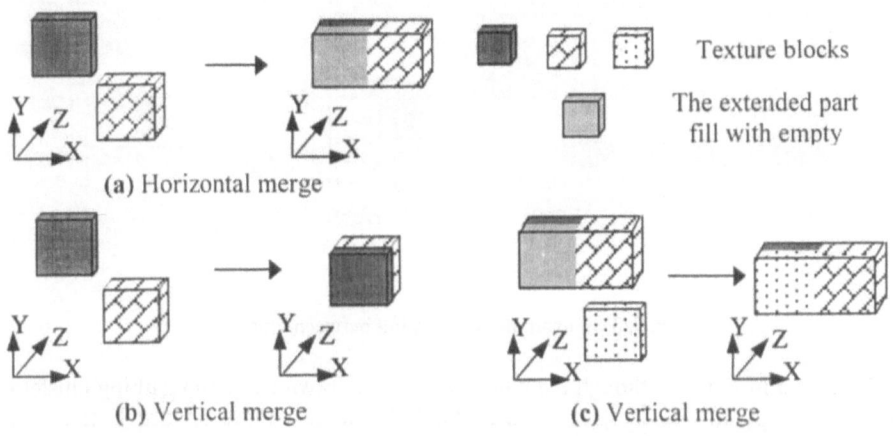

(a) Horizontal merge

(b) Vertical merge   (c) Vertical merge

Fig. 4. The horizontal and vertical merge operation

3.2.3.2 Merging in cluster
The texture blocks in a cluster are checked first. If the trimming axis of a data block is not Z, the texture block is rotated and reorganized in the texture buffer so that the trimming axis is changed to Z. Two kinds of merge operations are executed until no pair of blocks in the cluster can be merged. In horizontal merge, pairs of blocks will be merged according to the ascending order of their thickness difference. That is, the pair of texture blocks with the smallest thickness difference will be merged first.

3.2.3.3 Merging Between clusters
After the above merge operations, the texture blocks that have not been merged in the cluster may be merged further if the merging between clusters is considered. For parallel projection of the volume data, the texture clusters are traversed and rendered along the Z axis first, then along the Y axis, at last along the X axis. For each axis, whether ascending or descending order of the indices is used for traversing is

determined by the view direction. Hence, a sequence of clusters $\{B_{i,j,k} \mid 1 \leq k \leq N_z, N_z$ is the number of clusters along Z axis$\}$ are loaded into the texture memory sequentially. If two texture blocks from adjacent clusters along Z axis are merged, after loading the merged texture chunk into texture memory for rendering one cluster, the other texture block in the adjacent cluster is also resident in texture memory. Thus the latter block needs not to be loaded again when rendering the adjacent cluster.

In this step, for each row of clusters along Z direction (that is, the clusters with same X and Y indices), if there are texture blocks that have not been merged in pair of adjacent clusters, the two merge operations are executed for them. Fig. 5 shows two examples of the merge between clusters, where the cluster contains 4 texture blocks.

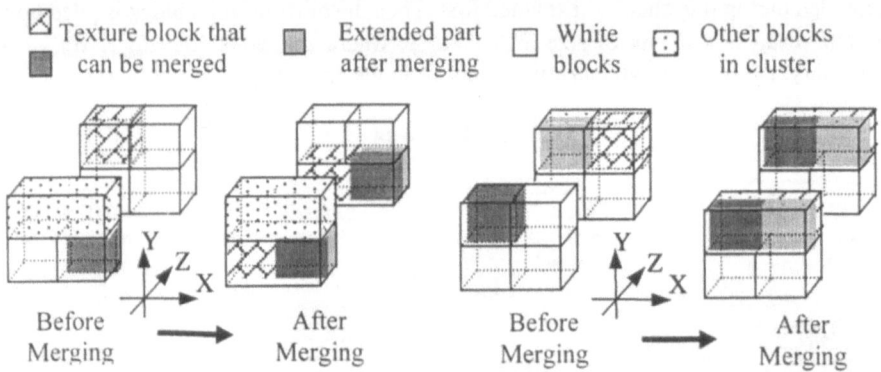

Fig. 5. Two examples of merging between clusters

Note that in Fig. 5, although the merged chunk is drawn twice in resulting clusters, only one copy of the merged texture chunk is stored in the texture buffer, and referenced by the pointers of block nodes in two clusters. This guarantees that the merged data can always be loaded and resident in texture memory for rendering the two clusters with different traversing orders.

3.2.3.4 Notes about the texture block merging

In the above merging procedure, the texture blocks are rotated, moved and merged in the texture buffer. In order to keep the correct texture mapping, the block position in the world space is reserved in the block node, while the texture coordinates of the block need to be updated accordingly. Thus during rendering, the trimmed blocks will "stay" in the original position in world space so that the volume blocks can be rendered correctly for a given view direction. A 2D example is illustrated in Fig. 6.

In addition, for a set of texture blocks, different orderings of the texture merging operations would lead to different merging results. For a special configuration of the texture hardware, texture blocks with different sizes should be loaded with the different speeds. Adjusting the order of the merging operation properly may produce texture chunks that lead to higher texture I/O performance.

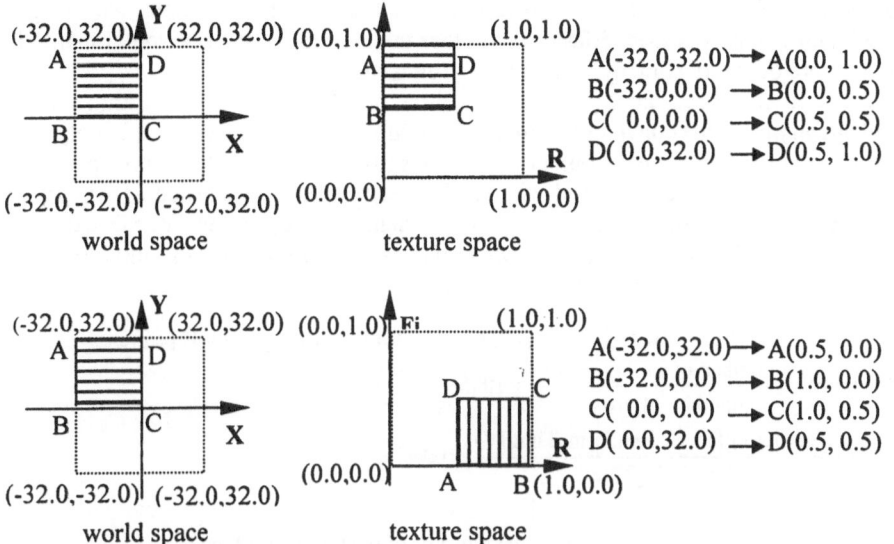

Fig. 6. The top figure shows a block ABCD and its corresponding data (i.e. texture) stored in the texture space. After moving and rotating the data in texture space (bottom figure), the texture coordinates of the block is updated accordingly, while the world coordinate is preserved So the block can be rendered with the same texture.

3.3 Volume Data Rendering

For any given view direction, the traverse order (ascending or descending) of the texture clusters along each axis is determined first. Then according to the viewing order, the texture clusters are traversed from back to front, and rendered cluster by cluster. If all blocks in a cluster are white blocks, the cluster is skipped. Otherwise, the texture chunks that do not reside in the texture cache are now loaded into the texture cache. The texture blocks in the cluster are also rendered from back to front. If a block is white, it is skipped without any further operation. Otherwise, the block is resampled by a set of parallel polygons from back to front. The bounding box of the object voxels in each texture block is used to clip the resampling polygons so that only the voxels inside the bounding box are resampled. After composing all resampling slices from back to front, the final image can be displayed on the screen.

4 Experimental Results

Our method has been implemented on SGI Octane/MXI, configured with R10000/195MHZ CPU, 128MB memory and 4MB texture memory. On this system the size of the texture page is 4k(32×64×2)[10] and the size of texture cache is *128×128×64*. Thus the size of the volume block is *64×64×64*. The texture cluster is composed of 2×2 texture blocks. Since in our platform the texture block with longer length along X axis is loaded faster than other texture blocks, the horizontal merge is used before the vertical merge in texture merging step.

We have compared our algorithm with three other algorithms. Algorithm I is an implementation of the conventional method [2]. In Algorithm II, the resample polygons are clipped by the envelope of object voxels. To show the effect of the merge operation, an implementation of our method without the merge operation is tested as Algorithm III. The full version of our algorithm is implemented as Algorithm IV. Six data sets from [6][8][13] are used for the test, where the size of the test image is 400×400. The experimental results are listed in Table I, where

- Data Processing Time = Time $_{\text{data classification}}$ +Time $_{\text{block trimming}}$ + Time $_{\text{block merging}}$

- Empty $= \dfrac{\text{number of empty voxl es}}{\text{number of voxels}} \times 100\%$

- $S = \dfrac{(\text{RenderingTimeI} - \text{RenderingTimeIV})}{\text{RenderingTimeI}} \times 100\%$

Table 1. The experiment results of the four algorithms

Data	Data Size	Empty (%)	Algorithm I Rendering Time (s)	Algorithm II Rendering Time (s)	Algorithm III Rendering Time (s)	Algorithm IV Rendering Time (s)	S(%)	Data Processing Time (s)
Head I	256×256×128	74.1	0.700	0.499	0.406	0.287	59.0	0.156
Head II	512×512×56	68.4	0.963	0.740	0.787	0.510	47.0	0.140
Body I	512×512×78	70.7	1.329	1.032	1.022	0.652	50.9	0.242
Frog	500×470×176	83.4	1.963	1.446	1.235	0.786	59.9	0.374
Foot	250×512×352	76.6 ʹ	2.552	2.056	1.578	0.978	61.7	0.535
Head III	512×512×256	67.0	3.575	3.166	2.792	1.604	55.1	2.624

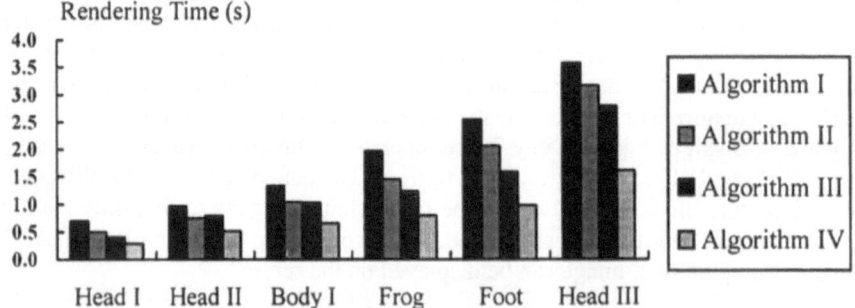

Fig. 7. The rendering time of four algorithms

As shown in Fig. 7, compared with the conventional method, about 40% to 60% rendering time can be saved by our method for the test data sets. Our method is also significantly faster than the Algorithm II and Algorithm III.

Fig.8 is the rendering result of the Head I data set by the conventional texture mapping hardware assisted volume rendering. Fig. 9 is the rendering result of the same data set by our new method. There is no difference between the visual quality of two images because the excluded and skipped empty voxels contribute nothing to the final image. Fig. 10 is the rendering result of the Frog data set by our algorithm. Fig. 11 is the rendering result of the Foot data set.

In our method, after the classification threshold is modified by the user, the volume blocks are classified again, and then only the texture blocks that contain object voxels are moved into the texture buffer for rendering. As listed in Table 1, although the volume data cannot be processed in real time, the time spent for processing is approximately on the same order of the rendering time.

5 Conclusion and Further Work

To summarize, we proposed an acceleration method to render large volume data with texture mapping hardware. Unlike the conventional method, our method uses a four-level volume-loading pipeline, through which the volume data can be processed efficiently before rendering. In the volume loading step, the empty voxels in the volume blocks are identified and trimmed off by the bounding boxes of the object voxels. Then the trimmed blocks are loaded into the texture buffer and become texture blocks. Finally, the texture blocks are merged into texture chunks to optimize the I/O performance of texture cache. During rendering, the resampling slices are clipped by the bounding box. As a result, both texture memory I/O and the texture resampling operations are reduced effectively. Experiments show that our method boosts the speed of rendering large volume data with texture mapping hardware approximately by a factor of 2. Moreover, the classification of the volume data is also accelerated by the min-max arrays of the volume blocks. Thus the raw volume data can be classified and processed quickly for the modified classification threshold.

In our method, if the volume data cannot be entirely stored in the main memory (e.g., the HEAD III data set), the frequent memory swapping operation would result in long preprocessing time. Fortunately, the volume data can be compressed and stored in the volume buffer. After the volume data is classified, only the texture blocks that contain object voxels are reconstructed from the compressed data on the fly and then moved to the texture buffer. In our further work different kinds of compression methods will be investigated to find the best compression scheme for our method. In addition, the possibility of using our method to render multiresolution volume data or the volume data on remote machine will also be explored in the future.

References

1 Akeley, K.,: RealityEngine Graphics. In Proceedings of ACM SIGGRAPH'93. (1993) 109-116

2 Cabral, B., Cam, N., and Foran. J,: Accelerated Volume Rendering and Tomographic Reconstruction Using Texture Mapping Hardware. ACM Symposium on Volume Visualization. (1994) 91-98

3 Cohen, D. and Shefer, Z.,: Proximity Clouds - An Acceleration Technique for 3D Grid Traversal. Technical Report FC93-01, Ben Gurion University of the Negev. (1993)

4 Gelder, A. V. and Kim, K.,: Direct Volume Rendering with Shading via 3D Textures. In Proceedings of ACM Symposium on Volume Visualization. (1996) 23-30

5 Lacroute, P. and Levoy, M.,: Fast Volume Rendering Using a Shear-Warp Factorization of the Viewing Transformation. In Proceedings of ACM SIGGRAPH'94. (1994) 451-458

6 LBL.,: Whole Frog Project. From http://www-itg.lbl.gov/. (1994)

7 Levoy, M.,: Efficient Ray Tracing of Volume Data. ACM Transactions on Graphics. Vol. 9, No. 3, (1990) 245-261

8 NLM.,: Http://www.nlm.nih.gov/research/visible/visible_human.html. (1997)

9 Silicon Graphics.: OpenGL on Silicon Graphics System. From http://trant.sgi.com/opengl/docs/docs.html. (1997)

10 Sobierajski, L. M., and Avila, R. S.,: A Hardware Acceleration Method for Volumetric Ray Tracing. In Proceedings of IEEE Visualization'95. (1995) 27-34

11 Westermann, R. and Ertl, T.,: Efficiently Using Graphics Hardware in Volume Rendering Applications. In Proceedings of ACM SIGGRAPH'98. (1998) 169-178

12 Yagel, R.,: Towards Real Time Volume Rendering. Proceedings of Graphicon'96. Vol. 1, July, (1996) 230-241

13 Zubal, I.G., Harrell, C.R., Smith E. O., Rattner, Z., Gindi, G. R. and Hoffer, P. B.,: Computerized Three-dimensional Segmented Human Anatomy. Medical Physics, Vol. 21, No. 2, (1994) 299-302

14 Zuiderveld, K., Koning, A. H. J., and Viergever, M. A.,: Acceleration of Ray Casting Using 3D Distance Transforms. In Proceedings of Visualization in Biomedical Computing 1992. October, (1992) 324-335,.

Visualization of Medical Data and Molecules

(Research Papers)

Real-Time Maximum Intensity Projection

Lukas Mroz, Andreas König and Eduard Gröller

Vienna University of Technology, Institute of Computer Graphics *

Abstract

Maximum Intensity Projection (MIP) is a volume rendering technique which is used to extract high-intensity structures from volumetric data. At each pixel the highest data value encountered along the corresponding viewing ray is determined. MIP is commonly used to extract vascular structures from medical MRI data sets (angiography). The usual way to compensate for the loss of spatial and occlusion information in MIP images is to view the data from different view points by rotating them. As the generation of MIP is usually non-interactive, this is done by calculating multiple images offline and playing them back as an animation.

In this paper a new algorithm is proposed which is capable of interactively generating Maximum Intensity Projection images using parallel projection and templates. Voxels of the data set which will never contribute to a MIP due to their neighborhood are removed during a preprocessing step. The remaining voxels are stored in a way which guarantees optimal cache coherency regardless of the viewing direction. For use on low-end hardware, a preview-mode is included which renders only more significant parts of the volume during user interaction. Furthermore we demonstrate the usability of our data structure for extensions of the MIP technique like *MIP with depth-shading* and *Local Maximum Intensity Projection (LMIP)*.

keywords: Volume Visualization, Maximum Intensity Projection, angiography

1 Introduction

The ability to depict blood vessels is of enormous importance for many medical imaging applications. CT and MRI scanners can be used to obtain volumetric data sets which allow the extraction of vascular structures. Especially data originating from MRI, which are most frequently used for this purpose, exhibit some properties which make the application of standard volume visualization techniques like ray casting [4] or iso-surface extraction [6] difficult. MRI data sets contain a significant amount of noise. Inhomogeneities in the sampled data make it difficult to extract surfaces of objects by specifying a single iso-value.

MIP exploits the fact, that within angiography data sets the data values of vascular structures are higher than the values of the surrounding tissue. By depicting the maximum data value seen through each pixel, the structure of the vessels contained in the data is captured. A straight forward method for calculating MIP is to perform ray

---

* Institute of Computer Graphics, Vienna University of Technology, Karlsplatz 13/186/2, A-1040 Vienna, Austria. email:{mroz, koenig, groeller}@cg.tuwien.ac.at

casting and search for the maximum sample value along the ray instead of the usual compositing process done in volume rendering. In contrast to direct volume rendering, no early ray termination is possible and the whole volume has to be processed. Depending on the quality requirements of the resulting image, different strategies for finding the maximum value along a ray can be used.

- **Analytical solution:** for each data cell which is intersected by the ray the maximum value encountered by the ray is calculated analytically. This is the most accurate but also computationally most expensive method [7].
- **Sampling and interpolation:** as usually done for ray casting, data values are sampled along the ray using trilinear interpolation. The cost of this approach depends on how many interpolations that do not affect the result can be avoided [7][9].
- **Nearest neighbor interpolation:** values of the data points closest to the ray are taken for maximum estimation. In combination with discrete ray traversal this is the fastest method. As no interpolation is performed, the voxel structure is visible in the resulting image as aliasing [1].

Recent algorithms for MIP employ a set of approaches for speeding up the rendering:

- **Ray traversal and interpolation optimization:** Sakas et al. [7] interpolate only if the maximum value of the examined cell is larger than the ray-maximum calculated so far. For additional speedup they use integer arithmetic for ray traversal and a cache-coherent volume storage scheme. Zuiderveld et al. [9] apply a similar technique to avoid trilinear interpolations. In addition, cells containing only background noise are not interpolated. For further speedup a low-resolution image containing lower-bound maximum estimations for each pixel is used. Cells with values below this bound can be skipped when the final image is generated. Finally a distance-volume is used to skip empty spaces.
- **Use graphics hardware:** Heidrich at al. [3] use conventional polygon rendering hardware to simulate MIP. Several iso-surfaces for different threshold values are extracted from the data set. Before rendering, the geometry is transformed in a way, that the depth of a polygon corresponds to the data value of its iso-surface. MIP is approximated by displaying the z-buffer as a range of gray values.
- **Splatting and shear warp:** Several approaches [1],[2] exploit the advantages of shear-warp rendering [5] to speed up MIP. Cai et al.[1] use an intermediate "worksheet" for compositing interpolated intensity contributions for projection of a single slice of the volume. The worksheet is then combined with the shear image to obtain the maxima. Several splatting modes with different speed/quality tradeoffs are available, run-length encoding and sorting of the encoded segments by value are used to achieve further speedup.

As a MIP contains no shading information, depth and occlusion information is lost (see Figure 1a). Structures with higher data value lying behind a lower valued object even appear to be in front of it. The most common way to ease the interpretation of such images is to animate the viewpoint while viewing. Another approach is to modulate the data values by their depth to achieve a kind of depth shading [3] (see Figure 1b). As the data values are modified before finding the maximum, MIP and depth shaded MIP of the same data may display different objects.

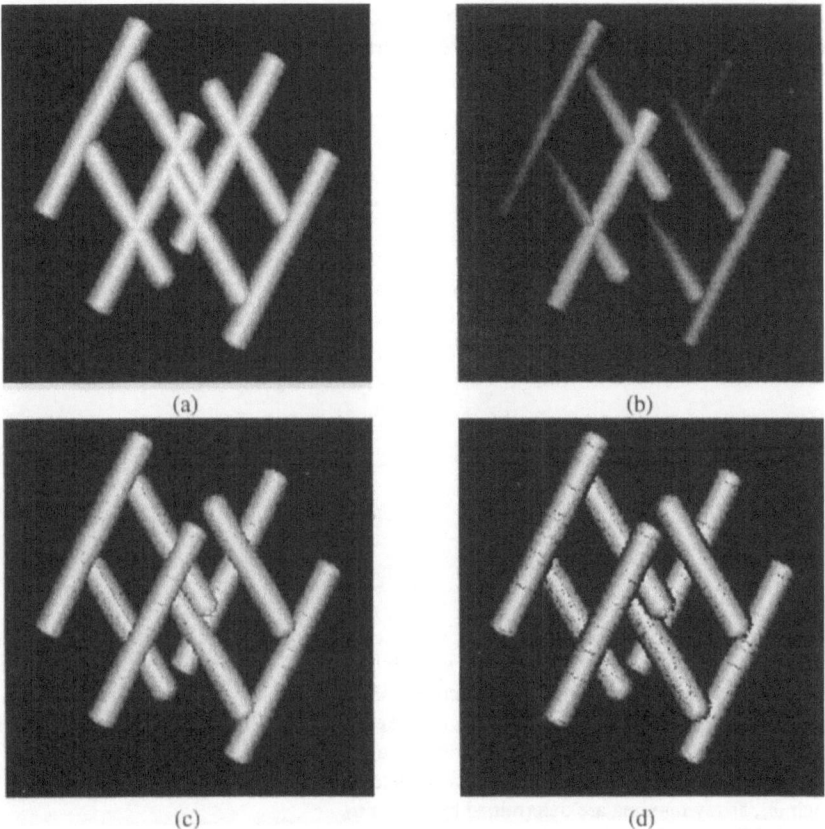

(a) (b)

(c) (d)

Fig. 1. a) MIP of a test data set containing soft bounded cylinders; b) Although depth shading provides some depth impression, the cylinders seem to intersect; c), d) While LMIP provides more information on the spatial structure, the results are sensitive to the threshold: c) high, d) low

Depth shading provides some hints on the spatial relation of objects. However it's performance is rather poor especially for tightly grouped structures. In such cases Closest Vessel Projection [7] or LMIP [8] can be used. As for MIP the volume is traversed along a ray, but the first local maximum which is above a user defined threshold is selected as the pixel color. If no value above the threshold is found along a ray, the global maximum along the ray is used for the pixel. With a high threshold value this method produces the same result as MIP, with a carefully selected one, less intense structures in front of more intense background are depicted, producing an effect similar to shading (see Figures 1c, d). As this method is very sensitive to the setting of the threshold, the ability to interactively tune this parameter is extremely important.

In the following we present a novel approach to generate MIP images. In Section 2 we discuss how to preprocess the volume data to eliminate voxels which will not contribute to a MIP. Furthermore we present a volume storage scheme which gets rid of overhead associated with skipping those voxels for rendering. In Section 3 the algorithm for generating MIP is presented. In Section 4 extensions of the algorithm for generating depth-shaded MIP and LMIP using our data structure are discussed.

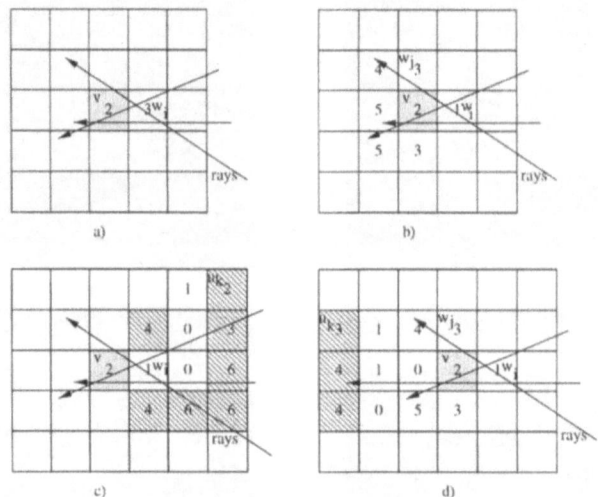

Fig. 2. Detection of voxels v which do not contribute to MIP. For simplicity the different cases are shown in 2D. a) v irrelevant for rays through w_i as $d(w_i) \geq d(v)$ b) v irrelevant for rays through w_i. If discrete 26-connected rays are tracked as in our algorithm, the two voxels with a value of 3 need not to be considered, as a ray through v and w can not pass through them. c) v irrelevant for rays through w_i as ray maxima are determined by voxels u_k. d) v irrelevant for rays through w_i, as ray maxima are determined by voxels u_k.

2 Preprocessing and Volume Storage

Most approaches to optimize volume traversal aim on excluding from the traversal and rendering process voxels which contain less-important information like low-valued background noise. In fact, in addition to this low-importance data, volumetric data sets usually contain a remarkable amount of voxels which never contribute to a MIP image. A voxel v is never visible and can be discarded if all possible rays through the voxel hit another voxel w with $d(w) \geq d(v)$ either before or after passing through v, where $d(v)$ is the data value at voxel v. This fact can be exploited when original voxel values are used for rendering using nearest neighbor interpolation. The hidden voxels can be identified by classifying all 26 neighboring voxels w_i according to the behavior of rays passing through w_i and v. If v does not contribute to the rays through any of it's neighbors, it can be removed. For efficiency reasons, this process is subdivided into four steps, first considering only the values of the direct neighbors w_i:

- If for a direct neighbor $d(w_i) \geq d(v)$, all rays through w_i and v will have at least the value $d(w_i)$, v is not needed (see Figure 2a).

- If for a direct neighbor $d(w_i) < d(v)$, but all possible viewing rays incoming from w_i through v hit direct neighbors w_j of v, with $min(d(w_j)) \geq d(v)$, all rays through w_i and v will have at least a value of $min(d(w_j))$. Thus v has no influence on rays through w_i (Figure 2b).
- If the first two tests fail for voxel w_i, the rays through v and w_i still may be influenced by more distant voxels u_j. If $min(d(u_j)) \geq d(v)$ and every ray hits some u_i, all rays through w_i and v are not influenced by v. (Figure 2c)
- Else, rays entering v through w_i and leaving through w_j may be blocked by distant voxels u_k with $min(d(u_k)) \geq d(v)$. In this case, v has also no influence on the maxima of the rays. (Figure 2d)

Voxel elimination based on the first two tests can be performed rapidly by accessing just direct neighbors of each voxel. The tracking of rays to more distant voxels can be implemented by recursively scanning the sub-volume containing all affected rays. Recursion in a certain direction is terminated if a voxel $d(u_j) \geq d(v)$ is found which blocks all rays in this direction. To avoid unnecessary recursive checks, voxels scheduled for removal are replaced by the minimum of their obscuring values (which is always $\geq$ their original value). To compensate for noise in the data and to increase the number of rejected voxels a user definable tolerance ε can be included into the comparison process, artificially lowering the value of each checked voxel v by ε. For voxel reduction rates using different elimination efforts and tolerance values please refer to Table 1.

data set	resolution	Fast, $\varepsilon=0\%$	Full, $\varepsilon=0\%$	Fast, $\varepsilon=1\%$	Full, $\varepsilon=1\%$
mr_angio	256x256x64	25%	47%	29%	53%
mr01	256x256x74	33%	46%	45%	55%
mr03	256x256x124	29%	58%	29%	58%

Table 1. Volume reduction rates. Fast optimization considers only direct neighbors, full optimization performs recursion up to a distance of 10 voxels. The results show the percentage of voxels removed. The fast optimization takes about 5s on a PII/333, a non-optimized implementation of the full preprocessing takes 3-4 minutes for the 256x256x74 data set.

Although distance volumes are an efficient way to skip large empty volume regions, they are less suited for volumes with large numbers of small empty spaces or even single empty voxels. In addition, several accesses to distance data stored in the empty space are usually required to skip a gap.

The data structure commonly used to store volume data is a 3D array which stores explicitly data values and encodes their position in space implicitly in the value's position in storage.

A way to completely get rid of unwanted voxels from a volume is to store the required voxels in an array of voxel-positions and to encode the value of the voxels implicitly into their position in this array (See Figure 3). The required sorting of all voxels according to their data value can be performed in linear time, as the limited range of possible data values allows to use histogram based sorting. The coordinates of a voxel within a volumetric dataset can be packed into a 32 bit integer, allowing the encoding

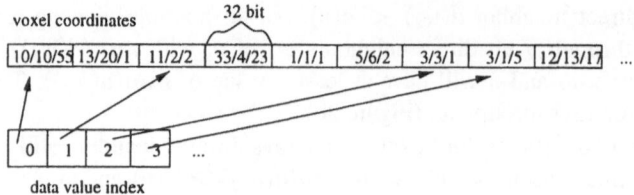

Fig. 3. Value-sorted array structure: All voxels are sorted according to data value and their position in space is stored in an array. An additional index points to the first voxel in the array for each possible data value

of volumes of up to 2048 x 2048 x 1024 voxels. A straight forward conversion of a 16 bit/value volume to a 32-bit position representation would double the memory cost. Omitting the voxels which have been marked by the preprocessing step as irrelevant leads to a factor of 1.0 to 1.5 in storage size compared to the original data. By storing voxels in a value-sorted array we gain several important advantages for MIP:

- The voxels can be splatted in the order of ascending data values as the array is traversed. Therefore **comparing** the value of the actual voxel with the screen content is **not necessary**. The value of the actual voxel is always $\geq$ the content of the screen, thus it's projection can be written into the image.
- As the array is traversed in the same way independent of the viewing direction, **optimum cache coherency** is always achieved.
- As blood vessels are represented by high data values, lower data values usually contain less important information. If an interactive display of the full data set is not possible on the given hardware, lower intensity values can be simply skipped (See Figure 4) to achieve interactivity during user interactions. The algorithm can be adjusted to display MIP at **any desired framerate**. The number of voxels which can be displayed at a specific frame rate can be easily derived automatically from a measurement of the number of voxels which can be rendered per second.

The voxel array structure does not store neighborhood information. It is therefore most useful for algorithms which work with voxels rather than cells. Acquiring the eight data values of a cell from the array without access to the original volume data would be a rather inefficient task.

3   Maximum Intensity Projection

As the voxels of the volume are processed in an arbitrary spatial order which is defined by the order within the value-sorted array, fast incremental techniques for projecting voxels along a ray can not be used. The parallel projection P of a point in 3D, however, can be expressed as the sum of independent projections of it's x, y, and z components.

$$\mathbf{P}\begin{pmatrix} x \\ y \\ z \end{pmatrix} = \mathbf{P}\begin{pmatrix} x \\ 0 \\ 0 \end{pmatrix} + \mathbf{P}\begin{pmatrix} 0 \\ y \\ 0 \end{pmatrix} + \mathbf{P}\begin{pmatrix} 0 \\ 0 \\ z \end{pmatrix}$$

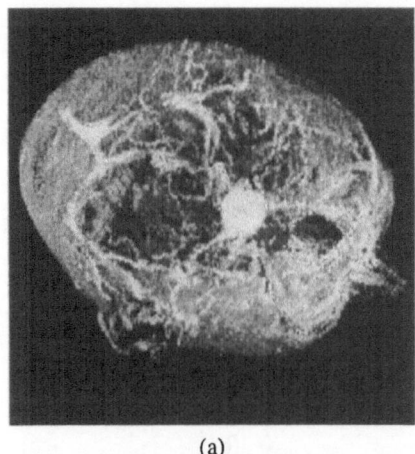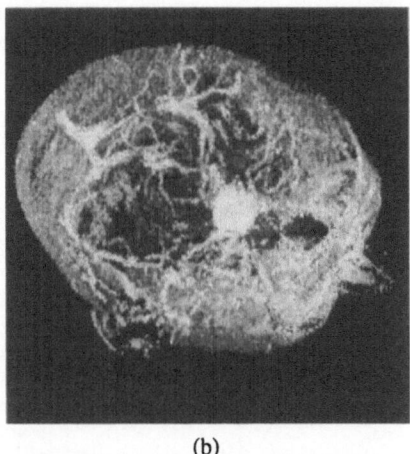

(a)   (b)

Fig. 4. a) MIP of the mr01 data set, 2.6M voxels. b) Interactive preview of the same data at 10 fps on a P233MMX with approximately 25% of the volume data displayed. Only slight differences are noticeable

The projections of each possible voxel coordinate along a coordinate axis can be precalculated and stored into a template. A combination of three templates xs,ys,zs, can be used to inexpensively calculate the position of each voxel's projection on the screen. For increased efficiency, the templates can immediately store offsets into the image buffer instead of x/y screen coordinates. The resulting rendering loop is simple and efficient:

```
xs=make_template(viewmatrix, volumesize_x);
do the same for ys, zs;
for intensity = 0 to max do
  color = gray[intensity];
  for i = first[intensity] to last[intensity] do
    voxpos = value_sorted_array[i];
    screen[xs[voxpos.x]+ys[voxpos.y]+zs[voxpos.z]]=color;
  end
end
```

Similar to other MIP algorithms this approach is also a tradeoff between speed and quality. As each voxel of the data set is projected onto exactly one pixel, the size of the resulting image depends on the size of the volume. Furthermore the composition of multiple integer templates for determining the position of the projection in combination with the projection of a voxel onto exactly one pixel may produce "holes" within the projection of the volume. Although clearly apparent in a static image, due to the high interactivity of this method these artifacts do not disturb the perception of the data during interactive work. The artifacts can be significantly reduced by using sub-pixel coordinates in x-direction. As the templates store offsets into the image buffer ($x + img\_width * y$) instead of x/y coordinates, subsampling in y-direction is not possible.

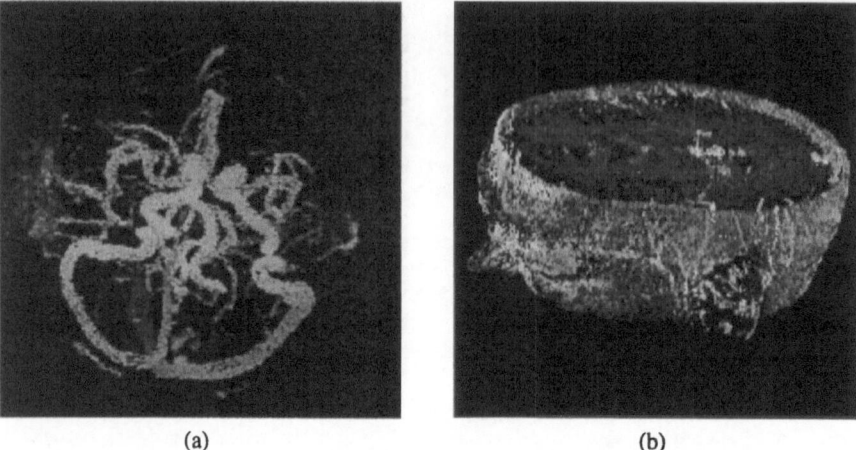

(a)   (b)

Fig. 5. a) MIP of the mr03 angiography data set with depth shading. b) LMIP of mr01

4 Extensions

The algorithm presented in the previous section can be easily extended to generate depth-shaded images (Figure 5). Similar to the screen position templates, three z-templates are generated to calculate the depth of each voxel's projection. The intensity value of each voxel is modulated by it's depth and written into the image buffer only if the value of the pixel at this position is smaller than the new value. As the depicted maximum values do not directly correspond to data values, applying this method to preprocessed volume data may produce incorrect results.

The second extension of the algorithm is capable of generating LMIP images providing the possibility to interactively adjust the threshold parameter. For generating a MIP image, the order of examining samples along a ray is not relevant. Straight forward LMIP requires the samples to be processed front to back along the viewing ray in order to find the first local maximum. Using z-templates for voxel depth calculation and two z-buffers per pixel allows to extract the closest local maximum from samples arriving in arbitrary order. The z[pix] buffer stores the depth of the currently visible voxel, while zb[pix] stores a "back-clipping" distance for each pixel. As the voxels are processed in order of ascending data value, all voxels below the LMIP-threshold can be first projected using the simple and fast MIP algorithm without z-calculation and local minimum tracking. Among all voxels above the threshold which are processed later, closest local maxima have to be found. At the beginning, z[] is initialized to contain the maximum possible distance. With z and v containing the depth and value of a projected voxel, the closest local maximum along a ray is found by

```
if (z<z[pix])
  screen[pix]=v; zb[pix]=z[pix]; z[pix]=z;
else if (z<zb[pix])
  screen[pix]=v; z[pix]=z;
```

The first condition detects voxels closer than the currently displayed voxel. As they have at least the same intensity as the currently displayed one, they are entered into the `screen` and `z[]` buffers, the back clipping plane is moved forward to the previous `z[]` position (Figure 6a). The second condition takes care of new voxels which are placed between the currently displayed voxel and the back clipping distance and are at least of the same intensity as the current one. In this case the `screen` and `z[]` are set to the new voxel. Voxels behind the back clipping distance are ignored (Figure 6b).

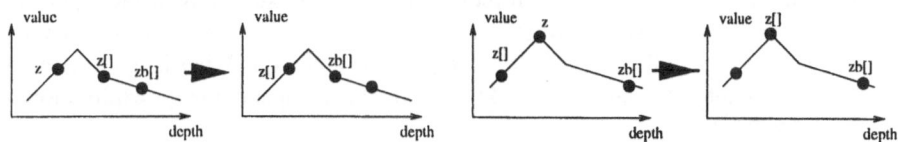

Fig. 6. Local Maximum Intensity Projection for non-sequential samples

5 Future Work

A shear-warp based projection could be used instead of templates with little impact on performance, with the advantage of removing most of the artifacts. Additional future work will concentrate on preprocessing and the use of cells instead of voxels in order to allow high-quality MIP rendering using the same sorted data structure.

6 Results

The volume preprocessing technique as well as all three rendering algorithms presented in this paper have been implemented into a Java applet (*http://www.cg.tuwien.ac.at/research/vis/vismed/RT-MIP/*). The rendering times in Table 2 have been measured on a PII/333 PC using the Java virtual machine of JDK 1.1.6 with a just in time compiler. A C++ version of the MIP rendering algorithm has shown an approximately 30% better performance than the Java version. For measurement a full preprocessing step with a tolerance of 1% has been applied to the data (See Table 1).

data set	MIP	depth-shaded	LMIP
mr_angio 256x256x64	198ms	275ms	274ms
mr01 256x256x74	212ms	279ms	359ms
mr03 256x256x124	343ms	457ms	388ms

Table 2. Frame rendering times for the different algorithms on a PII/333 PC. The timings for LMIP depend on the chosen threshold.

7 Conclusion

We have presented a new rendering algorithm for Maximum Intensity Projection. In combination with a preprocessing step which removes parts of the volume which do not contribute to a MIP image and a volume storage scheme which eliminates overhead for traversing empty regions of the volume, the algorithm achieves highly interactive frame rates. A preview mode for interactions on low-end hardware is provided basically "for free" by skipping voxels with low data values. Parallel projection is reduced to the addition of three values stored in lookup tables. Two extensions of the algorithm allow to generate depth-shaded MIP and LMIP at comparable speed. Although some artifacts are present in the image as no interpolation and a quick but jaggy projection method is used, the high interactivity of this method make it suitable for exploration of angiography data sets.

8 Acknowledgements

The work presented in this publication has been funded by the $V^{is}M^{ed}$ project (http://www.vismed.at). $V^{is}M^{ed}$ is supported by *Tiani Medgraph*, Vienna, http://www.tiani.com, and the *Forschungsförderungsfonds für die gewerbliche Wirtschaft*, Austria, http://www.telekom.at/fff/. The mr_angio data set is courtesy of Tiani Medgraph GesmbH, Vienna. The mr01 and mr03 data sets are available from the United Medical and Dental Schools (UMDS) Image Processing Group in London http://www-ipg.umds.ac.uk/archive/heads.html

References

1. W. Cai and G. Sakas. Maximum intensity projection using splatting in sheared object space. In *Proceedings EUROGRAPHICS '98*, pages C113–C124, 1998.
2. B. Csebfalvi, A. König, and E. Gröller. Fast maximum intensity projection using binary shear-warp factorization. Technical Report TR-186-2-98-27 at the Institute of Computer Graphics, Vienna University of Technology, 1998.
3. W. Heidrich, M. McCool, and J. Stevens. Interactive maximum projection volume rendering. In *Proceedings Visualization '95*, pages 11–18, 1995.
4. J. T. Kajiya. Ray tracing volume densities. In *Proceedings of ACM SIGGRAPH'84*, pages 165–174, 1984.
5. P. Lacroute and M. Levoy. Fast volume rendering using a shear-warp factorisation of the viewing transform. In *Proceedings of ACM SIGGRAPH'94*, pages 451–459, 1984.
6. W. E. Lorensen and H. E. Cline. Marching cubes: A high resolution 3d surface construction algorithm. In *Proceedings of ACM SIGGRAPH'87*, pages 163–189, 1987.
7. G. Sakas, M. Grimm, and A. Savopoulos. Optimized maximum intensity projection. In *Proceedings of 5th EUROGRAPHICS Workshop on Rendering Techniques*, pages 55–63, Dublin, Ireland, Italy, 1995.
8. Y. Sato, N. Shiraga, S. Nakajima, S. Tamura, and R. Kikinis. Lmip: Local maximum intensity projection - a new rendering method for vascular visualization. *Journal of Computer Assisted Tomography*, 22(6), 1998.
9. K. J. Zuiderveld, A. H. J. Koning, and M. A. Viergever. Techniques for speeding up high-quality perspective maximum intensity projection. *Pattern Recognition Letters*, 15:507–517, 1994.

Fast Volume Rotation
using Binary Shear-Warp Factorization

Balázs Csébfalvi

Department of Control Engineering and Information Technology,
Technical University of Budapest,
Budapest, Műegyetem rkp. 11, H-1111, HUNGARY
cseb@seeger.fsz.bme.hu

Abstract. This paper presents a fast volume rotation technique based
on binary shear-warp factorization. Unlike many acceleration algorithms
this method does not trade image quality for speed and does not require
any specialized hardware either. In order to skip precisely the empty
regions along the rays to be evaluated a binary volume is generated
indicating the locations of the transparent cells. This mask is rotated by
an incremental binary shear transformation, executing bitwise boolean
operations on integers storing the bits of the binary volume. The ray
casting is accelerated using the transformed mask and an appropriate
lookup-table technique for finding the first non-transparent cell along
each ray.

1 Introduction

Direct volume rendering is a flexible but computationally expensive technique
for visualization of 3D density arrays. Because of the huge number of voxels to
be processed the recent software-only acceleration methods are not fast enough
for interactive applications. Real-time frame rates can be achieved using large
multi-processor systems [1][4][10][13], but they are not widely used because of
their high costs. In the last two decades several accelerated volume-rendering
techniques have been proposed, which exploit the coherence of the data set.

Early methods use hierarchical data structures like pyramids, octrees or K-d
trees [2][6][11] to quickly traverse the transparent regions decreasing the number
of samples to be evaluated. Hierarchical data structures are used in homogeneity
acceleration techniques as well [2][14], which apply a simplified approximate
evaluation for the homogeneous regions.

Recent algorithms like distance transformation based methods [15][16] or
Lacroute's shear-warp factorization projection [5] concentrate on a more precise
skipping of empty ray segments instead of approximated evaluation of homogeneous regions. The main advantage of these techniques over hierarchical methods
is the applied encoding scheme, since the information about the empty cells, is
available with the same indices used for the volume data. Furthermore, there is
no additional computational cost for handling a hierarchical data structure during the ray traversal. Applying these acceleration methods the rendering time

can be reduced significantly, but the frame rates are still far from interactivity. Although there are surface oriented algorithms which provide real-time rotation, their limitation is the lack of the opacity manipulation [17][18].

2 The Algorithm

The software-only acceleration technique presented in this paper can also be considered as a surface oriented algorithm since interactive frame rates can be achieved if the opacity function is binary or near binary (the non-zero opacities are near one). On the other hand unlike the previous iso-surface methods it supports the high quality transfer function based rendering as well. In a practical application the surface rendering can be used as a fast preview, where the user can set the appropriate viewing angle interactively and afterwards the final image is rendered using the alpha-blending evaluation according to the selected transfer function.

2.1 Preprocessing

The input data of a direct volume rendering process is a spatial density function $f : R^3 \to R$ sampled at regular grid points, yielding a volume $V : Z^3 \to R$ of size $X \times Y \times Z$, where

$$V_{i,j,k} = f(x_i, y_j, z_k).$$

In the classification process, according to the density values different attributes like opacity or color are assigned to each voxel. The opacity function maps the volume V onto a classified volume $C : Z^3 \to [0,1]$ of the same size. In order to handle the empty cells efficiently many acceleration algorithms create a binary volume assigning zero to the transparent and one to the non-transparent cells, where a cell is considered transparent if all of its eight corner voxels have zero opacities. In our method the definition of a cell depends on the principal direction of the viewpoint. Without loss of generality, we assume that the principal direction is the z-axis. In this case, the proposed algorithm resamples the volume only in planes $z = z_k$, where $k = 0,1,2, ...$ $Z-1$. The density samples on these planes are computed from the densities of the four closest voxels using bilinear interpolation. The opacity of the sample is non-zero if at least one of the four voxels is opaque, thus the binary volume $B : Z^3 \to \{0,1\}$ of size $(X-1) \times (Y-1) \times Z$ is defined the following way:

$$B_{i,j,k} = \begin{cases} 1 \text{ if } C_{i,j,k} > 0 \text{ or} \\ \quad C_{i+1,j,k} > 0 \text{ or} \\ \quad C_{i,j+1,k} > 0 \text{ or} \\ \quad C_{i+1,j+1,k} > 0 \\ 0 \text{ otherwise.} \end{cases}$$

2.2 Ray Casting

The binary volume B can be stored in an integer array, where an integer represents a segment of a bit row parallel to the z-axis. In the special case, when the viewing direction is exactly the z-axis the volume can be rendered very efficiently by parallel ray casting, since the problem of finding the first non-transparent voxel hit by a ray can be reduced to the problem of finding the first non-zero bit inside an integer. The optimal solution to this problem is the direct addressing of a lookup table by the given integer which contains the position of the first non-zero bit for all the possible combinations. For a typical integer size like 32 bits it would require the allocation of a 2^{32} byte array which cannot be used in practice. Instead of this, the lookup table can store the offset of the first non-zero bit inside one byte for each byte combination and it can be addressed by the first non-zero byte of the given integer. Assuming that the most significant bit is the nearest one to the viewer the first non-zero byte can be determined by binary search. The following routine provides the position of the first non-zero bit, where the size of an integer is supposed to be 32 bits:

```
int Depth(int segment) {
  if(segment < 0x00010000) {
    if(segment < 0x00000100) return 24 + lut[segment];
    else return 16 + lut[segment >> 8];
  } else {
    if(segment < 0x01000000) return 8 + lut[segment >> 16];
    else return lut[segment >> 24];
  }
}
```

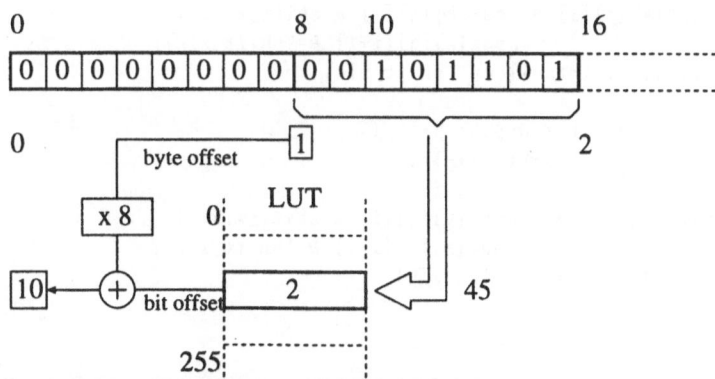

Fig. 1. An example for a LUT entry.

Usually one integer is not enough for storing a complete row of the binary volume B, thus the segments of the rows stored in integers have to be checked sequentially and the routine *Depth* is called only for the first non-zero integer.

2.3 Shear-Warp Factorization

The previous method works only for a special case but it can be extended to viewing directions, where the principal component is the coordinate z using binary shear-warp factorization. This transformation effectively moves the bits of the binary volume perpendicularly to the viewing direction. In an interactive volume rendering application the volume is required to be rotated continuously by small difference angles, in order to perceive the topology of the surfaces much better than in a static image. If the difference angle is small enough then there is no slice in the binary volume which has to be shifted by more than one bit. In this case, one shear operation can be performed very efficiently, since just bitwise operations need to be executed on neighboring integers. That is the reason why our method shears the binary volume B incrementally, applying a technique similar to the method proposed by Cohen-Or and Fleishman [1]. They used their so called incremental alignment algorithm in order to reduce the communication overhead in a large multi-processor architecture supporting shearing of volumes. Since some bits can be shifted out of the integer array storing the binary volume B, it has to be extended by $Z/2$ rows filled with zero values along the x-axis and along the y-axis as well in both directions. This extended array is defined as: int $mask[depth][height][width]$, where $depth = (Z + 31)$ div 32, $height = Y + Z$ and $width = X + Z$. The following routine demonstrates, how to execute one shear step in the left direction, processing 32 voxels in each step of the internal loop:

```
void ShearLeft() {
  int i, j, k;
  for(k = 0; k < depth/2; k++) {
    for(j = 0; j < height; j++) {
      for(i = 0; i < width-1; i++)
        mask[k][j][i] = mask[k][j][i] & shift_x[k] |
                        mask[k][j][i+1] & ~shift_x[k];
      mask[k][j][width-1] &= ~shift_x[k];
    }
  for(k = depth/2; k < depth; k++) {
    for(j = 0; j < height; j++) {
      for(i = width-1; i > 0; i--)
        mask[k][j][i] = mask[k][j][i] & shift_x[k] |
                        mask[k][j][i-1] & ~shift_x[k];
      mask[k][j][0] &= ~shift_x[k];
    }
}
```

For the sake of clarity this routine is not optimized, but it can be improved introducing local pointer variables in order to avoid unnecessary array addressing, and on the other hand only that part of the extended mask needs to be sheared which contains the non-zero bits representing the non-transparent cells. The integer array $shift\_x$ is defined as: int $shift\_x[depth]$ and it stores a binary vector of size Z indicating those z positions where the corresponding slices have to be shifted in the given shearing phase. There is also such an array denoted by *shift_y* for the y direction. In order to determine the offsets of the slices in different shearing phases, another two arrays are introduced for storing the real translations along the x-axis and the y-axis and they are defined as: *double trans_x[Z]* and *double trans_y[Z]* respectively. These translation arrays contain the x and y coordinates of those points, where the $z = z_k$ planes intersect the 3D line connecting the point p_0 (*trans_x[0]*,*trans_y[0]*,0) with point p_1 (*trans_x[Z-1]*,*trans_y[Z-1]*,Z-1). Initially, this line aligns to the z-axis, thus the translation arrays contain zeros. Before executing a binary shear operation the *shift* vectors are evaluated in advance according to the rotation direction. For example, when a clockwise rotation around the y-axis is needed, the point p_0 is translated by one along the x-axis into negative direction and p_1 is translated as well, but into positive direction. After this, the intersection points of the line connecting the new p_0 and p_1 and the planes $z = z_k$ are computed anew and the coordinates are stored in the translation arrays. The new binary shift vectors are determined according to these translation values. For example, the new *shift_x* array can be computed using the following routine:

```
void ComputeNewShiftX()
  {
    double x0 = trans_x[0] -= 1.0, x1 = trans_x[Z-1] += 1.0;
    int bit = 0x80000000, l = Z - 1;
    for(int z = 0; z < Z; z++) {
      double x = (x0 * (l - z) + x1 * z) / l;
      if(floor(x) != floor(trans_x[z])) shift_x[z/32] |= bit >> (z % 32);
      else shift_x[z/32] |= ~(bit >> (z % 32));
      trans_x[z] = x;
    }
  }
```

If the floors of the previous and the new translation values are not the same then the corresponding bit is set to one in the *shift_x* array, indicating that the associated slice needs to be shifted in the next binary shear operation. Since the difference between the old and new translations is the greatest in the plane $z = z_0$ (or $z = z_{Z-1}$) the difference cannot be greater than one in the intermediate z points, thus there is no slice which needs to be shifted with more than one bit.

2.4 Resampling

Using the transformed binary volume an intermediate image of size *width* × *height* is generated casting the rays from the grid points. Due to the shear transformation the *Depth* routine can be applied in the general case as well, since the segments of the rows perpendicular to the temporary image plane are stored in integers. The position of the first non-zero bit in a row determines the index i of the $z = z_i$ plane, where the first opaque sample is located along the corresponding ray. The accurate location of this sample is computed taken into account the exact translation values at the given depth z. In order to calculate the density in this sample point location, bilinear interpolation is used for the four closest voxels. Since the opacity of the sample is not necessarily one so the ray has to be traced further and evaluated according to the transfer function. If the volume contains internal empty regions (like a human skull) it makes sense to use the binary volume for evaluating the rest of the samples. First, the integer representing a z-row of bits is copied into a temporary variable, and whenever a non-zero bit has been processed it is set to zero, thus the routine *Depth* can be used again for finding the z position of the next non-transparent cell. Having the intermediate image generated it has to be warped in order to produce the final image, which is the parallel projection of the volume.

3 Extensions

The presented method can be extended to arbitrary viewing directions since a binary volume can be created for each principal direction. Applying an appropriate scaling for the slices perpendicular to the principal component, the algorithm can be used for perspective projection as well. The next two subsections describe two further improvements which could be useful in a practical implementation.

3.1 Rotation of Large Data Sets

Due to the incremental shear transformation the effective speed of the rotation could be low, especially processing larger volumes (256^3). Since most of the time is used for rendering, a possible solution to this problem is to render the volume after a couple of incremental shears. Increasing the size of the data set the ratio of the rendering and shearing times is getting lower, thus this strategy is not the best one. Another alternative is the introduction of super cells, which are square regions in the slices perpendicular to the principal direction. In the binary volume, one is assigned to the corresponding super cell if at least one voxel in it is opaque. Increasing the size of the super cell the shear transformation becomes faster but the rendering process slows down since the routine *Depth* does not necessarily return the exact depth only a lower bound, so the rays have to be traced further until having found a non-transparent sample. In the next section the performance is analyzed investigating the optimal cell size for data sets of different sizes.

3.2 Adaptive Thresholding

The primary limitation of the presented technique is that the volume has to be classified in advance in order to generate the binary volume, and afterwards the opacity function cannot be modified in a flexible way. Supposing that, the user wants to operate with a fixed number of transfer functions an appropriate density encoding scheme can be used to allow rapid switching between them. Each transfer function has a lower density threshold, where below this threshold zero opacity is assigned to the given sample. Assume that we want to use only three transfer functions. The lower density thresholds divide the density domain into intervals I_0, I_1, I_2, I_3 sorted in ascending order by their borders with increasing index. To each interval the two bit binary format of the corresponding index is assigned as a unique code. The code of a cell is defined as the code of the interval which contains the highest corner voxel density. The cell codes are stored in an integer array which is similar to the *mask* array but it contains two bits for each cell instead of one. This array can be sheared as well, but the bits of a code should always be moved at the same time in order to avoid the cutting of the codes. In the rendering phase the routine *Depth* must use the appropriate lookup table depending on the bit pattern to be searched for. Whenever the user changes the transfer function the variable *lut* has to be set to the address of the corresponding lookup table. This encoding scheme allows rapid access to the first non-transparent cell along the viewing ray independently from the selected transfer function. Let us take an example from the medical imaging practice, where only four materials (air, fat soft tissue, and bone) can be separated according to the *Hounsfield densities*[3][9]. In this case it makes sense to divide the density domain according to the lower density threshold of fat, soft tissue, and bone respectively. For example, having selected a transfer function which assigns zero opacities to the samples below the lower threshold of the soft tissue, only the codes 10 and 11 will be searched for in the binary volume, precisely skipping the transparent cells. In this case, the corresponding lookup table contains the bit offset of the first 10 or 11 pattern inside the given byte. The presented density encoding scheme does not affect significantly the performance and allows fast switching between the predefined transfer functions.

4 Implementation

The proposed fast rotation technique has been implemented in C++ and it has been tested on an SGI Indy workstation. Table 1 summarizes the running time measurements for a CT scan of a human head and Table 2 contains the test results for a higher resolution volume of the same data. The applied transfer function assigns high opacities to the voxels representing the bone thus rays terminate right after reaching the boundary of the skull (Figure 2).

cell size	shearing time	rendering time	frame rate
1	0.019 sec	0.114 sec	6.87 Hz
2	0.005 sec	0.107 sec	8.64 Hz
3	0.002 sec	0.118 sec	8.11 Hz
4	0.001 sec	0.156 sec	6.26 Hz

Table 1. Test results for the CT head of size $128 \times 128 \times 113$.

Note that, the optimal super cell size is not necessarily the one which the highest frame rate belongs to, since with larger super cell size the effective rotation speed is higher. In order to rotate the volume continuously the cell size

cell size	shearing time	rendering time	frame rate
1	0.160 sec	0.590 sec	1.21 Hz
2	0.040 sec	0.492 sec	1.81 Hz
3	0.017 sec	0.535 sec	1.78 Hz
4	0.009 sec	0.709 sec	1.37 Hz

Table 2. Test results for the CT head of size $256 \times 256 \times 225$.

must be set small and higher rotation speed can be achieved by setting larger cell size producing approximately the same frame rates.

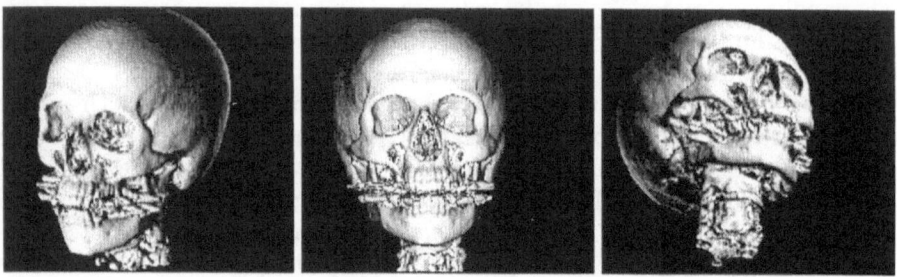

Fig. 2. Interactive rotation using fast iso-surface rendering.

Using transfer functions which assign low opacity values to different tissues the rendering time increases drastically, since after skipping the empty regions the alpha-blending evaluation of the semi-transparent segments is computationally very expensive. Although setting larger super cell size higher rotation speed can be achieved in the fast previewing phase the high quality rendering slows down since the binary volume contains less precise information about the transparent cells. Table 3 and Table 4 show the average rendering times for the low and high resolution data sets respectively using three different transfer functions as demonstrated in Figure 3.

cell size	transfer function A	transfer function B	transfer function C
1	0.36 sec	0.19 sec	0.18 sec
2	0.39 sec	0.21 sec	0.21 sec
3	0.43 sec	0.25 sec	0.24 sec
4	0.51 sec	0.32 sec	0.31 sec

Table 3. Rendering times for the volume of size $128 \times 128 \times 113$.

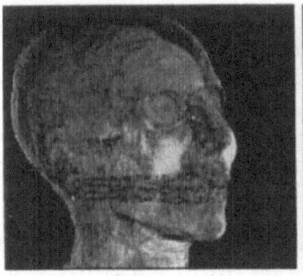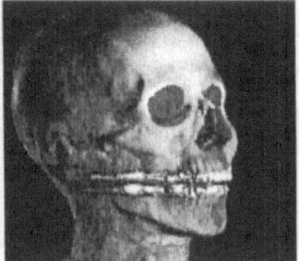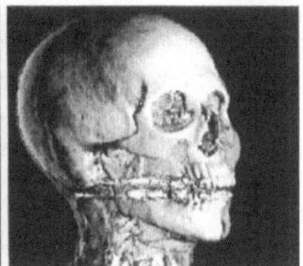

Fig. 3. Alpha-blending rendering using different transfer functions.

cell size	transfer function A	transfer function B	transfer function C
1	1.97 sec	1.02 sec	0.99 sec
2	1.93 sec	1.01 sec	0.96 sec
3	2.08 sec	1.09 sec	1.08 sec
4	2.44 sec	1.45 sec	1.46 sec

Table 4. Rendering times for the volume of size $256 \times 256 \times 225$.

5   Conclusion

In this paper a fast volume rotation technique has been presented which provides interactive frame rates without using any specialized hardware support. Real-time rotation can be achieved using binary or near binary opacity function, when rays terminate right after reaching an opaque surface. In this sense the proposed technique is a surface oriented algorithm but unlike other interactive iso-surface methods it significantly speeds up the classical transfer function based ray casting. Because of the precise skipping of empty regions it is approximately as efficient as the classical shear-warp algorithm based on run-length encoding. In a practical implementation it can be applied as a fast previewer rendering the iso-surface defined by the lower density threshold of the selected transfer function, where the viewing direction can be set interactively, and the final image is rendered using the alpha-blending evaluation. The effective rotation speed can be increased by setting a larger super cell size and applying the proposed adaptive thresholding extension the user can switch rapidly between predefined transfer functions.

Acknowledgements

This work has been funded by the V^{is}M^{ed} project (http://www.vismed.at). V^{is}M^{ed} is supported by *Tiani Medgraph*, Vienna (http://www.tiani.com), and by the *Forschungsförderungsfond für die gewerbliche Wirtschaft*. This paper has also been supported by the National Scientific Research Fund (OTKA ref.No.:

154

F 015884) and the Austrian-Hungarian Action Fund (ref.No.: 29ö4 and 32öu9). The CT scan was obtained from the Chapel Hill Volume Rendering Test Dataset. The data was taken on the General Electric CT scanner and provided courtesy of North Carolina Memorial Hospital.

References

1. Daniel Cohen-Or and Shachar Fleishman. An incremental alignment algorithm for parallel volume rendering. *Computer Graphics Forum (EUROGRAPHICS '95 Proceedings)*, pages 123–133, 1995.
2. John Denskin and Pat Hanrahan. Fast algorithms for volume ray tracing. *Workshop on Volume Visualization*, pages 91–98, 1992.
3. R.A. Drebin, L. Carpenter and P. Hanrahan. Volume rendering. *Computer Graphics (SIGGRAPH '88 Proceedings)*, 22:65–74, 1988.
4. Jürgen Hesser, Reinhard Männer, Günter Knittel, Wolfgang Strasser, Hanspeter Pfister and Arie Kaufman. Three architectures for volume rendering. *Computer Graphics Forum (EUROGRAPHICS '95 Proceedings)*, pages 111–122, 1995.
5. Philippe Lacroute and Marc Levoy. Fast volume rendering using a shear-warp factorization of the viewing transformation. *Computer Graphics (SIGGRAPH '94 Proceedings)*, pages 451–457, 1994.
6. David Laur and Pat Hanrahan. Hierarchical splatting: A progressive refinement algorithm for volume rendering. *Computer Graphics (SIGGRAPH '91 Proceedings)*, pages 285–288, 1991.
7. Marc Levoy. Display of surfaces from ct data. *IEEE Computer Graphics and Application*, 8:29–37, 1988.
8. Marc Levoy. Efficient ray tracing of volume data. *ATG*, 9(3):245–261, 1990.
9. Derek R. Ney, Elliot K. Fishman, Donna Magid and Marc Levoy. Computed tomography data: Principles and techniques. *IEEE Computer Graphics and Application*, 8, 1988.
10. Peter Schröder and Gordon Stoll. Data parallel volume rendering as line drawing. *Workshop on Volume Visualization*, pages 25–32, 1992.
11. K.R. Subramanian and Donald S. Fussell. Applying space subdivision techniques to volume rendering. *IEEE Visualization '90*, pages 150–159, 1990.
12. L. Szirmay-Kalos (editor). *Theory of Three Dimensional Computer Graphics*. Akadémia Kiadó, Budapest, 1995.
13. Guy Vézina, Peter A. Fletcher and Philip K. Robertson. Volume rendering on the maspar mp-1. *Workshop on Volume Visualization*, pages 3–8, 1992.
14. Jason Freund and Kenneth Sloan. Accelerated volume rendering using homogeneous region encoding. *IEEE Visualization '97*, pages 191–196, 1997.
15. D. Cohen and Z. Shefer. Proximity clouds - an acceleration technique for 3D grid traversal. *TR FC93-01, Ben Gurion University, Israel*, 1993.
16. K. Zuiderveld, A. Koning, Viergever and A. Max. Acceleration of ray casting using 3D distance transformation. *Visualization in Biomedical Computing*, pages 324–335, 1992.
17. Jae-jeong Choi and Yeong Gil Shin. Efficient Image-Based Rendering of Volume Data. *TR http://cglab.snu.ac.kr/ jjchoi/ibr.html, Seoul National University, Korea*, 1998.
18. Björn Gudmundsson and Michael Randén. Incremental generation of projections of CT-volumes. *In First Conf. on Visualization in Biomedical Computing, Atlanta*, 1990.

VIVENDI - A Virtual Ventricle Endoscopy System for Virtual Medicine

Dirk Bartz[1] and Martin Skalej[2]

[1] WSI/GRIS, University of Tübingen,
Auf der Morgenstelle 10/C9,
D72076 Tübingen, Germany
Email: bartz@gris.uni-tuebingen.de
[2] Department of Neuroradiology, University Hospital Tübingen
Hoppe-Seyler-Str. 3
D72076 Tübingen, Germany

Abstract. Virtual Medicine is an emerging and challenging field in Computer Graphics. Numerous visualization methods are used to model and render data of different modalities.

In this paper, we present a new endoscopy system for virtual medicine. The main purpose of this system is to provide support for the planning of complicated endoscopic interventions inside of the ventricular system of the human brain. Although, our current focus is on ventricle endoscopy, this system is applicable to other areas as well.

In order to achieve interactive framerates on workstations with medium graphics performance, we apply an efficient implementation of a basic algorithm for general visibility queries.

Keywords: Virtual Medicine, Virtual Environments, Surgical Assist Systems.

1 Introduction

Minimally-invasive neurosurgical procedures are of increasing importance in neurosurgery. In comparison to commonly used surgical techniques, less brain tissue is damaged. Furthermore, minimally-invasive procedures have less deleterious effects on the patient. On the other hand, these procedures lack fast access in case of serious complications, such as strong bleeding. Therefore, careful planning and realization of this procedure is essential, in order to avoid such complications. This problem aggravates, because handling and control of these endoscopes is very difficult, mainly due to limited flexibility of and limited field of view through the endoscope (which provides only a very limited orientation), and the sensitive nature of the brain tissue.

To optimize the success of these interventions, an improved planning and training environment is required. Consequently, we developed a virtual ventricle endoscopy system in cooperation with the Department of Neuroradiology of the university hospital at Tübingen.

Traditional planning of endoscopic interventions in neurosurgery is based on the thorough examination of slice images of a MRI scan. We add two additional stages to this planning pipeline; a pre-processing stage and the virtual endoscopy of the ventricular system (Figure 1).

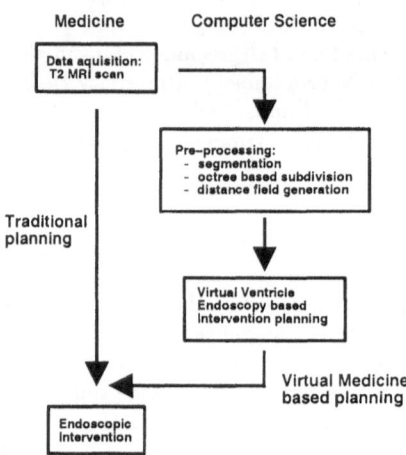

Fig. 1. Flow of virtual endoscopy-based intervention planning.

We will discuss these two stages in this paper, which is organized as follows: We start with a brief survey of methods for virtual medicine. In Section 3, we outline the basic medical problems that indicate the need for ventricle endoscopy. In Section 4, we present our system for planning endoscopic interventions inside the ventricular system. Finally, in Section 5, we state a conclusion and give perspectives to future work.

2 Related Work

There has been quite some work in the field of virtual medicine. In this Section, we present a brief discussion of some of the related work. Among the first is [17], where methods of Computer-Aided Geometric Design (CAGD) were applied for planning of cranial surgery.

In [13, 10, 9], Finite Element Methods were used to construct models of tissue of the human body. These models were used to simulate deformations of the tissue to predict the outcome of plastic or craniofacial surgery.
Computer-based anatomical atlases were introduced by Höhne et al. [5]. Based

on CT and MRI volume data, methods of artificial intelligence and volume rendering were combined.

Recently, Serra et al. introduced a framework for stereo-tactic frame surgery [16], using a set of computer-based tools and a 3D-output device, similar to the virtual table paradigm.

One of the major recent research topics in virtual medicine is virtual endoscopy. Frequently, the proposed methods generate off-line animations of virtual cameras simulating an endoscopic session through various hollow organs. In 1994, Geiger and Kikinis proposed using Finite Element Methods to specify a path of the camera [3]. Similar approaches of automatic generation of the camera path were investigated by Vining et al. [18], Lorensen et al. [11], and Hong [6]. In contrast, Rubin et al. manually specify a key-frame interpolated camera path [14].

Hong et al. proposed a new navigation method, implementing the guided-navigation paradigms [7]. By combining distance fields and kinematic rules, an intuitive scheme for navigating inside the human colon was developed. Furthermore, a customized visibility algorithm was proposed in order to reduce the number of surface polygons of the inner surface of the colon to a feasible size. While this system provided a fast and intuitive handling of the virtual endoscope, it required high-end SGI InfiniteReality graphics for interactive framerates.

3 Ventricle Endoscopy

In minimally-invasive neurosurgery of the brain, existing cavities can be used as a preformed path for movements of the endoscope, without destruction of brain tissue. Our focus is on the ventricular system of the human brain, in which the brain liquor (cerebrospinal fluid or CSF) is produced and resorbed (Figure 2 (a)). To access the ventricles, a hole is drilled through the skull and a tube is placed through this hole into the ventricular system. Thereafter, the endoscope is introduced through the tube, which is used as a stable guide for the endoscope.

Because of the water-like optical property of the CSF - which fills the ventricular system, viewing of the surrounding tissue is possible. Movement of the endoscope - guided by video-control via the small field of view of the endoscope - is limited by the tube and the surrounding tissue (Figure 2 (b)). Micro-instruments, introduced through an additional canal inside the endoscope, can then be used to perform the actual minimally-invasive procedure, e.g. removing accessible mass lesions.

Due to respiration and other metabolistic activity, the CSF flows through the cavities inside the human brain. In some cases, the drain inside the ventricles - the aqueduct (Figure 2 (a), B) - is blocked by occlusion or stenosis. This causes a serious disturbance of the natural flow of the CSF, which frequently leads to a dangerous increase of pressure inside the skull and can damage the brain severely. The clinical picture of this hydrocephalus is one of the major indications

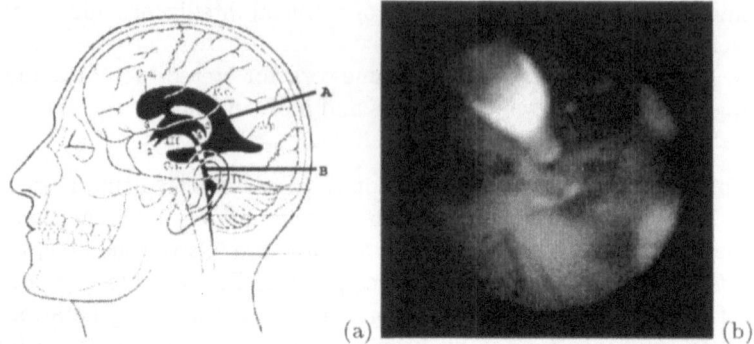

Fig. 2. Ventricular system of the human head: (a) A ventricles, B aqueduct; (b) view through endoscope.

for a minimally-invasive intervention in the ventricular system, where a bypass is realized by perforating the floor of the third ventricle. However, the limited view and orientation through-out the intervention increases the necessary time of the intervention and consequently, the inherent risks of serious complications. To overcome these drawbacks, we propose the use of a virtual endoscopy system to improve the planning of and orientation during this procedure.

4   VIVENDI - Virtual Ventricle Endoscopy

In this Section, we discuss the elements of the VIVENDI system for virtual ventricle endoscopy. In some parts, VIVENDI follows the VICON system for interactive colonoscopy [7]. However, due to the different anatomical topology, we use a different subdivision and visibility scheme. Therefore, the only common method is the guided-navigation system, which is already discussed in detail in [7].

4.1   System Architecture

The endoscopy system itself consist of two stages: pre-process and interactive virtual endoscopy. The pre-processing stage is responsible for the generation of numerous auxiliary data, which is later used during the interactive virtual endoscopy. It is organized in three major steps, which are outlined in Figure 3. In the first step, we segment the voxels classified as inside of the ventricles using a 3D region growing algorithm starting from an interactively specified seed point. Subsequently, the default path of our virtual camera is generated using Dijkstra's minimum path algorithm [7], using the seed point and an additionally specified target point as start and end point of the camera path. Thereafter, we

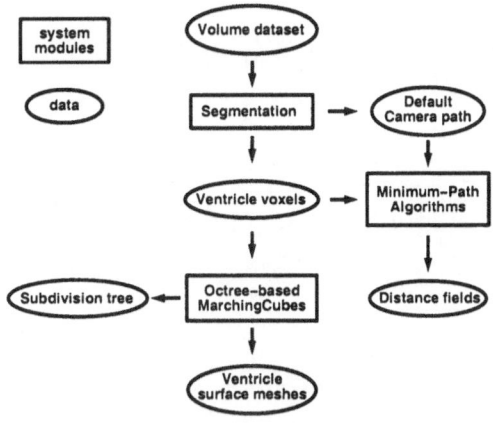

Fig. 3. Pre-process flow.

extract the isosurfaces of the ventricular system, using an octree-based parallel implementation of the Marching Cubes algorithm [1]. The size of the leaf blocks of the octree depends on a specified granularity value of volume cells which intersect with the isosurface[1]. Based on this octree representation, a subdivision of the extracted isosurface is generated, where the isosurface geometry associated with an octree leaf block is considered as one subdivision entity. Finally, three distance fields are computed, implementing a collision avoidance scheme and the motion of the virtual camera for guided-navigation [7] towards a target point.

For interactive colonoscopy, complete pre-processing time took up to ten hours. Most of the time was spent on generating the default camera path (or skeleton of the colon) and on the generation of the subdivision along this skeleton. However, this time can be greatly reduced by using improved data structures and algorithms. Replacing FIFO-queue based priority queues of the first pre-processing step by Fibonacci heaps [12], we could reduce the algorithmic time complexity, hence reducing the time comsumption significantly.

For the previous skeleton-based subdivision step, a back-tracking algorithm was used, in order to optimize the size of the respective subdivision cells. However, this was a computational expensive approach, which frequently lead to a processing time of several hours. Due to the shape of the ventricular system, a tube-like subdivision is not available. Instead, we used a generalized octree-based subdivision scheme where computational costs are only a fraction of the skeleton-based subdivision. In total, we reduced the pre-processing time down to approximately 15 minutes.

---

[1] We call these cells relevant cells and the respective granularity value *relevant cell load* or RCL.

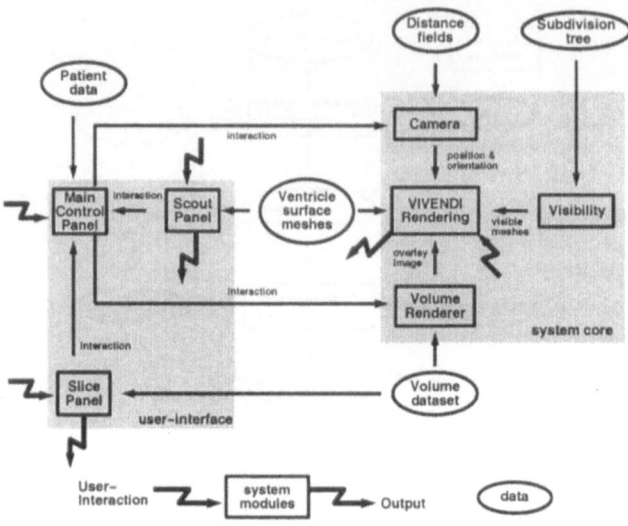

Fig. 4. VIVENDI control flow.

The interactive endoscopy stage of the VIVENDI system is built from an user-interface and system core (Figure 4). The centerpiece of the latter is the VIVENDI render system (module VIVENDI Rendering), which is responsible for the OpenGL rendering of the geometry of the organ (Fig. 5). In order to reduce its complexity, a visibility culling algorithm is applied (module Visibility, see Section 4.2 for details). User-interaction (for navigation and measurement) via the rendering area of the VIVENDI rendering is bypassed to the Main Control Panel. Position and orientation of the virtual camera are provided by the guided-navigation system (module Camera, see [7] for details). If the user initiates direct volume rendering (module Volume Renderer), the generated images are overlaid with the polygonal rendering of VIVENDI Rendering.

On the left hand side of Figure 4, the user-interface (see Fig. 5) is organized around the Main Control Panel, which provides control over camera navigation, volume rendering, animation generation, and other general parameters of the system. The Main Control Panel communicates with the Slice Panel, which provides the traditional three orthogonal slices of the volume dataset in different resolutions. The Scout Panel provides a fully rotatable 3D overview of the surface of the ventricular system. Furthermore, it administrates the position markers and controls the multiple camera paths (Section 4.3). If a position marker is used as a jump mark, it provides the virtual camera with the new position and orientation via the Main Control Panel.

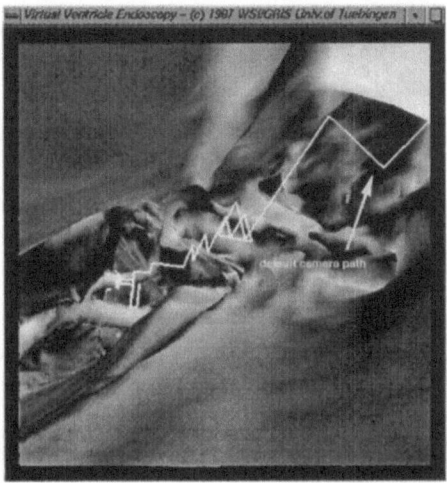

Fig. 5. Endoscopic view.

4.2   Visibility

Research on algorithms for visibility queries is one of the most active areas in Computer Graphics. Consequently, several algorithms for occlusion culling have been proposed. Most notable are the hierarchical Zbuffer (HZB) approach [4], hierarchical occlusion maps (HOM) [19], and the aggregate cull rectangle algorithm (ACR) [7]. Considering the structure of the problem and necessary interactivity, none of these algorithms are suited for virtual ventricle endoscopy. The HZB-algorithm lacks efficiency for interactive rendering of medium sized datasets (up to 1M triangles) on current graphics computers. In contrast, the HOM-algorithm requires a dedicated scene subdivision into separated objects in order to get a good occluder pre-selection, which is necessary for high performance. Finally, the ACR-algorithm provides a good interactive performance, based on a tube-like topology — like the human colon. However, this topology is not available in the ventricular system. For these reasons, we chose view-frustum culling [2] for our application.

After subdividing the ventricle dataset into octree blocks in a pre-process step, each octree block of the subdivision tree above a given granularity of relevant cells (RCL) is tested for intersection with the view-frustum. If the octree block is completely within the view-frustum, all its child blocks are within the view-frustum as well, hence they are potentially visible. If the octree block is only partially inside the view-frustum, we proceed with its child blocks. If this block has no further children — it has reached the specified polygon load —, all its polygons are considered potentially visible. Finally, if the octree block is completely outside the view-frustum, all its child blocks and associated polygons

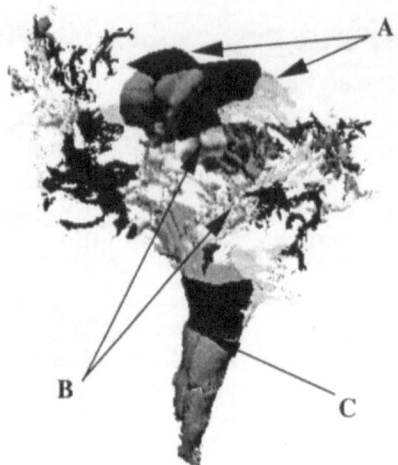

Fig. 6. Octree Subdivision of Ventricular system - different blocks are marked with different grey-levels. Note that the segmentation does not only select voxels of the ventricular system, but all connected CSF-filled cavities. A: Lateral Ventricles, B: 3rd and 4th Ventricles, C: Central Canal.

are not visible and therefore, not rendered.

We exploit the *OpenGL selection mode* to implement our culling algorithm [8]. The whole view-frustum is used as the sensitive area for selection. We render the named (with an ID) octree blocks in selection mode, which generates no contribution to the framebuffer. If any geometry of the octree blocks intersects with the view-frustum, the selection mode returns a *hit report*, including the IDs associated with these blocks and their minimum z-values. If a block completely contains the view-frustum, this simple test would fail. Therefore, it is necessary to explicitly check before-hand for this case.

Recently, Hewlett-Packard introduced their fx-series of graphics systems. A speciality of this system is the support for occlusion culling with the *HP Occlusion Culling Flag* [15]. We rendered the depth-sorted (according to their minimum z-value) potentially visible octree blocks using this functionallity on a HP B180/fx4 graphics workstation. In a special *occlusion mode*, the HP flag detects if a rendered geometry would change the content of the z-buffer of the framebuffer. By checking this flag, we can determine if the octree block is occluded.

Table 1 shows the trade-off between visibility test overhead and rendering performance. The first subdivision into 120 blocks achieves the best framerate. However, the framerate of the second subdivision is only 0.4 frames/second (fps) lower, although the number of rendered polygons is 15% larger. Furthermore, Table 1 show the results of view-frustum culling with and without occlusion culling on a HP B-class fx4 workstation using a 180 MHz PA-7300 LC CPU and

RCL	total #blocks	% blocks visible	#polygons visible	cull rate [%]	framerate [fps]
View-frustum culling only					
1000	120	52	185,834	52.4	7.6
2000	59	27	220,414	43.7	7.14
4000	28	16	248,618	36.5	7.12
View-frustum and occlusion culling					
1000	120	45	27,121	93.1	19.4
2000	59	13	50,054	87.2	17.8
4000	28	13	83,941	78.6	15.5

Table 1. Average performance of view-frustum culling of a ventricular system with more than 391K polygons.

the HP Occlusion Culling flag. Overall, a culling performance of up to 93% of the model and framerates up to 19.4 fps on the HP were achieved. However, in some areas with low occlusion, occlusion culling does not increase the framerate. In fact, it might reduce the framerate if the occlusion culling overhead exceeds the benefits of culling polygons of the model. Although we only presented results for one dataset, other datasets showed a similar performance.

4.3 Multiple Camera Paths

Segmentation and navigation depend heavily on the start point of the camera path (the seed point of the segmentation) and the camera path itself respectively. However, not all areas of interest are connected by voxels classified as inside. This is due to partial volume effects, lack of resolution or obstruction of narrow areas. Furthermore, some examinations require different camera paths in order to reach different locations within the ventricular system. For these cases, VIVENDI supports multiple camera paths, which combines the models of different areas of interest of one volume dataset into a joined model. Each single model is pre-processed individually - providing its own camera path (and reconstructed isosurface if necessary) - and finally combined into the joined model with the respective number of camera paths. Alternatively, this functionality can be used to generate multiple camera paths in a single model[2].

For the actual virtual endoscopic examination, all individual models of the joined model which are not the current one, are considered not visible. To change to one of the other camera paths, automatic pre-defined blue model markers at the ends of the camera paths on the scout panel can be selected (Figure 7). In the course of changing to another camera path, the distance fields are changed too. This is necessary to reduce the look-up time of the binary tree on the compressed representation of the distance fields.

---

[2] In this case, the single model is not simply copied, but referenced for each additional camera path.

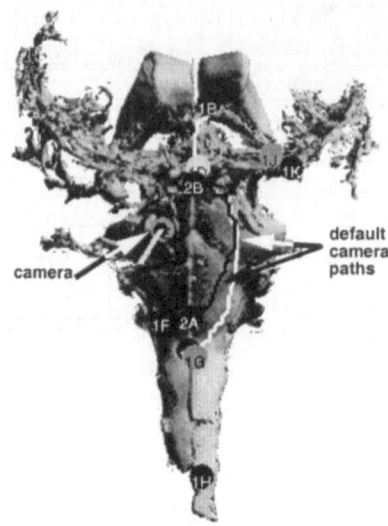

Fig. 7. Snapshot from the Scout Panel: Two different camera paths are combined into one model. Blue (dark) markers are pre-defined model markers; red, green, and yellow (different shades of grey) markers are user-defined markers.

5   Conclusion and Future Work

In this paper, we presented a virtual endoscopy system, applied to virtual endoscopy of the ventricular system of the human brain. Starting from a virtual colonoscopy system, we replaced a visibility algorithm for tube-like objects by a basic view-frustum culling method, due to the non-tube topology of the ventricular system. Furthermore, we improved and adapted the pre-processing steps of our application. Pre-processing time of several hours was reduced to a few minutes. Finally, we introduced multiple camera paths - combining and handling of several camera paths in one model - and a new user-interface to adapt to the needs of our partners at the hospital.

Our system is suitable to many applications for virtual endoscopy, although the presented application is virtual ventricle endoscopy to support planning of endoscopic interventions. Therefore, one of the future focuses will be on further exploration of other applications of virtual endoscopy, especially inside the human head.

Early feedback from the Department of Neurosurgery suggests a high usability and acceptance for this clinical application. Nevertheless, further clinical studies are on the agenda of future work.

For our measurements, we used a HP B180/fx4 workstation. However, measurements of smaller ventricular system models suggest that computer systems with less rendering performance and without hardware support for occlusion culling, i.e., a SGI O_2 or a PC graphics card, have sufficient graphics performance if powerful standard OpenGL-based occlusion culling algorithms are used[8]. Consequently, a major research focus will be on occlusion culling methods.

Acknowledgments

This work has been supported by the MedWis program of the German Federal Ministry for Education, Science, Research and Technology, and by hardware of the Hewlett-Packard Workstation Systems Lab., Ft. Collins, CO. Early implementations were derived from a prototype of the VICON system of the Center of Visual Computing at Stony Brook.

Datasets were provided by the Department of Neuroradiology of the University Hospital Tübingen. Special thanks to Dorothea Welte, Barbara Kortmann, Rupert Kolb, and Mechthild Uesbeck of the Department of Neuroradiology, Frank Duffner of the Department of Neurosurgery, and Hendrik-Jan van Veen of the Max-Planck-Instiute for Biological Cybernetics. Last but not least, we thank Michael Doggett and Michael Meißner for proof-reading.

References

1. D. Bartz, R. Grosso, T. Ertl, and W. Straßer. Parallel Construction and Isosurface Extraction of Recursive Tree Structures. In *Proc. of WSCG*, volume III, 1998.
2. B. Garlick, D. Baum, and J. Winget. Interactive Viewing of Large Geometric Databases Using Multiprocessor Graphics Workstations. In *ACM SIGGRAPH'90 course notes: Parallel Algorithms and Architectures for 3D Image Generation*, 1990.
3. B. Geiger and R. Kikinis. Simulation of Endoscopy. In *AAAI Spring Symposium Series: Application of Computer Vision in Medical Image Processing*, pages 138–140, 1994.
4. N. Greene, M. Kass, and G. Miller. Hierarchical Z-Buffer Visibility. In *Proc. of ACM SIGGRAPH*, pages 231–238, 1993.
5. K. Höhne, M. Bomans, M. Riemer, R. Schubert, and U. Tiede. A 3D Anatomical Atlas Based on a Volume Model. *IEEE Computer Graphics Applications*, 12:72–78, 1992.
6. L. Hong, A. Kaufman, Y. Wei, A. Viswambharan, M. Wax, and Z. Liang. 3D Virtual Colonoscopy. In *IEEE Symposium on Biomedical Visualization*, pages 26–32, 1995.
7. L. Hong, S. Muraki, A. Kaufman, D. Bartz, and T. He. Virtual Voyage: Interactive Navigation in the Human Colon. In *Proc. of ACM SIGGRAPH*, pages 27–34, 1997.
8. T. Hüttner, M. Meißner, and D. Bartz. OpenGL-assisted Visibility Queries of Large Polygonal Models. Technical Report WSI-98-6, ISSN 0946-3852, Dept. of Computer Science (WSI), University of Tübingen, 1998.
9. E. Keeve, S. Girod, P. Pfeifle, and B. Girod. Anatomy-based Facial Tissue Modelling Using the Finite Element Method. In *Proc. of IEEE Visualization*, pages 21–28, 1996.

166

10. R. Koch, M. Gross, F. Carls, D. Büren, G. Fankhauser, and Y. Parish. Simulating Facial Surgery Using Finite Element Methods. In *Proc. of ACM SIGGRAPH*, pages 421–428, 1996.

11. W. Lorensen, F. Jolesz, and R. Kikinis. The Exploration of Cross-Sectional Data with a Virtual Endoscope. In R. Satava and K. Morgan, editors, *Interactive Technology and New Medical Paradigms for Health Care*, pages 221–230. 1995.

12. K. Mehlhorn, S. Näher, and C. Uhrig. The LEDA Platform for Combinatorial and Geometric Computing. In *24th International Colloquium on Automata, Languages, and Programming (ICALP-97)*, volume LNCS 1256, pages 7–16, 1997.

13. S. Pieper. *CAPS: Computer Aided Plastic Surgery*. PhD thesis, MIT, 1992.

14. G. Rubin, C. Beaulieu, V. Argiro, H. Ringl, A. Norbash, J. Feller, M. Dake, R. Jeffrey, and S. Napel. Perspective Volume Rendering of CT and MR Images: Application for Endoscopic Imaging. In *Radiology*, volume 199, pages 321–330, 1994.

15. N. Scott, D. Olsen, and E. Gannett. An Overview of the VISUALIZE fx Graphics Accelerator Hardware. *The Hewlett-Packard Journal*, (May):28–34, 1998.

16. L. Serra, W. Nowinski, T. Poston, N. Hern, L. Meng, C. Guan, and P. Pillay. The Brain Bench: Virtual Tools for Stereotactic Frame Surgery. In *Medical Image Analysis*, volume 1(4), pages 317–329, 1997.

17. M. Vannier, J. Marsh, and O. Warren. Three Dimensional Computer Graphics for Craniofacial Surgical Planning and Evaluation. In *Proc. of ACM SIGGRAPH*, pages 263–273, 1983.

18. D. Vining, D. Gelfand, R. Bechtold, E. Scharling, E. Grishaw, and R. Shifrin. Technical Feasibility of Colon Imaging with Helical CT and Virtual Reality. In *Annual Meeting of American Roentgen Society*, page 104, 1994.

19. H. Zhang, D. Manocha, T. Hudson, and K. Hoff. Visibility Culling Using Hierarchical Occlusion Maps. In *Proc. of ACM SIGGRAPH*, pages 77–88, 1997.

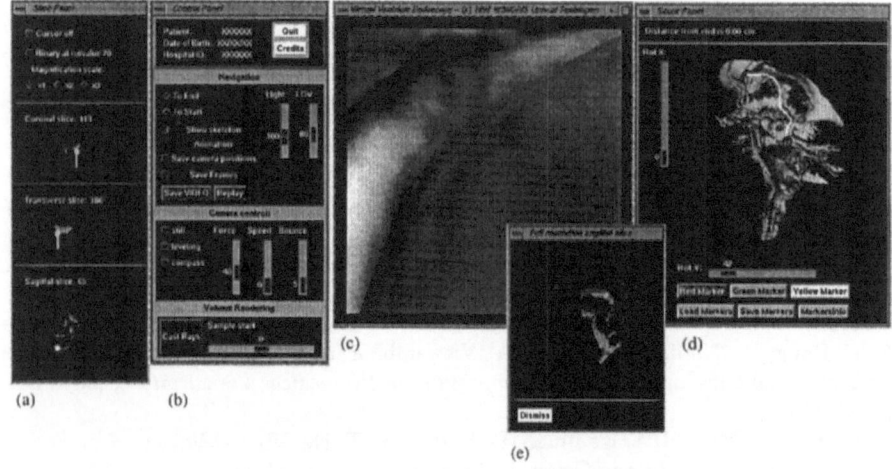

Fig. 8. User-interface of the virtual ventriculoscopy system: (a) Slice Panel, (b) Control Panel, (c) Main View, (d) Scout Panel, and (e) Full Resolution Sagittal Slice Panel.

Editors' Note: see Appendix, p. 324 for colored figures of this paper

A Client-side Approach towards Platform Independent Molecular Visualization over the World Wide Web

Michael Bender, Hans Hagen, and Axel Seck

bender|hagen|seck@informatik.uni-kl.de
Computer Graphics Group, Department of Computer Science,
University of Kaiserslautern

Abstract. A web-based, entirely platform independent Molecular Visualization System has been developed using state of the art Internet programming techniques. This system offers the visualization of various molecular models, molecular surfaces and molecular properties which can be displayed at the same time. The system itself has been developed using the Java programming language, which allows flexible and platform independent use, perfect integration with the World Wide Web and due to its object-oriented structure easy extension and maintenance. All necessary calculations, e.g. the calculation of a Richards' Contact Surface or an Isosurface, take place on the client's side exploiting the computational power of modern desktop workstations and personal computers.

1 Introduction

From the 70s on, computer molecular representations began to reduce the common limitations of physical molecular models, namely size restriction due to weight and mechanical stability of the models themselves. While the first generations of computers permitted only two-dimensional illustrations for the reason of low computational and graphical power, the development in the last twenty years, led quickly to the kind of scientific visualization we are currently used to.

Commercial and public domain programs offer a variety of molecular representations from simple Stick or Ball-and-Stick Models to advanced surface models like the Van-der-Waals Surface Model or the Richards' Contact Surface Model, allowing easy mapping of molecular properties onto the surfaces. On high-end graphics workstations, e.g. an Onyx2, it is not worth mentioning visualizing hundreds of molecules at the same time, displaying animations in real-time or even computing dynamic interactions with the user in real-time.

Two aspects of applications for Molecular Visualization, not yet considered, are platform independence and – directly related to this – Molecular Visualization over the World Wide Web. These are two main topics this work is focused on. Since highly-optimized computational algorithms and high-speed graphical output depend strongly on the specific underlying system architecture, our approach includes the use of standard methods and libraries to achieve our goal:

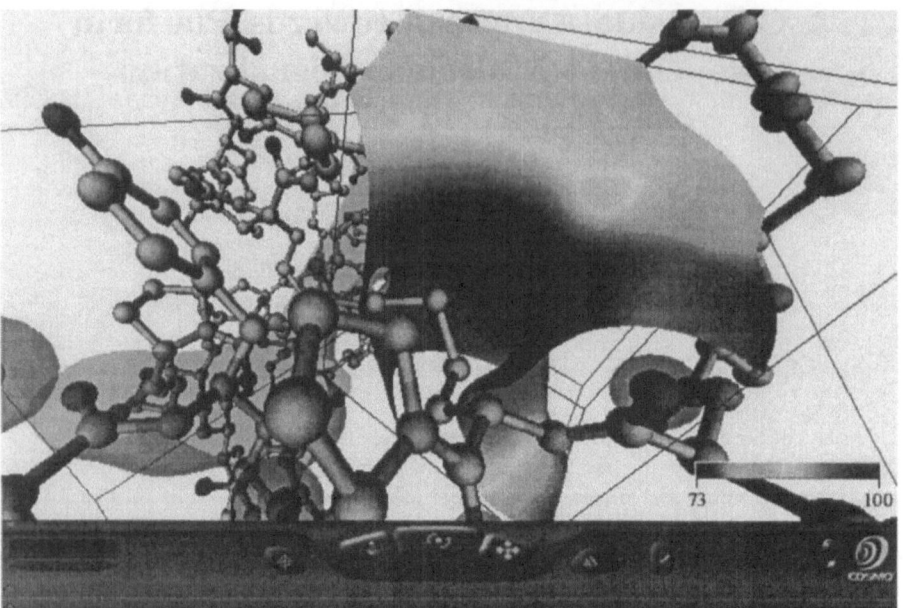

Fig. 1. Plant seed protein *(Ball-and-Stick Model, different Isosurfaces)*

An entirely platform independent web-based tool for displaying molecules and their chemical and physical properties. The molecule data as well as the application itself, which is completely written in Java, are loaded directly by a client from the World Wide Web via an Internet Browser and the application is then executed by the client doing all necessary calculations for visualization.

The rest of this paper is organized as follows. In Chap. 2 we briefly review some of the previous work that has been done in web-based Molecular Visualization. Chapter 3 gives a short description of common molecular models, corresponding computation algorithms and molecular properties. In Chap. 4 we outline our approach and Chap. 5 presents our results. Finally, we give our conclusions in Chap. 6 and point out promising future research.

2   Previous and Related Work

Within the scope of the World Wide Web special description languages for chemical objects apart from HTML have been defined (e.g. CML[1]). Moreover, for several molecule data formats (e.g. PDP, XYZ, Alchemie) extensions of the MIME format exist, requesting the use of special Browser plug-ins for visualization (e.g. Chime[2]). For the present the most common description format to distribute three-dimensional data over the World Wide Web is certainly VRML,

---

[1] the Chemical Markup Language, see http://www.venus.co.uk/OMF/.

[2] see http://www.mdli.com/download/chime/.

which may be replaced by something like Java3D in future. [1] describes a general WWW-based visualization service where the user enters some data on the client which is transferred to the server calculating the VRML model which is then retransmitted to the client for display. Each shape changing interaction by the user starts another cycle. The "Virtual Reality Modelling Language in Chemistry"[3] fits to this description. A complex in-house chemical information system based on Internet programming techniques with selected tasks shifted to the client's side is described in [2]. [3] presents a Java applet for interactive 3D visualization on the Web which makes use of the Marching Cubes Algorithm on the client's side. In [4] a progressive approach for transmitting Isosurfaces to an applet is applied to avoid Internet bottlenecks. The Molda System [5] is a client-side molecular modeling and graphics system using Java and VRML and offering the visualization of standard molecular models.

What all these systems have in common is that they emphasize the server's role (clients are pure display clients) or that they do not exploit the clients' today's abilities.

3 Molecular Models and Properties

3.1 Standard Models

The most common standard models in Molecular Visualization are Wireframe Models, Stick Models, Ball-and-Stick Models and space-filling Corey-Pauling-Koltun (CPK) Models (see Fig. 2 for examples).

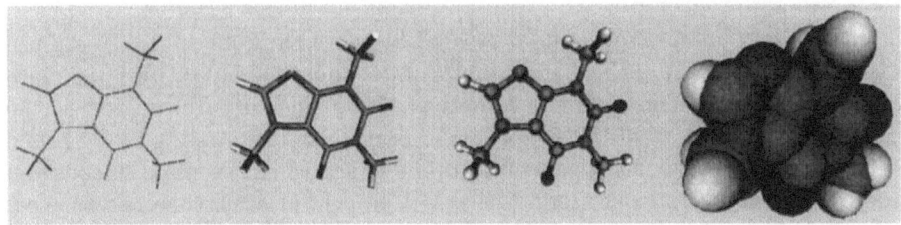

Fig. 2. Standard molecular models *(from left to right: Wireframe, Stick, Ball-and-Stick and CPK Model of caffeine)*

In *Wireframe Models* bonds are represented by lines while atoms are not regarded. *Stick Models* extend this idea by representing the bonds by cylinders of a chosen diameter. *Ball-and-Stick* Models include atoms represented by spheres which diameters are a chosen fraction of the atoms' Van-der-Waals radii. In *CPK Models* bonds are invisible with atoms visualized as spheres using their Van-der-Waals radii. All standard models choose their colors according to the CPK standard colors.

---

[3] see http://www.pc.chemie.tu-darmstadt.de/vrml/.

3.2   Surface Models

Computerized visualizations allow the display of various molecular surfaces, which depend on positions and radii of the atoms forming the displayed molecule and on chemical or physical properties. In addition, different molecular properties can be mapped onto these surfaces. A comparison of common molecular surface definitions can be seen in Fig. 3.

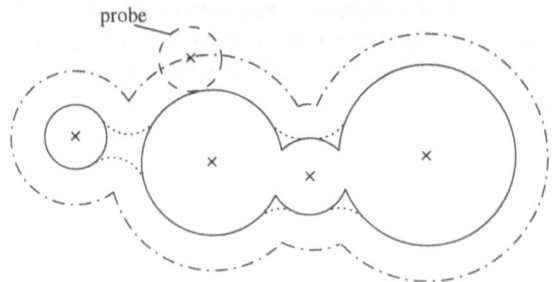

Fig. 3. Comparison of different molecular surface definitions   *(solid: Van-der-Waals Surface, dot: Richards' Contact Surface, dash-dot: Richards' Accessible Surface)*

The *Van-der-Waals Surface* is the simplest molecular surface. It consists of the visible part of the union of all atoms composing the molecule displayed with their Van-der-Waals radii.

Richards' Contact Surface [6] is created by rolling a probe sphere of a given radius, usually 1.4 Å (sphere including a water molecule H_2O), over the Van-der-Waals Surface. The surface is composed of two kinds of surface patches: the part of the Van-der-Waals Surface of each atom which is accessible to the probe sphere and the inward facing part of the probe sphere when it is simultaneously in contact with more than one atom. Richards' Contact Surface is constructed using Conolly's Algorithm [7, 8] combining an analytical computation of the different surface patches with a following triangulation step of a chosen accuracy. The previously described Van-der-Waals Surface can be treated as a special case of Richards' Contact Surface using a probe sphere of infinitesimal radius.

Richards' Accessible Surface [6] is constructed in the same way as Richards' Contact Surface, except that the surface itself is composed of the positions reached by the probe's center.

Isosurfaces represent all spatial positions with constant values of a chosen scalar molecular property. A very popular way to compute an Isosurface is the Marching Cubes Algorithm [9]. After a grid based subdivision of the given volume of density values this algorithm merges local triangular approximations to an approximation of the entire Isosurface. Of course, Isosurfaces depend on the calculation method used for the selected molecular property.

3.3 Molecular Properties

There are a lot of different molecular properties which can be mapped onto the previously described molecular surfaces or which can be used to define Isosurfaces. Molecular orbitals are generally represented by Isosurfaces. Curvature can describe the curvature of a given molecular surface, the gradient of a selected scalar property or the value (of the gradient) of a non-scalar molecular property. Color mapping of interaction energy allows the chemist to rapidly and intuitively locate the regions corresponding to favorable interactions. Hydrophobicity represents the molecule's aversion against water molecules. It could be approximated using the tables found in [10]. A simple approximation of the molecule's electrostatic potential V is given by

$$V(r) = \sum_{i=1}^{n} \frac{q_i}{|r - r_i|} \tag{1}$$

where r is the selected position, r_i the atoms' centers and q_i the atoms' electronegativities.

4 Our Approach

Our goal was the design of a web-based Molecular Visualization system meeting the following requirements: 1. The system structure has to allow easy extension and exchange of system components. 2. The construction of the system has to provide a way to load the client's components dynamically over the Internet. 3. The system should not be affected by Internet or server bottlenecks. 4. The system has to operate platform independent but it should make use of resources offered by the client machine (e.g. OpenGL hardware acceleration). 5. The chemical data to visualize has to be transferred via the Web in a common standardized data format. 6. The system should offer advanced molecular surface models with real-time interaction rates. 7. The Graphical User Interface has to provide a straightforward and standardized user control.

The stated requirements lead quickly to the use of the Java programming language guaranteeing perfect integration with the World Wide Web and excellent platform independent use. Additionally, the object-oriented structure of Java permits the split of the system into interacting components and allows easy maintenance. Besides this, the ability to produce standardized Graphical User Interfaces is one of Java's inherent properties. All remaining requirements will be the contents of the following paragraphs explaining the system's structure and its interacting components with their functions.

4.1 Environment

The entire system has been developed using Java version 1.1. For the reason of Java3D not yet running reliably enough for our tasks we have chosen VRML 2.0

as an intermediate solution. The display of the VRML models is done by Silicon Graphics' CosmoPlayer, which is available as a Browser plug-in for Windows based systems, for SGIs and for Macintosh platforms. For the communication with the VRML Browser the External Authoring Interface (EAI) is used. We apply Netscape's Navigator as our favorite Web Browser.

On the server's side arbitrary http servers providing a Common Gateway Interface can be used – we are applying Apache and Roxen Challenger servers.

We use the Brookhaven Protein Data Bank (PDB) Format[4] as a standardized molecule data format. This is no restriction with Babel[5] offering on-line translation between different data formats.

4.2 System Structure

The design of our system follows the client-server approach but with essential tasks dedicated to the clients. Figure 4 shows a simplification of the system's design.

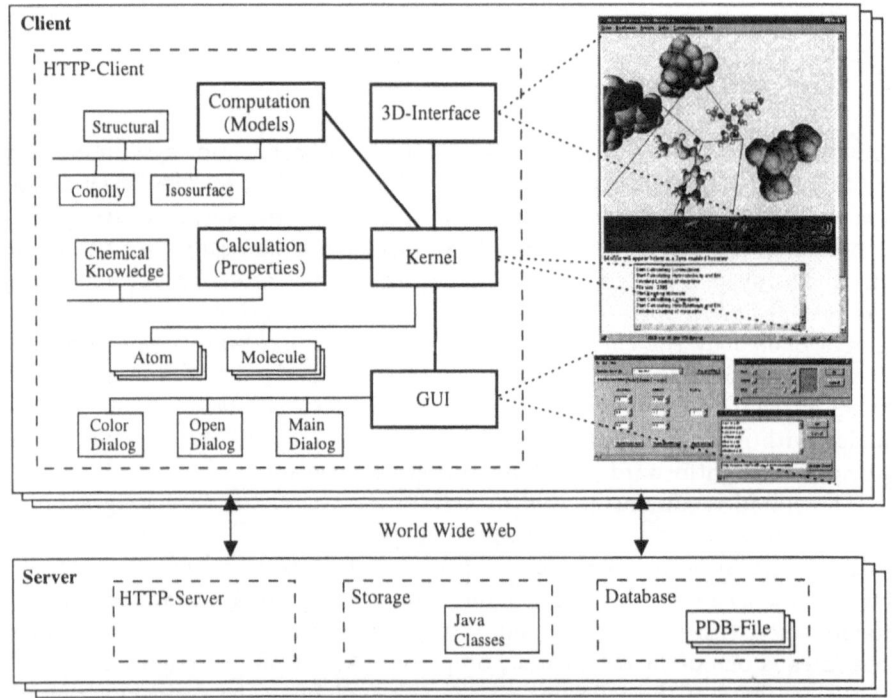

Fig. 4. System structure

---

[4] see http://pdb.pdb.bnl.gov/.
[5] see http://www.chemoinformatics.com/free/babel.html.

The server's side The server machine runs a http server daemon which can access a local database of chemical atomic coordinate files in PDB format. The http server has to provide a list of PDB files and it has to be able to send each PDB file in a compressed format to a client. The client's Java classes are stored on a server to take advantage of dynamic Java class loading.

The client's side The client machine has to run a Java enabled http client, usually a Web Browser. The client's part of our system consists of five main components each corresponding to one or more Java packages, but we will disregard single Java classes and the Java class hierarchy in our descriptions.

The *kernel* corresponds to the Java applet which actually is the most important part of the Java program on the client's side controlling the main thread and establishing all connections with the user's Web Browser and with the 3D-interface. The kernel initializes the other modules and it is responsible for all internal and Internet communication. The client's part of the Molecular Visualization system can handle several molecules with individual visualization parameters and molecular models at the same time – one of them is always labeled as the active molecule. For this purpose each molecule is stored separately, including all information concerning it (atoms, current molecular models, visualization parameters). Operations (add, delete, duplicate) on this internal database for molecules are also handled by the kernel component. It is important to point out, that all data concerning visualization is stored in a native format to become independent of the currently used 3D-interface and its data format. The kernel displays all kind of status information (molecule name, communication status, computation and calculation progress, memory usage, et cetera) in a console window below the 3D-interface in the user's Web Browser.

The *computation module* is responsible for the production of visualization-ready molecular models which can be passed through the kernel to the 3D-interface. While the computation of a standard model is a simple generation of drawing primitives the computation of an advanced surface model requires time and memory consuming algorithms. We have implemented Conolly's Algorithm to compute Richards' Contact Surface. The accuracy in the triangulation step is determined by a user-controlled error criterion. Isosurfaces are computed with the Marching Cubes Algorithm. In addition to a resolution vector (n_x, n_y, n_z) used for the subdivision of the regular grid the user can limit the algorithm's processing to a bounding box. This is extremely useful to examine surface details. The computation module is also responsible for the mapping of molecular properties onto surfaces models.

The *calculation module* offers algorithms approximating molecular electrostatic potential and hydrophobicity at any given spatial position. Currently both approximations work with weighted sums so the calculation time is proportional to the number of atoms in a molecule.

The *3D-interface* is responsible for the display of the molecular models. It provides drawing primitives like lines, spheres, cylinders or triangle meshes representing complex surface shapes. The primitives' properties, especially color, are used to map molecular properties. Besides navigation (rotating, moving, zooming) with the mouse, our current 3D-interface offers simple user interaction with callback functions. The 3D-interface is supposed to exploit the client's graphical power to maximize the visualization quality, for example frame and interaction rate. Silicon Graphics' CosmoPlayer for instance is able to use the support of hardware accelerated OpenGL graphics boards on Windows based systems. This component encapsulates all accesses to the three-dimensional in- and output, so it can easily be adapted to Java3D in future.

The *GUI* is responsible for all user interactions. Its dialog windows offer various possibilities to start up actions (e.g. calculations) and to change parameters.

Networking A client can receive PDB data from several servers on the Internet and on the other hand one server may offer its service to many clients on the Web. Internet connections are only kept for short times when they are really needed. This is the case for the dynamic Java class loading, for the choice of a molecule from a server's browsable list and for the download of a molecule's PDB description. Java classes as well as molecular descriptions are both transferred over the Web in a compressed format to avoid bottlenecks.

5 Results

To examine our system's capabilities - in particular its computational and its rendering performance - it has been tested on different computer platforms (Windows and SGI based systems with different hardware configurations) with several molecules of different structure and complexity. As an example we present some results achieved on a today's typical client system, a medium equipped Windows based PC System with a 350MHz Pentium II processor, 128MB main memory and a hardware accelerated OpenGL graphics board – currently such a configuration can be obtained for less than $US 1,000. The speed analysis in Table 1 lists the measured computation time for the model, the resulting triangle count to be displayed and the frame rate achieved. The numbers indicate a minor importance of speed restrictions introduced by the Java programming language. Note that the Ball-and-Stick and the CPK Model do not use triangle primitives.

Table 1. Measured computation time (sec), triangle count and frame rate

	Ball-and-Stick		CPK		Richards' Contact			Isosurface		
ethanol	0.0 –	30.3	0.0 –	29.9	1.559	2,668	20.5	5.815	22,690	6.1
aspirin	0.0 –	20.1	0.0 –	19.1	4.592	10,922	11.9	6.423	22,492	6.2
morphine	0.0 –	9.5	0.0 –	8.2	14.942	34,098	6.2	46.387	18,436	9.2
strychnine	0.0 –	8.4	0.0 –	7.0	19.311	43,502	4.6	55.739	18,414	8.7

In the upper left picture of Fig. 5 (see Appendix) a Ball-and-Stick Model of ribonucleic acid is combined with an Isosurface (yellow, transparent) determining the value 140 of electrostatic potential. In the upper right picture hydrophobicity is mapped onto a Richards' Contact Surface of vitamin-b1 while the Isosurface (grey, transparent) depicts a hydrophobicity value of 0.7. The same Isosurface can be seen in the Stick Model of vitamin-b1 with electrostatic potential mapped onto its surface. A second Isosurface (green) determines the value 2.0 of hydrophobicity. The restriction of Isosurface calculations to a bounding box is very useful for complex molecules, for example the plant seed protein in Fig. 1 (see Appendix). Here electrostatic potential is mapped onto an Isosurface depicting a hydrophobicity value of -1.7 while a second Isosurface (purple, transparent) determines a hydrophobicity value of -2.2. Both Isosurfaces are restricted to a chosen volume respectively.

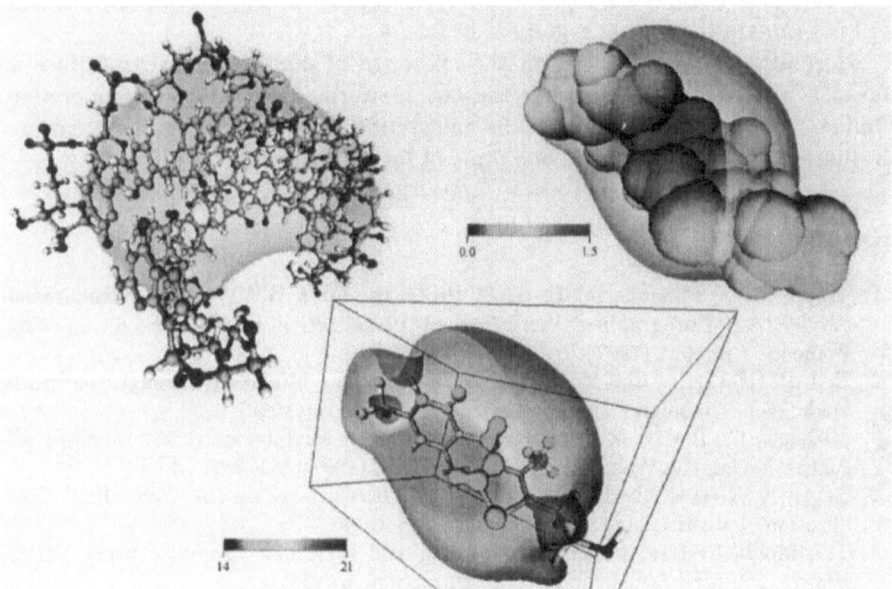

Fig. 5. Ribonucleic acid *(Ball-and-Stick Model, Isosurface)*, vitamin-b1 *(Richards' Contact Surface, Isosurface)*, vitamin-b1 *(Stick Model, different Isosurfaces)*

6   Conclusion and Future Work

Our idea of an efficient, client-side Molecular Visualization system offers much more interaction to the user than standard web-based client-server systems. Although the use of the Java programming language implies a reduced execution speed in some areas, Java leads to a flexible program design allowing easy extension and providing excellent portability. In fact, according to our performance tests computational and graphical power is nothing to be really concerned about any longer. Furthermore the users are offered several parameters to regulate the trade-off between visualization quality and time. Our system offers the visualization of various molecular models and properties which can be displayed at the same time. By placing all necessary calculations on the client's side our approach exploits the computational power of modern desktop workstations and personal computers. This also reduces the needed Internet bandwidth, only a few kilobytes of data containing the molecule description are transferred – the users become more independent of server or Internet bottlenecks.

With regard to the still decreasing costs of computational power and the further development of the World Wide Web, this powerful approach promises new possibilities in teaching, tele-working, Web publishing or even drug design. Thinking about the new generation of Network Computers (NCs), offering hardware Java virtual machines, for many tasks of web-based (Molecular) Visualization, run time reasons calling for specific applications and hardware platforms will not remain the killing argument in future.

Currently we are working on the exchange of our present 3D-interface to Java3D. This will finally remove the last platform dependence of our system. Online editing of molecules and the calculation of the following client-side re-evaluation of the molecules is one topic of future research.

References

1. Trapp, J. C., Pagendarm, H.-G.: A Prototype for a WWW-based Visualization Service. 8th Eurographics Workshop on Visualization in Scientific Computing, Boulogne sur Mer, (1997).
2. Ertl, P. (Novartis Crop Protection AG): Molecular Modelling through the World Wide Web. Chemistry and the Internet Conference, (1998).
3. Michaels, C., Bailey, M.: VizWiz: A Java Applet for Interactive 3D Scientific Visualization on the Web. IEEE Visualization, , (1997) 261–267.
4. Engel, K., Grosso, R., Ertl, T.: Progressive Iso-Surfaces on the Web. IEEE Visualization, Late Breaking Hot Topics, (1998) 37–40.
5. The Molda System: Molecular Modeling and Molecular Graphics using VRML Viewer. See http://cssj.chem.sci.hiroshima-u.ac.jp/molda/.
6. Richards, F. M.: Areas, Volumes, Packing, and Protein Structure. Annual reviews of Biophysics and Bioengineering, Vol. 6 (1977) 151–176.
7. Conolly, M. L.: Analytical Molecular Surface Calculation. Journal of Applied Crystallography, Vol. 16 (1983) 548–558.
8. Conolly, M. L.: Molecular Surface Triangulation. Journal of Applied Crystallography, Vol. 18 (1985) 499–505.
9. Lorensen, W. e., Cline, H. E.: Marching Cubes: A High Resolution 3D Surface Construction Algorithm. Computer Graphics, Vol. 21 (1987) 163–169.
10. Viswanadhan, V. N., Ghose, A. K., Revankar, G. R., Robins, R. K.: Atomic Physiochemical Parameters for Three-Dimensional Structure-Directed Quantitative Structure-Activity Relationships IV. Additional Parameters for Hydropobic and Dispersive Interactions. Journal of Chemical Information and Computer Science, Vol. 29 (1989) 163–172.

Editors' Note: see Appendix, p. 325 for colored figures of this paper

Geometry, Grids, and Systems

(Research Papers)

Geodesic Flow on Polyhedral Surfaces

Konrad Polthier and Markus Schmies

Technische Universität Berlin
{polthier, schmies}@math.tu-berlin.de

Abstract. On a curved surface the front of a point wave evolves in concentric circles which start to overlap and branch after a certain time. This evolution is described by the geodesic flow and helps us to understand the geometry of surfaces. In this paper we compute the evolution of distance circles on polyhedral surfaces and develop a method to visualize the set of circles, their overlapping, branching, and their temporal evolution simultaneously. We consider the evolution as an interfering wave on the surface, and extend isometric texture maps to efficiently handle the branching and overlapping of the wave.

1 Introduction

The study of geodesics and their behaviour under variations helps us to understand the geometry of curved surfaces. In this paper we try to compute and visualize aspects of the geodesic flow and related terms like injectivity radius and conjugate points, and extend these notions to polyhedral surfaces. By avoiding a formal definition right now some of these terms can be immediately related with the evolution of a wave front. For example, consider the front of a point wave on a surface evolving with unit speed. Then the time of the first hit upon itself is equal to the injectivity radius, unless a previous branching of the front occurs – the branch point is called conjugate point.

We solve two problems: First, to compute the evolution of the front of a point wave on a polyhedral surface. At each time the front is a topological circle on the surface, which may overlap and have singular points resulting from previous branchings of the front. In a numerical step, each point of the front is moved a constant distance in orthogonal direction to the curve, i.e. a constant distance along the geodesic normal to the curve in this point. In section 3 we employ the concept of straightest geodesics on polyhedral surfaces to give a thorough definition of the evolution on polyhedral surface and describe its numerics.

The second problem is to visualize the evolution consisting of a set of concentric, overlapping, and branched circles on a curved surface. Directly drawing such circles as curves would lead to overpainting of previous circles, due to the overlapping, unless transparency is added. As indicated above, the picture of a point wave allows a better visualization of many geometric characteristics as easily perceived phenomena, which includes the overlapping and the temporal evolution. For the visualization of the evolving wave we use isometric texture maps to avoid metric distortions between texture space and the surface, and we

 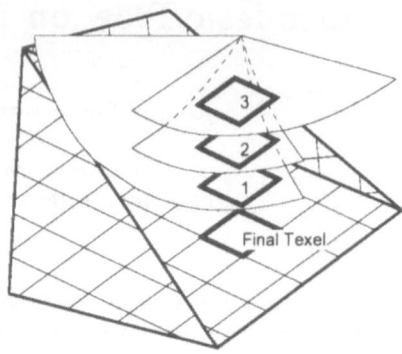

Fig. 1. Point waves on surfaces develop singularities at so-called conjugate points where the wave branches. Right figure shows a stack of abstract texels of the branched texture map, each corresponding to one layer of the wave.

extend these maps to include the multiple covering of some parts of the surface by different layers of the wave. We define branched texture maps which combine the notion of global texture maps that cover the whole surface, and local texture maps, which cover certain polygons. Isometric texture maps on surfaces are used e.g. in the line integral convolution technique in flow visualization [2].

The numerics and visualization ideas in this paper easily extend to other applications besides the geodesic flow. For example, the evolution and interference of other wave fronts over flat or curved surfaces. More abstract even the visualization of a homotopy of a curve, i.e. a one-parameter deformation, may be visualized using the interpretation as an evolving wave front. Waves have been studied in computer graphics from different aspects. In the animation [5] Max used bump mapping to perturb the surface of water for simulating waves viewed from a distance. Fournier and Reeves [4] explicitly model waves using parametric surfaces which allow simulation of detail structure such as waves curling over. The present paper describes ideas used in the video Geodesics and Waves [7].

2   Circles on Surfaces and the Geodesic Flow

A *point wave* in the euclidean plane starts at an origin p and evolves in concentric circles around p. In a particle model all particles of the wave front move with constant unit speed along radial rays away from the origin p if we neglect surface tension. At each time t the outer wave front forms a distance circle $\sigma(t)$ with center p and radius t.

For the construction of concentric circles on curved surfaces we use a similar picture. Particles of a wave front move along geodesic rays emanating from the origin p, therefore the circle at radius t consists of all points at distance t along a geodesic ray from p. It is one purpose of this paper to compute, study, and visualize such distance circles. We start with a review of geodesic curves, their properties, and their extension to polyhedral surfaces.

2.1 Evolution of Distance Circles and Point Waves

Geodesic curves on smooth surfaces are characterized by two equivalent properties, either as locally shortest curves or as straightest curves. This equivalence fails on polyhedral surfaces and leads to different concepts of locally shortest [1][3] and straightest geodesics [6].

Definition 1. *On a smooth surface M a curve γ is a* geodesic curve *if it is not curved within the surface. Formally, a smooth curve $\gamma : [a, b] \to M$ with tangent vector γ', $|\gamma'| = 1$, and surface normal N is* geodesic *if γ'' is parallel to N.*

Since this definition is equivalent to a 2nd-order ordinary differential equation, geodesics are uniquely determined by an initial point and an initial direction. This property allows particles with an initial impulse to move on surfaces along geodesics if there is no tangential acceleration.

Geodesics carry information about the underlying geometry of the surface. We briefly mention two terms, cut locus and conjugate points, since both have a strong relation to distance circles and point waves.

2.2 Cut Locus and Conjugate Points

The set of geodesics emanating from a given point p is conveniently described by the exponential map.

Definition 2. *The* exponential map *at a point p on a smooth surface M associates to each tangent vector v in the tangent space T_pM a point on a geodesic γ through $\gamma(0) = p$ with initial direction $\gamma'(0) = v$ as follows:*

$$\begin{aligned} \exp_p &: T_pM \to M \\ \exp_p(v) &= \gamma(1) \end{aligned} \qquad (1)$$

Here $\gamma(1)$ is a point on γ at distance $|v|$ from p since γ runs at speed $|v|$.

The exponential map maps small circles around 0 in the tangent space T_pM to *distance circles* around p on M, i.e. to circular curves on M where all points have the same distance to p. For a given vector $v \in T_pM$ the radial lines $rv \subset T_pM$, $r \in \Re$, are mapped isometrically to a geodesic ray γ.

In this formalism it is straight forward It is now easy to describe distance circles around a point p on a surface M. Let $V(0) \subset T_pM$ be a small neighbourhood of $0 \in T_pM$, then its image under the exponential map is a neighbourhood $U(p) := \exp_p V(0)$ of p. Using polar coordinates (r, φ) in $V(0)$ then each vector $v := r \cdot (\cos \varphi, \sin \varphi)$ is uniquely determined by its coordinates (r, φ). Its image under the exponential map given by $\gamma_\varphi(r) := \exp_p(r \cdot (\cos \varphi, \sin \varphi))$ induces a polar coordinate system on $U(p)$.

Using the above definitions, the particle model of a wave front in the plane extends immediately to curved surfaces M. The particles of a wave front start

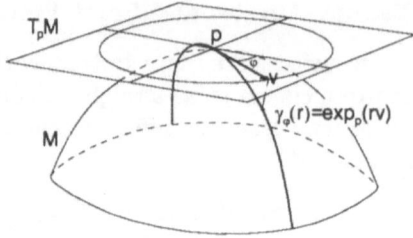

Fig. 2. Exponential map of geodesics emanating at a point.

at $p \in M$ and move with constant speed along geodesics rays emanating at p. If we normalize the speed to 1 then the wave front at a time t is a distance circle δ_t with radius t given by

$$\delta_t : [0, 2\pi] \to M$$
$$\delta_t(\varphi) := \gamma_\varphi(t) \tag{2}$$

In contrast to the euclidean case, the wave front on a curved surface will usually self intersect after some time t_0. There are two possible reasons for the intersection. First of all, if the surface is not simply connected and has a handle like a torus, then for each point p there exist two emanating geodesic rays, γ_{φ_1} and γ_{φ_2}, with $\varphi_1 \neq \varphi_2$ and a time t_0 such that $\gamma_{\varphi_1}(t_0) = \gamma_{\varphi_2}(t_0) =: q$ (which go around a handle of the surface). At q the wave front hits upon itself and starts to interfere. The time of the first hit $t_0 = \min_q dist(p, q)$ is called the *injectivity radius* at p. A second type of intersection occurs at so-called conjugate points of p. At conjugate points $q = \gamma_{\varphi_0}(r_0)$ the differential $\nabla \exp_p$ does not have maximal rank, i.e. $\partial/\partial\varphi \exp_p(r_0, \varphi_0) = 0$. Here, the wave front branches and nearby geodesics intersect shortly behind the conjugate point. The branching occurs in the form of a swallow's tail, see figures 3 and 7. For $t > t_0$ the polar coordinates fail to be a coordinate chart.

2.3 Review of Straightest Geodesics on Polyhedral Surfaces

Straightest geodesics are introduced in [6] to solve the initial value problem for geodesics on polyhedral surfaces. Since this property is essential for tracing particles we recall the basic definition:

Definition 3. *Let M be a polyhedral surface. A polygonal curve γ on M is a straightest geodesic if for each point $p \in \gamma$ the left and right curve angles θ_l and θ_r at p are equal, see figure 8.*

A straightest geodesic in the interior of a face is locally a straight line, and across an edge it has equal angles on opposite sides. The definition of straightest geodesics on faces and through edges is identical to the concept of shortest geodesics, but at vertices the concepts differ.

The most important property of straightest geodesics is the unique solvability of the initial value property which is not available for shortest geodesics.

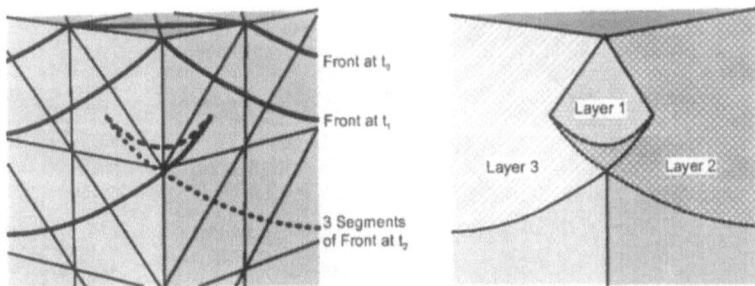

Fig. 3. Front of a point wave branches at conjugates points in the form of a swallow's tail. Behind the vertex of a cube the wave splits in two layers, and a third new layer is generated at the conjugate point. All three layers start to interfere.

Theorem 1 (Discrete Initial Value Problem). *Let M be a polyhedral surface and $p \in M$ a point with polyhedral tangent vector $v \in T_pM$. Then there exists a unique straightest geodesic γ with*

$$\gamma(0) = p \tag{3}$$
$$\gamma'(0) = v,$$

and the geodesic extends to the boundary of M.

3 Computing Discrete Distance Circles

For the computation of distance circles at a point p on a polyhedral surface we start with a topological polygonal circle $\sigma(0)$ such that all of its vertices lie at p. Each vertex $q \in \sigma(0)$ has a unit tangent vector associated to it, and therefore the circle $\sigma(0)$ at time $t_0 = 0$ is completely described by a set of pairs $(q_{i,0}, v_{i,0})$, $i = 1, .., n$. For the numerics, it is essential to distribute the tangent vectors equally spaced in angular direction since they determine the geodesic along which the particles $q_{i,0}$ will move.

In the numerical iteration step from time t_j to t_{j+1}, the circle $\sigma(t_{j+1})$ is obtained by computing the set of vertices and tangent vectors

$$q_{i,j+1} = \gamma_{(q_{i,j},v_{i,j})}(1) \atop v_{i,j+1} = \dot{\gamma}_{(q_{i,j},v_{i,j})}(1), i = 1, .., n, \tag{4}$$

where $\gamma_{(q_{i,j},v_{i,j})}$ is the straightest geodesic starting at $\gamma_{(q_{i,j},v_{i,j})}(0) = q_{i,j}$ with initial direction $\dot{\gamma}_{(q_{i,j},v_{i,j})}(0) = v_{i,j}$. (Compare with figure 4.) It should be noted that $\gamma_{(q_{i,j},v_{i,j})}(0) = \gamma_{(p,v_{i,0})}(t_j)$, i.e. for fixed j all points $q_{i,j}$ lie on a distance circle with distance t_j to p.

Equation 4 is essentially the computation of a segment of a straightest geodesic for all vertices on the outer circle $\sigma(t_j)$. On the other hand, the distance between adjacent vertices on the same circle grows exponentially with the radius.

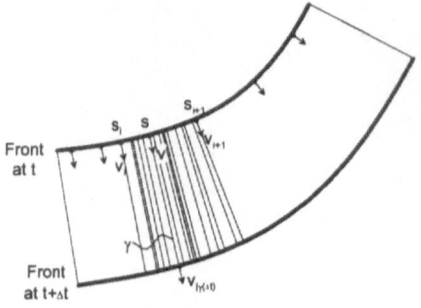 

Fig. 4. Set of distance circles with direction of movement (right). Left, the front at time $t + \Delta t$ is generated from the front at time t by computing geodesics with length Δt.

Therefore each timestep includes a refinement and coarsening step to maintain nearly constant distance between adjacent points on each circle. For the insertion of new vertices, say between $q_{i,j}$ and $q_{i+1,j}$, we connect both points by a geodesic segment and insert a new vertex on this geodesic. In fact, for a fixed t_j, the circle $\sigma(t_j)$ is piecewise geodesic - a natural generalization of piecewise linear.

When a curve reaches a conjugate point it starts to form a swallow's tail with sharp edges. This does not irritate the algorithm since each vertex on the curve still has a vector attached which uniquely determines its further movement. We remark that on a polyhedral surface each positively curved vertex is a conjugate point for all points in a small neighborhood. The resulting branching is studied on the cube in figures 3 and 9. When approximating a smooth geometry with a polyhedral surface, we suppress this type of local branching related to the discretization in favor of the global branching related to the shape of the smooth surface by using a reasonably fine mesh to distinguish between both types of branching.

A direct visualization of the set of circles gives reasonable results only for a small number of circles, see figure 4. In the following section we interpret the set of concentric circles as an evolving wave and use a resolution independent visualization based on texture map techniques.

4 Branched Coverings and Textures

4.1 Isometric Texture Maps

Texture maps of a 2-dimensional texture domain onto a general surface are faced with the problem that the texture images are metrically distorted. These principle difficulties can be avoided in the case of piecewise linear triangulated surfaces where one can choose the texture triangles isometric (up to scaling) to their corresponding surface triangles. Then each texel is isometric to its image on the surface and no distortion of the texture image occurs. Additionally, this

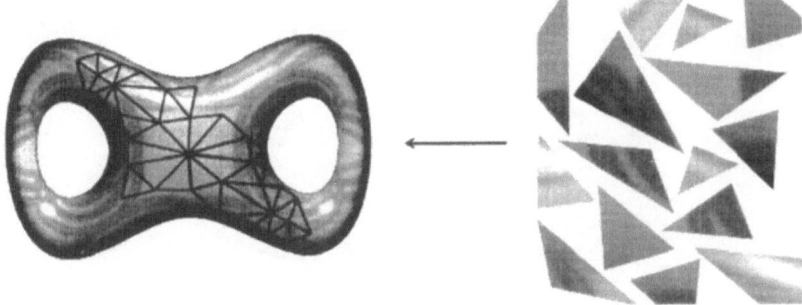

Fig. 5. Locally isometric texture maps allow 2d-texturing of arbitrary surfaces without distortion effects. Corresponding triangles in the texture domain and on the surface are required to be similar up to scaling.

concept allows texture maps on arbitrary triangulated surfaces, as shown in figure 5. It is implemented in animation systems like Softimage and employed e.g. in the context of LIC [2].

A major problem of this method arises from the rasterization of the texture domain since adjacent triangles on the surfaces are not adjacent in texture space. The texels along the common edge must be synchronized to avoid aliasing artefact. The use of bilinear texture interpolation even requires synchronization of texels lying outside the triangle as indicated in figure 6. An extrapolation scheme would introduce new discontinuities, therefore we prefer the following direct scheme: let T be a triangle on the polyhedral surface and ϕ_T the isometric map from T to the texture domain. If the point $p \in T$ is hit and assigned a color c, then the corresponding texel $\tau_T(p)$ in $\phi_T(T)$ is assigned the new value

$$F(\tau_T(p)) := \frac{F(\tau_T(p)) + c(p)}{\#Hits(\tau_T(p)) + 1}.$$

If $\phi_T(p)$ is close to the boundary of $\phi_T(T)$ then we color the texture image of the adjacent triangle T_1 in the corresponding texel $\tau_{T_1}(p)$ too. Let $(b_1, b_2 b_3)$ be the barycentric coordinates of p in the triangle $T = (v_1, v_2, v_3)$. The position $\phi_{T_1}(p)$ near $\phi_{T_1}(T_1)$, and therefore the texel $\tau_{T_1}(p)$, is easily computed using the triangle $(\phi_{T_1}(v_1), \phi_{T_1}(v_2), \phi_{T_1}(v_3))$ using

$$\phi_{T_1}(p) = \sum_{j=1}^{3} b_j \phi_{T_1}(v_j), \tag{5}$$

see figure 6. To assign multiple texels efficiently we calculate all occuring values $\phi_T(v)$ in a preprocessing step and later only use equation 5.

4.2 Branched Texture Maps

Two-dimensional textures on surfaces are usually given as a texture map from a two-dimensional image onto the surface. One distinguishes between global

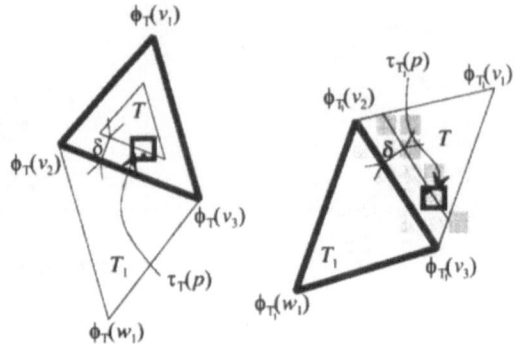

Fig. 6. Texel assignment for isometric texture maps.

texture maps covering the complete surface and local texture maps covering subsets of the surface. A surface may have multiple textures of each kind. During the rendering process, the final texture of a surface is computed by blending the textures associated to a point. The application of local textures requires the explicit specification of a domain on a surface where the local texture shall be applied, and it is therefore conceptually different from the use of global texture maps.

The approach of *branched texture maps* combines global and local textures in one concept and avoids the specification of subdomains for local textures in terms of regions on the surface. We start by covering the complete surface with a set of base texels, e.g. such a covering may be an isometric texture map as discussed in the previous section. At rendering time, these texels will contain the final texture. We associate to each base texel a stack of abstract texels. Each element of the stack corresponds to a texture layer covering a local region of the surface, and two different entries correspond to two different layers. For example, a surface which has two global textures associated would have a stack of constant height 2 at every point on the surface, and in the case of one global and one local texture the height would change between 1 and 2.

In contrast to the use of global and local texture maps, we do not allocate a new global or local texture image for each new layer. Instead whenever the wave fronts reaches a point on the surface, the corresponding texel stack is adjusted only at this base texel if necessary. Figure 1 shows some layers and the corresponding stack at a base texel.

5 Dynamic Computation of a Wave Texture

We divide the simulation of the wave in two major computational steps: first, the computation of the evolution of the wave front, which consists of geometric problems described in section 3 and leads to a static set of concentric circles, i.e. a set of wave fronts. Second, the simulation of the actual flow by animating the set of wave fronts. The animation is not done on the original set of fronts but on the level of textures. From the set of fronts, i.e. a set of geometric curves on the surface, we produce a single branched texture map which associates to each base texel of the surface a stack of abstract texels. The final animation of the wave is obtained from the single branched texture map by imposing a period function, but without any new computation of the wave evolution.

The separation of the numerical step and the use of branched texture maps makes the animation of moving waves a very cheap computation. To produce static pictures it is sufficient to color the base texels directly.

5.1 Generating the Branched Texture Map

The set of wave fronts $\sigma(t_j), j = 1, 2, \ldots$ computed in section 3 are a discretization of the exponential map from $T_p M$ to M and is now translated into a branched texture map. First, we construct an isometric texture map covering M once with so-called base texels and associate to each base texel of M an empty stack of abstract texels, as shown in figure 1. The height of each stack is not known in advance and will vary from base texel to base texel. Now we analyze the set of curves $\sigma(t_j)$, and whenever the front has flowed over a base texel we add a new abstract texel to the stack of this base texel. Each abstract texel is essentially of the following structure:

struct {float α, t; AbstractTexel next} AbstractTexel.

The task of generating the stack is simple if the maximal distance between two successive curves is smaller than half the diameter of the smallest texel on the surface, which can be easily controlled for isometric texture maps. Here we sample each wavefront which carries the necessary information about angle α and time t. To obtain smooth results, it is essential to hit texels more frequently, say 4-8 times, and store average values $(\bar{\alpha}, \bar{t})$.

Each sheet of the wave hitting a given base texel corresponds to exactly one abstract texel above the base texel. A serious problem is the detection of the sheet corresponding to the current hit. For polyhedral geometries, we avoid this problem by letting the front detect branch points from the vertex curvature and split. This allows us to assign to each front segment a unique level identifying the sheet.

But, when approximating smooth surfaces, we need to distinguish between the branch points of the smooth geometry and those induced by polyhedral vertices. In this case, we let each base texel reconstruct the necessary information for each circle, resp. geodesic, from the time and angle of the current hit. Let m_T be the midpoint of a base texel T with edge size δ and let each abstract texel have stored average values $(\bar{\alpha}, \bar{t})$. Assume a circle $\sigma(\alpha, t)$ hits the base texel at a point q corresponding to an angle α, then q belongs to the same layer of the abstract texel if

$$dist_M(m_T, q) \approx \sqrt{\frac{d}{d\alpha}\sigma(\alpha, t)^2 \left(\alpha - \bar{\alpha}\right)^2 + (t - \bar{t})^2} \leq \frac{\delta}{\sqrt{2}}$$

for a threshold δ depending on the discretization of the flow.

In practice, we have a lower resolution in time direction and compute fewer wave fronts with distance of more than a few triangle diameters, and interpolate between successive fronts as indicated in the left image in figure 4.

6 Summary and Acknowledgments

As an application of the concept of straightest geodesics on polyhedral surfaces, we have computed concentric circles around a given point. Considering the set of circles as an evolving wave front by animating the radius offers a natural visualization approach of the circle homotopy, where geometric properties like injectivity radius and conjugate points are easily perceived as properties of the evolving wave. For the visualization we define branched texture maps, which extend isometric texture maps to (partially) multiply covered surfaces. The visualization methods presented in this paper may be applied to other problems such as propagation of general waves and homotopies of curves.

The authors appreciate the cooperation with Martin Steffens and Christian Teitzel during the production of the video. We thank the anonymous referees for helpful comments.

References

1. A. D. Aleksandrov and V. A. Zalgaller. *Intrinsic Geometry of Surfaces*, volume 15 of *Translation of Mathematical Monographs*. AMS, 1967.
2. H. Battke, D. Stalling, and H.-C. Hege. Fast line integral convolution for arbitrary surfaces in 3d. In H.-C. Hege and K. Polthier, editors, *Visualization and Mathematics*, pages 181–195. Springer Verlag, Heidelberg, 1997.
3. E. Dijkstra. A note on two problems in connection with graphs. *Numer. Math.*, 1:269–271, 1959.
4. A. Fournier and W. Reeves. A simple model of ocean waves. *ACM Siggraph 86*, pages 75–84, 1986.
5. N. Max. Carla's island. ACM Siggraph 81 Video Review, 1981. Animation.
6. K. Polthier and M. Schmies. Straightest geodesics on polyhedral surfaces. In H.-C. Hege and K. Polthier, editors, *Mathematical Visualization*, pages 135–150. Springer Verlag, Heidelberg, 1998.
7. K. Polthier, M. Schmies, M. Steffens, and C. Teitzel. Video on geodesics and waves. Siggraph'97 Video Review, 1997.

Editors' Note: see Appendix, p. 326 f. for colored figures of this paper

On Simulated Annealing and the Construction of Linear Spline Approximations for Scattered Data

Oliver Kreylos[1,2] and Bernd Hamann[1]

[1] Center for Image Processing and Integrated Computing (CIPIC), Department of
Computer Science, University of California, Davis, CA 95616-8562, USA
[2] Institut für Betriebs- und Dialogsysteme, Fakultät für Informatik, Universität
Karlsruhe (TH), 76128 Karlsruhe, Germany

Abstract. We describe a method to create optimal linear spline approximations to arbitrary functions of one or two variables, given as scattered
data without known connectivity. We start with an initial approximation
consisting of a fixed number of vertices and improve this approximation
by choosing different vertices, governed by a simulated annealing algorithm. In the case of one variable, the approximation is defined by line
segments; in the case of two variables, the vertices are connected to define a Delaunay triangulation of the selected subset of sites in the plane.
In a second version of this algorithm, specifically designed for the bivariate case, we choose vertex sets and also change the triangulation to
achieve both optimal vertex placement and optimal triangulation. We
then create a hierarchy of linear spline approximations, each one being
a superset of all lower-resolution ones.

1 Introduction

In several applications one is concerned with the representation of complex geometries or complex physical phenomena at multiple levels of resolution. In the
context of computer graphics and scientific visualization, so-called *multiresolution methods* are crucial for the analysis of very large numerical data sets [1–5].
Examples include high-resolution terrain data (digital elevation maps) and high-
resolution, three-dimensional imaging data (e. g., magnetic resonance imaging
data).

We present an approach for the construction of multi resolution representations of very large scattered data sets using an iterative optimization algorithm
and the principle of *simulated annealing* [9–12]. Our goal is the computation of
several optimal linear spline approximations to a given scattered data set.

We assume that the given data sets are samples of a real function of one or two
variables, with the samples randomly distributed in the function's domain and no
known connectivity between them. Each individual linear spline approximation
is defined by its control points and, in the case of multivariate functions, by the
way these points are connected to form a triangulation. We only place control
points at given sample positions and only use the supplied function values at
those positions.

1.1 Visualizing Large Data Sets

To create a hierarchy of approximations to a given scattered data set we choose N_k vertices from the set at each hierarchy level k. We ensure that the set of vertices of any hierarchy level $j < k$ is a subset of level k's vertex set. After having decided which vertices to select for a hierarchy level k, that level's vertices are connected in an appropriate way to form a linear spline's control mesh. An example of such a hierarchy in the univariate case is shown in Fig. 1.

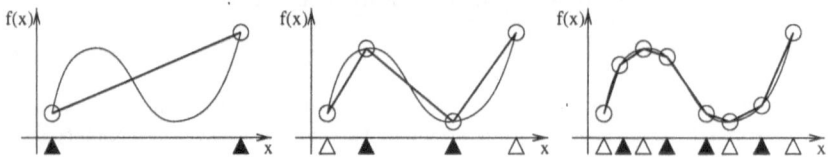

Fig. 1. A hierarchy of approximations in the univariate case. New vertices are inserted at the sites marked by solid triangles.

When representing high-resolution data sets with low-resolution linear spline approximations, one has to be careful where to place the spline's control points and how to connect them in order to achieve a faithful representation of the data set, see Fig. 2.

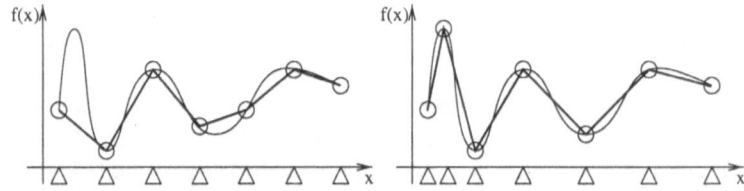

Fig. 2. Uniform vs. optimal control point placement for univariate data.

If the number of vertices for an approximation level is prescribed, one has to address two problems:

1. Which vertices should one choose for the approximation, i.e., how should one create the vertex placement?
2. How should one connect the chosen vertices, i.e., how should one create the connectivity?

In the special case of a function of one variable, we only have to address the first problem, since in the univariate case the connectivity is defined by the chosen sites' numerical order.

1.2 Finding Optimal Approximations

Our approach to finding an optimal linear spline approximation for a given, fixed number of vertices N_k is based on an iterative optimization algorithm. First, we create an initial configuration, then we improve this configuration by changing its vertex placement and its connectivity in every step. We judge a configuration's quality by its L^2 distance from the scattered data set. Since this optimization problem is high-dimensional and generally involves local minima in abundance, the algorithm of simulated annealing is well suited to construct "good" linear spline approximations [12,9].

Simulated annealing is an iterative method that applies random changes to the current configuration and accepts a step depending on the resulting change of the error measure and a value called "temperature." This value determines the probability of accepting a step that increased the error measure: The higher the temperature, the higher the probability of accepting a bad step. The so-called "annealing schedule" determines how fast the temperature is decreased during the iteration.

In the case of two or more variables the quality of a configuration depends on both vertex placement and connectivity. There are two different ways to proceed:

1. One can ignore the optimization of the connectivity by enforcing a fixed type of connectivity throughout the iteration process; in the bivariate case, an obvious candidate is the Delaunay triangulation [6]. Under this constraint the algorithm can proceed exactly as in the univariate case.
2. One can attempt to optimize both parts of the configuration in parallel. For example, before each step one could randomly decide to either move a vertex or *swap* a common edge of two adjacent triangles.

2 The Optimization Algorithm

We now describe the individual steps of our algorithm. Algorithm 1 is a high-level description. The subsequent sections describe the important steps in more detail.

Algorithm 1: Optimal linear spline approximation.

```
Create initial configuration (vertex placement and connectivity);
Determine initial temperature and create annealing schedule;
While iteration is not finished {
  Change current configuration;
  Calculate change in error measure;
  Undo iteration if rejected by simulated annealing; }
Return current configuration;
```

2.1   Creating an Initial Configuration

Our approximations are defined over the original sites' convex hull. In the univariate case, we cover the convex hull by choosing the leftmost and the rightmost original vertices and distribute the rest of vertices uniformly between them. In the multivariate cases cover the convex hull by always selecting all non-interior vertices; then we choose the rest of vertices randomly from the original data set. In the bivariate case, we define the initial connectivity by a Delaunay triangulation of the initial vertices' sites.

2.2   Creating an Annealing Schedule

A reasonable heuristic to define the initial temperature is to apply some steps of the iteration scheme and to define the initial temperature in a way that the annealing algorithm initially accepts an "expected bad" step with a probability of one half. Next, we lower the temperature in steps, leaving it constant for a fixed number of iterations and scaling it by a fixed factor afterwards.

2.3   Changing the Current Configuration

The simulated annealing algorithm's core is its iteration step. In principle, one can use any method to change the current configuration, but we have found out that the "split" approach, shown in Algorithm 2, works very well.

Algorithm 2: Changing the current configuration.

```
if(acceptWithProbability(moveProbability)) { /* move a vertex */
  Choose an interior vertex v;
  Estimate v's contribution vE to the error measure;
  if(vE < localMovementFactor * E)
    Move v globally;
  else
    Move v locally;
  if(moveProbability == 1) /* Vertex movements only? */
    Restore Delaunay property; }
else { /* swap an edge */
  Choose a swappable edge e;
  Swap edge e; }
```

The constant *moveProbability* is used to control the behaviour of the optimization process for bivariate functions. If this constant's value is one, the algorithm moves a vertex in every step, and after each vertex movement the current triangulation is updated to satisfy the Delaunay property. In the other case the algorithm can either move a vertex or swap an edge, thereby optimizing both vertex placement and triangulation simultaneously.

Estimating a Vertex' Error Contribution. To estimate how much the removal of an interior vertex v would increase the current error measure, we estimate the "volume" of v's platelet: We construct an approximating least squares hyperplane H for all vertices surrounding v. Then we calculate h as v's ordinate-direction distance from H and A as the area of v's platelet, see Fig. 3. We define the error contribution as $\sqrt{A \cdot h^2/2}$ in the univariate case and as $\sqrt{A \cdot h^2/3}$ in the bivariate case, to ensure that the ratio of a vertex' error contribution and the used L^2 error measure is scale-invariant.

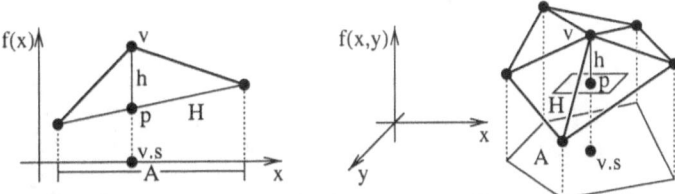

Fig. 3. Estimating a vertex' contribution to the error measure.

Global Vertex Movement. If v's error contribution is smaller than a constant *localMovementFactor* times the current error measure E, we assume that v is currently located in a "flat" region of the function and should be moved away from this region. We move v *globally* to a randomly chosen new site not already being part of the current configuration. By doing this we assure that vertices get driven away from nearly flat regions of a function in early stages of the iteration. Figure 4 shows the process of actually moving a vertex globally in detail.

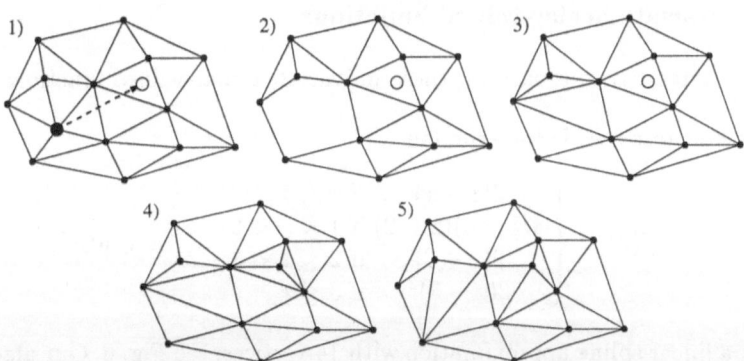

Fig. 4. Moving a vertex globally in the bivariate case. 1) initial state; 2) removing the vertex; 3) filling the hole; 4) inserting new vertex; 5) restoring the Delaunay property (only if the connectivity is ignored during optimization).

Local Vertex Movement. When a vertex' error contribution is larger than *localMovementFactor · E*, we assume it is currently located in an "important," high-curvature region of the target function, and we attempt to find a better site for this vertex by moving it *locally* to a new, unoccupied site in its platelet. To move a vertex locally, we "slide" the vertex on the line from its old to its new site, dragging the edges connecting it to all surrounding vertices along. Whenever a surrounding simplex becomes degenerate during the vertex' motion, we swap one edge of the affected simplex before moving the vertex any further, see Fig. 5.

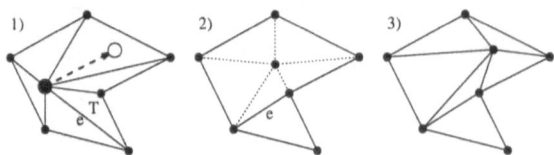

Fig. 5. Moving a vertex locally in the bivariate case: 1) initial state; 2) swapping edge *e* to prevent triangle *T* from becoming degenerate; 3) resulting state.

3   Examples and Results

In this section we show some of the experiments we did to evaluate our algorithm's behaviour. We begin with univariate and bivariate scalar-valued functions, and we then use a slight variation of the algorithm, where the Euclidian distance between vectors is used to calculate the L^2 error measure, to approximate bivariate vector-valued functions.

3.1   Univariate Scalar-valued Functions

We have tested our approach for these univariate scalar-valued functions:

1. The first test case is the function

$$
f(x) = \begin{cases} 2(1-x) & \text{if } x < 1 \\ 4(1-x)(x-2) & \text{if } 1 \leq x < 2 \\ 2(x-2) & \text{if } 2 \leq x < 3 \\ 2(x-3)^2 & \text{if } 3 \leq x \end{cases} \quad , \quad x \in [0,4],
$$

and a linear spline approximation with 14 vertices, see Fig. 6. Our algorithm finds a very good approximation, although the function has discontinuities in both the zeroeth and first derivatives. In the two quadratic sections [1, 2] and [3, 4] the sites are uniformly distributed; we thus assume that the resulting approximation is globally optimal.

2. The second test case is the function

$$f(x) = 4 \sum_{n=0}^{3} \frac{\sin((2n+1)x)}{2n+1} \quad , \quad x \in [0, 4\pi],$$

the fourth-order Fourier approximation of a square wave, and a linear spline approximation with ten vertices, see Fig. 7. The number of vertices is too small to capture all details of the function, but the algorithm still finds a very good approximation.

3. The third test case is the same function as in the second, but this time using a linear spline approximation with 30 vertices, see Fig. 8. Now all the function's important features are present in the approximation.

3.2 Bivariate Scalar-valued Functions

We have tested our approach for these bivariate scalar-valued functions:

4. The fourth test case is the function

$$f(x,y) = 2 \sum_{i=0}^{2} \sum_{j=0}^{2} \frac{\sin((2i+1)x)}{2i+1} \cdot \frac{\sin((2j+1)y)}{2j+1} \quad , \quad x, y \in [0, 2\pi],$$

the third-order Fourier approximation of a bivariate square wave, and a linear spline approximation with 50 vertices and a general triangulation, see Fig. 9. The number of vertices is too small to capture all details of the function, but the algorithm still finds a decent approximation.

5. The fifth test case is the same function as in the fourth, but this time using a linear spline approximation with 250 vertices and a general triangulation, see Fig. 10. Due to the high number of vertices the iteration takes much longer to converge, but captures all details of the target function.

6. The sixth test case is a scattered data set consisting of 37,594 vertices, resulting from a laser scan of a Ski-Doo hood and a linear spline approximation with 1,000 vertices and a general triangulation, see Fig. 11. This case shows that our algorithm can be used in surface reconstruction, as long as the source data can be interpreted as a bivariate, scalar valued function.

3.3 Bivariate Vector-valued Functions

We have applied our method to these RGB color image data sets:

7. The seventh test case is a photograph of the Golden Gate Bridge in San Francisco, resampled to a resolution of 329×222 pixels, see Fig. 12, and a linear spline approximation with 400 vertices and a general triangulation, see Fig. 13. The RGB image is interpreted as a bivariate vector-valued function, defined by samples positioned at the pixels' centers.

8. The eigth test case is the same function as in the seventh, but this time approximated by a linear spline with 1,600 vertices and a general triangulation, see Fig. 14. The resulting linear spline is a superset of the result of experiment seven as defined in Sect. 1.1. It is hard to see in these low-quality reproductions, but the approximation is very close to the original image.

196

4 Conclusions and Future Work

In this paper we presented a method to calculate optimal linear spline approximations to functions defined by scattered data, using an iterative optimization technique governed by the simulated annealing algorithm. Our method is a generalization of the data-dependent triangulation method discussed by Schumaker [9]. We have demonstrated that our method performs well for univariate and bivariate scalar-valued functions. Furthermore, we have found that our algorithm approximates RGB images very well, even when using only a small number of vertices. Our technique provides an interesting alternative way to transform images to a storage-efficient, resolution-independent representation.

The main areas for future research are the generalization of our algorithm to functions of three and more variables and the application of our method to image and video compression. If one treats video data as time-varying bivariate vector-valued functions, and exploits the strong frame coherence of video streams especially in tele-conferencing, our algorithm might lead to a real-time video compression method for this kind of video streams.

5 Acknowledgements

This work was supported by grants and contracts awarded to the University of California, Davis, including the National Science Foundation under contract ACI 9624034 (CAREER Award), the Office of Naval Research under contract N00014-97-1-0222, the Army Research Office under contract ARO 36598-MA-RIP, the NASA Ames Research Center through an NRA award under contract NAG2-1216, the Lawrence Livermore National Laboratory through an ASCI ASAP Level-2 under contract W-7405-ENG-48 (and B335358, B347878), and the North Atlantic Treaty Organization (NATO) under contract CRG.971628. We also acknowledge the support of Silicon Graphics, Inc., and thank all members of the Visualization Thrust at the Center for Image Processing and Integrated Computing (CIPIC) at the University of California, Davis.

References

1. Bonneau, G.-P., Hahmann, S. and Nielson, G. M., *BLaC-wavelets: A multiresolution analysis with non-nested spaces*, in Yagel, R. and Nielson, G. M., eds., Visualization '96 (1996), IEEE Computer Society Press, Los Alamitos, CA, pp. 43–48
2. Eck, M., DeRose, A. D., Duchamp, T., Hoppe, H., Lounsbery, M. and Stuetzle, W., *Multiresolution analysis of arbitrary meshes*, in Cook, R., ed., Proc. SIGGRAPH 1995, ACM Press, New York, NY, pp. 173–182
3. Gieng, T. S., Hamann, B., Joy, K. I., Schussman, G. L. and Trotts, I. J., *Constructing hierarchies for triangle meshes*, in IEEE Transactions on Visualization and Computer Graphics 4(2) (1998), pp. 145–161
4. Hamann, B., *A data reduction scheme for triangulated surfaces*, in Computer Aided Geometric Design 11(2) (1994), pp. 197–214

5. Hamann, B., Jordan, B. J. and Wiley, D. A., *On a construction of a hierarchy of best linear spline approximations using repeated bisection*, in IEEE Transactions on Visualization and Computer Graphics 4(4) (1998)
6. de Berg, M., van Kreveld, M., Overmars, M. and Schwarzkopf, O., *Computational Geometry* (1990), Springer-Verlag, New York, NY
7. Edelsbrunner, H., *Algorithms in Combinatorial Geometry* (1987), Springer-Verlag, New York, NY
8. Preparata, F. P., Shamos, M. I., *Computational Geometry*, third printing (1990), Springer-Verlag, New York, NY
9. Schumaker, L. L. *Computing Optimal Triangulations Using Simulated Annealing*, in Computer Aided Geometric Design 10 (1993), pp. 329–345
10. Nielson, G. M., *Scattered data modeling*, in IEEE Computer Graphics and Applications 13(1) (1993), pp. 60–70
11. Press, W. H., Teukolsky, S. A., Vetterling, W. T., and Flannery, B. P. *Numerical Recipes in C*, 2nd ed. (1992), Cambridge University Press, Cambridge, MA
12. Metropolis, N., Rosenbluth, A., Rosenbluth, M., Teller, A., and Teller, E., in Journal of Chemical Physics 21 (1953), pp. 1087–1092

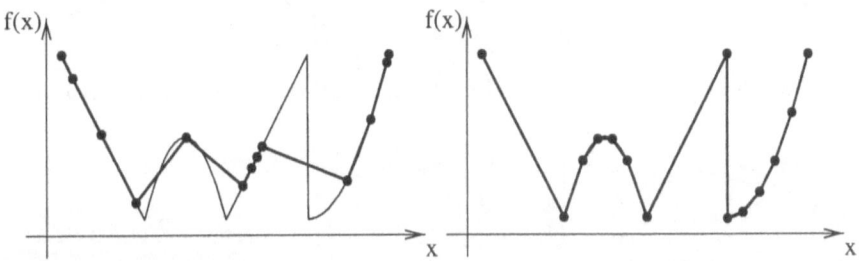

Fig. 6. First experiment. Left: initial vertex placement; right: final vertex placement.

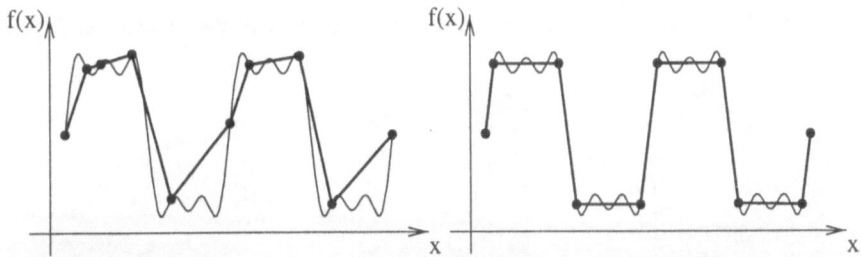

Fig. 7. Second experiment. Left: initial vertex placement; right: final vertex placement.

198

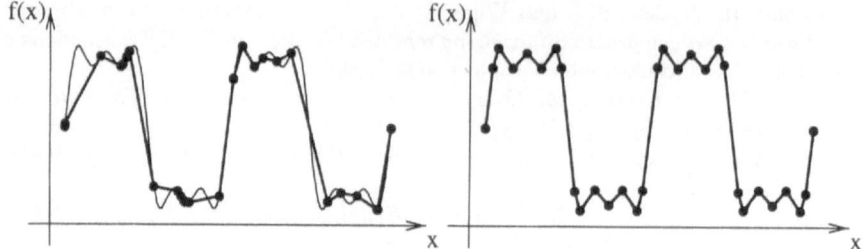

Fig. 8. Third experiment. Left: initial vertex placement; right: final vertex placement.

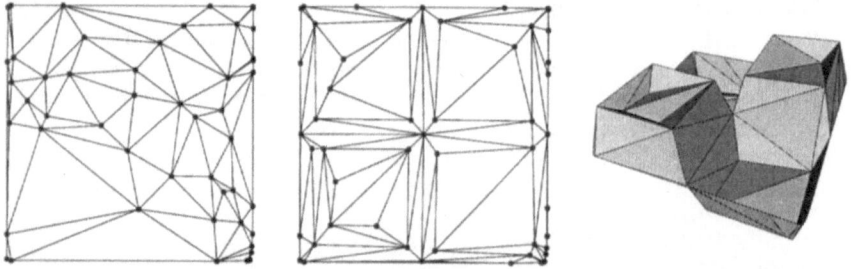

Fig. 9. Fourth experiment. Initial and final configurations and flat-shaded rendering.

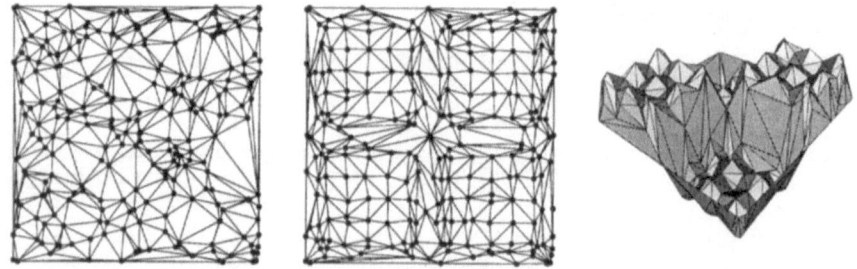

Fig. 10. Fifth experiment. Initial and final configurations and flat-shaded rendering.

Fig. 11. Sixth experiment. Initial and final configurations and flat-shaded rendering.

Editors' Note: see Appendix, p. 328 for colored figure of this paper

A Comparison of Error Indicators for Multilevel Visualization on Nested Grids

Thomas Gerstner, Martin Rumpf, and Ulrich Weikard

Department for Applied Mathematics, University of Bonn, Germany

Abstract. Multiresolution visualization methods have recently become an indispensable ingredient of real time interactive post processing. Here local error indicators serve as criteria where to refine the data representation on the physical domain. In this article we give an overview on different types of error measurement on nested grids and compare them for selected applications in 2D as well as in 3D. Furthermore, it is pointed out that a certain saturation of the considered error indicator plays an important role in multilevel visualization and can be reused for the evaluation of data bounds in hierarchical searching or for a multilevel backface culling of isosurfaces.

1 Introduction

A variety of multiresolution visualization methods has been designed to serve as tools for interactive visualization of large data sets [3, 9, 12, 20]. Here the local resolution of the generated visual objects, such as 2D graphs, or isosurfaces and color shaded slices in 3D, depends on error indicators which measure the error due to a locally coarser approximation of the data.

Different approaches have been presented to solve the outstanding continuity problem, i.e. to avoid cracks in adaptive isosurfaces. In the Delaunay approach by Cignoni et al. [4] and in the nested mesh method by Grosso et al. [10] the successive remeshing during the refinement guarantees continuity. Alternatively, Shekhar et al. [21] rule out hanging nodes by inserting additional points on faces with a transition from finer to coarser elements due to an adaptive stopping criterion.

We apply the method of adaptive projection on nested grids, which has been described in earlier publications. For the general concept we refer to [18]. Implementational aspects are especially described in [17]. The core of our approach is identical to the method of Zhou et al. [24]. In 3D it can be regarded as a generalization of the techniques presented by Livnat et al. [15] and in [8].

In this paper we give a detailed comparison of error indicators and the performance of corresponding multilevel methods. Here we do not focus on the methodology itself but on the indicators, their effect on cost reduction, and their relation to the actual error in a corresponding norm. Therefore, especially to simplify the exposition, we confine ourselves to simplicial grids generated by bisection, which are well known from adaptive numerical methods [1, 19]. In explicit, we deal with the recursive bisection of [13, 16].

Let us point out that there are other, more general approaches especially for surfaces by De–Floriani et al. [6] and Hamann and Chen [11] which also apply to non nested grid hierarchies, but with a different focus concerning the field of applicability.

2 A general multilevel algorithm on nested grids

We confine ourselves here to hierarchical simplicial grids which carry a piecewise linear data function. Let us consider a family of nested, conforming, simplicial meshes $\{\mathcal{T}^l\}_{0 \leq l \leq l_{\max}}$ in two or three dimensions. We denote by $h(T)$, $h(e)$ the diameter, respectively the length of an edge e of a simplex $T \in \mathcal{T}^l$. Furthermore, $\mathcal{N}(T)$, $\mathcal{N}(\mathcal{T}^l)$ denote nodal sets of single simplices, respectively entire triangulations.

The simplices, triangles in 2D, respectively tetrahedra in 3D are assumed to be refined by recursive bisection. For a simplex T, the midpoint of a predestined edge $e_{\text{ref}}(T)$ is thereby picked up as a new node $x_{\text{ref}}(T)$, and the simplex is cut at the edge, respectively face, $F_{\text{ref}}(T)$ spanned by $x_{\text{ref}}(T)$ and the nodes of T, which are not endpoints of the refinement edge $e_{\text{ref}}(T)$, into two child simplices $\mathcal{C}(T) = \{T_{\mathcal{C}}^1, T_{\mathcal{C}}^2\}$. A simple alternating scheme for the refinement edge e_{ref} [1, 13] guarantees the conformity of the resulting grids. Finally, let U^l denote the piecewise linear function on $\mathcal{T}^l$ uniquely described by the data values on the corresponding nodes.

The multilevel algorithm is based on a depth first traversal of the grid hierarchy. On every simplex we check for a stopping criterion. If it is true we stop and visualize locally. Otherwise, we recursively proceed on the child set $\mathcal{C}(T)$.

If we stop on a specific simplex T and refine another simplex $\tilde{T}$ which shares the refinement edge with T, i.e. $e_{\text{ref}}(T) = e_{\text{ref}}(\tilde{T})$, an inconsistency occurs at the hanging node x_{ref}. This leads to jumps in the color intensity or cracks in the isosurface. In the case of general nested grids we can apply adaptive projection operators to ensure consistency [18]. Here, we simply have to ensure that, whenever a simplex is refined, all the simplices sharing its refinement edge – in 2D only the one triangle opposite to T at the edge $e_{\text{ref}}(T)$ – are refined as well.

This can be achieved by defining error indicators $\eta(x)$ on the grid nodes and choosing $\eta(x_{\text{ref}}(T)) < \varepsilon$ as stopping criterion on a simplex for some user prescribed threshold value ε. Since all nodes, except those on the coarsest level, are refinement nodes $x_{\text{ref}}(T)$ on a refinement edge $e_{\text{ref}}(T)$, the indicator value $\eta(x)$ measures the error on those simplices sharing the edge. Therefore, the recursive traversal would stop not only on T but – if visited – on all other simplices sharing the refinement edge if their common stopping criterion is true.

However, for an arbitrary error indicator it might still occur that, although $\eta(x_{\text{ref}}(T)) < \varepsilon$, $\eta(x_{\text{ref}}(\hat{T})) \geq \varepsilon$ on some descendant $\hat{T}$ whose refinement node $x_{\text{ref}}(\hat{T})$ is located on the boundary of T. The adjacent tetrahedron will possibly be visited and then refined, whereas on T the stopping criterion already holds. To avoid this we assume the following saturation condition on the error indicator (for a generalization compare [18]):

Saturation Condition:
$\eta(x_{ref}(T)) > \eta(x_{ref}(T_C))$ *for all* $T \in \mathcal{T}^l$ *with* $l < l_{max}$ *and* $T_C \in \mathcal{C}(T)$.

An error indicator η is called admissible, if it fulfills the saturation condition. Otherwise, it can easily be adjusted in a preroll step (cf. section 3). The adaptive algorithm can be sketched in pseudo code as follows

```
Inspect(T) {
    if SimplexIsOfInterest(T)
        if C(T) ≠ ∅ ∧ η(x_ref(T)) ≥ ε
            { Inspect(T_C^1); Inspect(T_C^2); }
        else Extract(T);
}
```

where the function *SimplexIsOfInterest()* checks whether the simplex is a candidate for some local rendering or not. For example, in 3D slicing the cutting plane has to intersect the simplex T. In multilevel isosurface extraction this function checks whether the isosurface intersects the current simplex. At the end of the next section we show how this function can be implemented efficiently.

3　An overview of error measurement

In this section we will discuss several principle techniques of error measurement. The starting point will be some actual local error measure on the grid hierarchy. The local resolution and the visual impression of the numerical data is closely related to the specific type of error measurement applied in the adaptive traversal of the tree structure.

Choices for the error metric

Let $\eta^*(x)$ be a measure on nodes x, which weights the effect of stopping for some local rendering already on a simplex T with $x = x_{ref}(T)$ instead of traversing the locally finest grid level. Furthermore, let us denote by $S(x)$ the support of the piecewise linear base function corresponding to the node x. Then, given a fine grid data function U and a coarse grid function U^l on level l with $T \in \mathcal{T}^l$, we assume $\eta^*(x)$ to be the distance between U and U^l measured locally on $S(x)$ by some metric $d_{S(x)}$, i.e.

$$\eta^*(x) = d_{S(x)}(U, U^l)$$

Let us consider several widely used metrics:

- We can choose some local norm of the difference functions such as

$$\eta^*(x) := \|U - U^l\|_{p, S(x)},$$

where $\| \cdot \|_{p, S(x)}$ is the usual L^p norm for $p \in [1, \infty]$ restricted to the domain $S(x)$. Due to Hölder's inequality the error indicators obviously become sharper for increasing values of p.

– Instead of function values we can consider derivatives and define

$$\eta^*(x):=\|\nabla U - \nabla U'\|_{p,S(x)} \, .$$

In general the resulting error measurement is sharper then the one based on function values. By some worst case analysis based on inverse estimates we obtain the estimate

$$\|\nabla U - \nabla U'\|_{p,S(x)} \le C\, h_{\min}^{-1} \|U - U'\|_{p,S(x)}$$

where $h_{\min}:=\min_{T\in\mathcal{T}_{l-1}, T\subset S(x)} h(T)$. This estimate is asymptotically sharp on fine grid levels for a function U, which is the interpolation of some smooth function. Frequently, the norm of the gradient is taken as an error indicator. This is questionable, because rendering is "linearly exact" and therefore refinement in areas of uniformly large gradient norms does not improve the graphical representation.

– Third – a smooth graphical representation in mind – we may be interested in measuring the geometric smoothness of the approximation independently of the true function values. A possible measure is a discrete curvature quantity. For surfaces this should be related to the absolute curvature $\kappa = \sqrt{\kappa_1^2 + \kappa_2^2}$ where the κ_i are the principle curvature terms. As clearly indicated in the case of minimal surfaces with vanishing mean curvature or cylinders with vanishing Gaussian curvature, mean or Gaussian curvature discretization does not make sense in terms of general error control.

– A fourth choice of a suitable measure is closely related to geometric shapes [14]. In our simple case of a scalar function U a suitable approach is to compare the graphs of U, respectively U' on $S(x)$. If dist$(\cdot, \cdot)$ is a geometric distance metric on graphs, we are lead to $\eta^*(x):=\text{dist}(graph(U), graph(U'))$. For flat graphs this error indicator only slightly differs from measuring the difference of the function values.

Furthermore, the viewing direction and distance may enter the error metric [15], or the error measurement may depend on the distance to a specific region of interest [2, 5, 18]. We will here restrict ourselves to the basic error norms and discrete curvature measurement.

Hierarchical error measurement

Usually, an error measurement which locally compares coarse grid functions with the functions on the finest grid is expensive to evaluate even in a preprocessing step. We will apply an often used simplification, which only compares data on the current grid level to data on the next finer grid level. We will denote the corresponding one level look ahead error indicator by $\eta(x)$. However, the saturation condition as a minimum precondition to guarantee continuity of the adaptive projection may fail for η.

– **Hierarchical offset error indicators:** In analogy to the norm of the difference function we can consider the hierarchical offset function U_δ defined on a tetrahedon as

$$U_\delta|_T = U_l|_T - U_{l-1}|_T$$

The values of U_δ on $\mathcal{N}^l \setminus \mathcal{N}^{l-1}$ are related to the original data values by the following recursive formula

$$U(x_{ref}(T)) = \frac{U(x_1) + U(x_2)}{2} + U_\delta(x_{ref}(T))$$

where x_1 and x_2 are the end points of the edge corresponding to $x_{ref}(T)$ on a simplex T. For smooth data, i.e. $U(x) = u(x)$ for all nodes x with $u \in C^2$, $|U_\delta(x_{ref}(T))| = O(h(T)^2)$, which implies the saturation condition holds asymptotically on grids $\mathcal{T}^l$ for l sufficiently large. Let us emphasize that the handling of the U_δ–values would therefore allow an economical δ–compression of the data. The original values can easily be retrieved during the recursive tree traversal. Now, we define the hierarchical L^∞ error indicator

$$\eta_\infty(x) := |U_\delta(x)|.$$

Instead of the L^∞ norm we can analogously consider different integral norms applied to the difference function which corresponds to a new node. Using lumped mass integration we obtain

$$\eta_p(x) = \frac{1}{3} \left(\sum_{T,\, x \in \mathcal{N}(T)} |T| \right)^{\frac{1}{p}} |U_\delta(x)|$$

for $1 \le p < \infty$. Decreasing p leads to an earlier stopping of the tree traversal on simplices of small size.

– **Gradient type error indicator:** Instead of measuring the one level error with respect to function values, we can consider the error of the function gradient. We thus define

$$\eta_{1,p}(x) := \begin{cases} \left(\displaystyle\sum_{T,\, x \in \mathcal{N}(T)} |T| \right)^{\frac{1}{p}} \|\nabla U_\delta|_T\| & \text{for } 1 \le p < \infty, \\ \displaystyle\max_{T,\, x \in \mathcal{N}(T)} \|\nabla U_\delta|_T\| & \text{for } p = \infty. \end{cases}$$

The evaluation of these error indicators takes some effort in the precomputing step. If we replace simplices by simplex refinement edges without modifying the scaling we gain at least for $p = \infty$

$$\eta_{N,e} := \frac{2\, |U_\delta(x_{ref}(T))|}{h(e_{ref}(T))}.$$

- **Discrete curvature type indicators:** With a focus on an isosurface's geometric shape, we will now consider some kind of curvature estimation. We ask for a discrete curvature quantity which locally measures the quality of the data approximation from the perspective of the visual appearance [18]. In isosurface images consisting of linear patches we can easily recognize folds at surface edges. In each tetrahedron the data gradient ∇U^l is always perpendicular to an isosurface. Therefore, at any face F the normal component of the jump of the normalized gradient, denoted by $[\frac{\nabla U^l}{|\nabla U^l|}]_F$, locally measures the fold in the data function. Here the jump operator $[\cdot]_F$ is defined as the difference of the argument on both sides of the face. This jump obviously serves as a well–founded graphical error criterion and motivates the definition

$$\eta_N(x_{ref}(T)) := \left[\frac{\nabla U^l}{|\nabla U^l|}\right]_{F_{ref}(T)}.$$

We can apply the simplification of the previous indicator here as well and denote the resulting error indicator by $\eta_{N,e}$.

Ensuring the saturation condition

As pointed out above, the hierarchical error indicators do not fulfill the saturation condition. We can overcome this drawback by defining a modified error indicator $\bar{\eta}$, which is defined as the minimal saturated error indicator larger or equal to η. This definition is constructive in the sense that in a bottom up, breadth first traversal of the grid, we can blow up these error indicator values. In pseudo code this blow up mechanism looks as follows:

```
for (l = l_max − 1 ; l ≥ 0 ; l − − )
    for all T ∈ 𝒯^l and x = x_ref(T)
        η̄(x) = max{ max   η̄(x_ref(T_C)), η(x)};
                 T_C∈𝒞(T)
```

Let us emphasize that a depth first traversal of the hierarchy in the adjustment procedure would not be sufficient. If the error indicators are adjusted in this way the continuity problems are solved automatically.

Recursive blowup

Alternatively, we can ensure saturation of an indicator η by recursively defining:

$$\eta^+(x) = \eta(x) + \max_{T_C \in \mathcal{C}(T)} \eta^+(x_{ref}(T)) \qquad \text{for } x = x_{ref}(T).$$

On the finest grid level, where $C(T) = \emptyset$, we simply set $\eta^+(x) = \eta(x)$. The different error measures are obviously related to each other by $\eta \leq \bar{\eta} \leq \eta^+$.

Furthermore, we obtain $\eta^*_\infty \le \eta^+_\infty$ and $\eta^*_{1,\infty} \le \eta^+_{1,\infty}$ due to the triangle inequality. The indicator η^+, although the largest one derived from the original indicator η^*, and thus the weakest, can have other desirable properties. For instance, an easy computation of min/max-values for isosurface extraction or criteria for multilevel backface culling are possible, which is demonstrated next.

On the one hand, we are able to compute a bound $\beta_0(T)$ for second order off-set terms of the data function on a simplex $T \in \mathcal{T}'$, i.e. the difference of the true function and its linear approximation. This can be applied in the implementation of the *SimplexIsOfInterest()*-function. We obtain

$$\min_{x \in T} U' - \beta_0(T) \le U \le \max_{x \in T} U' + \beta_0(T).$$

The *SimplexIsOfInterest()* routine corresponding to the extraction of an iso-surface for the isovalue c can be written in pseudo code:

```
SimplexIsOfInterest(T) {
    if min_{x∈N(T)} U(x)−β₀(T) ≤ c ≤ max_{x∈N(T)} U(x)+β₀(T)
        return true;
    else
        return false;
}
```

In the hierarchical offset case and for the choice η^+_∞ we can define

$$\beta_0(T) = \begin{array}{ll} \frac{1}{2}\eta^+_\infty(x_{ref}(T)), & \text{for hierarchical offset indicators} \\ h(T)\eta^+_{1,\infty}(x_{ref}(T)), & \text{for gradient type indicators} \end{array}$$

In both cases, the expensive storing of min/max–values as discussed in [23] can be avoided.

On the other hand, we may check – based on coarse grid simplices – whether all polygons extracted by the algorithm will be backfaces. Let $N' = \frac{\nabla U'}{\|\nabla U'\|}$ denote the normal of some triangle of the final isosurface triangulation on the simplex $T \in \mathcal{T}'$, and V the viewing vector from the object to the eye (we confine ourselves here to parallel projection). If $N' \cdot V \ge 0$, the triangle is faced towards the viewer. Otherwise it does not need to be drawn. We obtain a significant acceleration of our isosurface algorithm, if on a much coarser grid level we recognize simplices containing only isosurface triangles which are faced away from the viewer so that we are already able to stop the local traversal at this level. If $\beta_N(T)$ is a bound of the modification of N' in $T \in \mathcal{T}'$, we obtain the multilevel backface test

$$N' \cdot V + \beta_N(T) \le 0, \qquad \text{with } \beta_N(T):=\eta^+{}_N(x_{ref}(T))$$

for the discrete curvature type error indicator $\eta^+{}_N$.

Skipping the normalization and considering instead a bound $\beta_1(T)$ which measures the possible offset in $\|\nabla U\|$, we alternatively obtain the rejection criterion

$$\nabla U' \cdot V + \beta_1(T) \le 0, \qquad \text{with } \beta_1(x) = \eta^+_{1,\infty}(x_{ref}(T)).$$

It can easily be seen that, on average, while arbitrarily rotating the object, we save up to one half of the computing time for an isosurface.

4 A quantitative comparison

Up to now we have analyzed qualitative aspects of different error indicators. In what follows, let us focus on a detailed quantitative discussion. Therefore, we study certain test problems in 2D as well as in 3D.

Test data sets

In 2D we pick up different examples from different classes of data sets. On the one hand, we choose a typical measurement data set, which represents a geographical map, originally sampled on a 257^2 regular grid, which we afterwards cover with a hierarchical triangular grid (see Appendix). It consists of regions with a significant roughness and other areas which are almost planar.

On the other hand, we apply multilevel visualization to a typical numerical data set already computed on a triangular grid hierarchy. It is characterized by smooth, less steep areas which alternate with thin transition zones where the data function is rather steep. Nevertheless, the frequencies are, on average, much more damped in the latter data set, and the numerical data set is much smoother than the geographical map. Here we consider a timestep of a Cahn-Hilliard simulation on the same 257^2 regular grid (see Appendix). It represents the density of an alloy after quenching (rapidly cooling), which leads to phase separation [7, 22].

In the 3D case we consider isosurface extraction and color slicing (see Appendix). Here the well known 129^3 bucky ball data set serves as an example. Like the Cahn-Hilliard data set it contains smooth areas in the interior of the molecule and steep areas in the vicinity of the carbon atoms.

Measures of cost, quality and efficiency

The cost of the visualization method is mainly controlled by the number of visited grid cells in the recursive traversal. We suppose that a suitable graphics hardware guarantees a fast processing and final rendering of graphic primitives on the adaptively finest grid levels, so that the CPU and not the graphics hardware is the bottleneck. Since error indicators come along with different ranges of indicator values on the grid nodes, we ensure comparability by normalizing the maximal indicator value to 1.

An alternative measure of the cost would be the number of rendered primitives. Not surprisingly both measures are closely related and therefore it does not really matter which one we choose. In our experiments there is at most a ratio of logarithmic size with respect to the maximal depth of the grid hierarchy. The following results are based on the visited-cell-count cost measure.

The crucial measure of the quality of an adaptive projection in visualization is the visual impression of the rendered image. However, this is impossible to quantify. So in order to get a comparable notion of the quality we chose the reciprocal of the corresponding global norm of the difference between the adaptively extracted function and the function on the finest grid. In this context the efficiency E of an error indicator is the quotient of quality and cost and would thus be

$$E_\eta(U, \varepsilon) = \frac{1}{k \cdot \|P_\eta U - U\|}$$

where k is the number of visited cells used for the adaptive projection $P_\eta U$.

Results

Fig. 1 and 2 compare results obtained for the different classes of error indicators. The scaling on the y axes is logarithmic. For the geographical data the different characteristics of the hierarchical offset error indicators compared to the error indicators based on derivatives are striking.

The smoother numerical data show a similar behaviour. Not surprisingly, the graphs for $\bar\eta_N$ and $\bar\eta_{N,e}$ are especially for the geographical data set nearly the same. Therefore, the simplification incorporated in $\bar\eta_{N,e}$ seems to be admissible and as $\bar\eta_{N,e}$ is easier to calculate, it is more favourable for practicable purposes than $\bar\eta_N$. As also can be expected, the indicators $\bar\eta_1$ and $\bar\eta_2$ are – in comparison to $\bar\eta_\infty$ – rather similar.

The efficiency of these indicators is depicted in Fig. 3. It becomes clear that $\bar\eta_\infty$ is less efficient than $\bar\eta_1$ and $\bar\eta_2$. In the case of the geographical data set and also for not too high threshold values in the case of the numerical data, the qualities of the three error indicators differ only slightly. So the main reason for the low efficiency of $\bar\eta_\infty$ is that even for high threshold values a large number of cells is visited. In the 3D-case the results are similar.

Finally, we compare the different methods for ensuring the saturation condition in case of the η_∞-indicator. In our experiments the differences in smoothness between the geographical and the numerical data are clearly visible in the characteristic if $\bar\eta_\infty$ is used. However, these differences are lost for η_∞^+. This is also true for other indicators as for example η_2^+ compared to $\bar\eta_2$. Hence, an application of an η^+-type saturated indicator is only reasonable if the advantages concerning min/max-bounds or backface culling are exploited.

Visual Impression

We also want to show that for reasonable threshold values the visual impression of the original and adaptively projected images are rather close (see Appendix). For the geographical map, the adaptive image consists of 13666 patches whereas the original image has a size roughly ten times larger (131072 patches). Additionally, we show extracted isosurfaces of the bucky ball data set with 128709 and 590018 triangles, respectively (see Appendix). In all these figures we used the $\bar\eta_\infty$ error indicator.

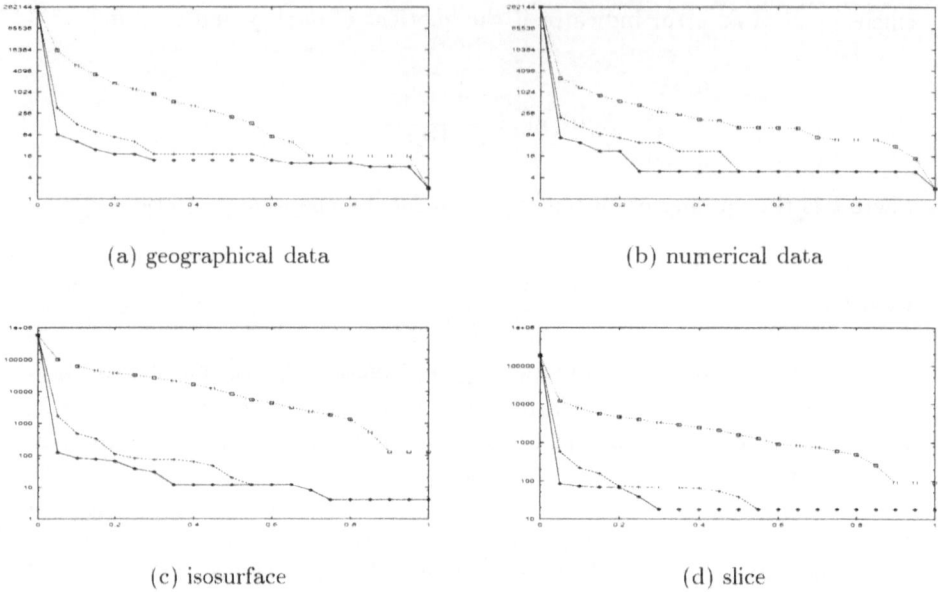

(a) geographical data   (b) numerical data

(c) isosurface   (d) slice

Fig. 1. Error indicators $\bar{\eta}_1$ (dashed), $\bar{\eta}_2$ (solid) and $\bar{\eta}_\infty$ (dotted) based on the different local L^1, L^2 and L^∞ norms of the hierarchical offset are compared, concerning the count of visited simplices for varying threshold values.

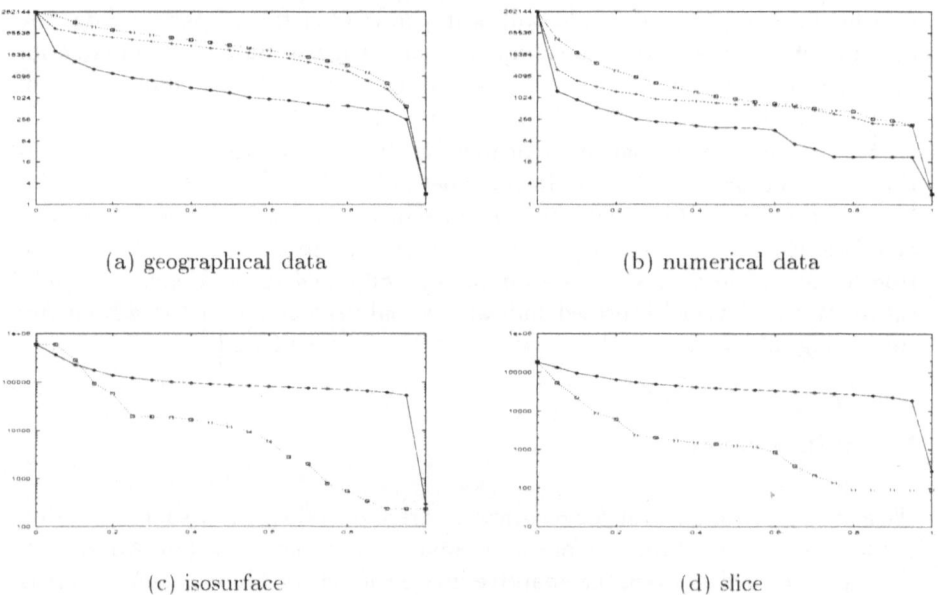

(a) geographical data   (b) numerical data

(c) isosurface   (d) slice

Fig. 2. The visited cell count is compared for the error indicators $\bar{\eta}_{1,\infty}$ (solid), $\bar{\eta}_N$ (dotted), and $\bar{\eta}_{N,e}$ (dashed, only 2D) respectively.

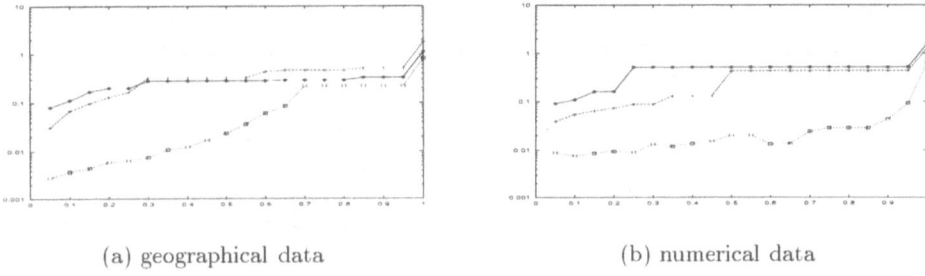

(a) geographical data   (b) numerical data

Fig. 3. Efficiency as function of threshold value for local $\bar{\eta}_\infty$-error (dotted), $\bar{\eta}_1$-error (solid) and local $\bar{\eta}_2$-error (dashed)

5 Concluding remarks

In this paper we have considered several error indicators which are used in multiresolutional visualization and compared their quantitative as well as their qualitative properties. We have specifically looked at local norms of difference functions, differences of gradients and discrete curvature measures. We have employed the saturation condition as an important prerequisite for interactive visualization since it solves the continuity problem.

We have shown how this condition can be fulfilled by a blowup of η. We have thereby defined the minimal saturated hierarchical error indicator $\bar{\eta}$, which indeed gives very good results concerning triangle count vs. global error.

On the other hand, by a slight alteration of the blowup mechanism, we have defined the error indicators η^+, which have a slightly worse efficiency but other desirable properties. For instance, based on the error indicator data bounds on simplices can be computed which then serve as stopping criteria for multiresolutional isosurface extraction. We have also shown how gradient type error indicators allow multilevel backface culling. In a series of numerical experiments on application data the different error indicators have been compared and their mutual advantages have been outlined.

References

1. E. Bänsch. Local mesh refinement in 2 and 3 dimensions. *IMPACT of Computing in Science and Engineering*, 3:181–191, 1991.
2. E. Bier, M. Stone, K. Pier, W. Buxton, and T. DeRose. Toolglass and magic lenses: the see–through interface. In *Proceedings of SIGGRAPH '93 (Anaheim, CA, August 1-6). In Computer Graphics Proceedings, Annual Conference series, ACM SIGGRAPH*, pages 73–80, 1993.
3. A. Certain, J. Popović, T. DeRose, T. Duchamp, D. Salesin, and W. Stuetzle. Interactive multiresolution surface viewing. In *SIGGRAPH 96 Conference Proceedings*, pages 91–98, 1996.

4. P. Cignoni, L. De Floriani, C. Montoni, E. Puppo, and R. Scopigno. Multiresolution modeling and visualization of volume data based on simplicial complexes. In *1994 Symposium on Volume Visualization*, pages 19–26, 1994.

5. P. Cignoni, C. Montani, and R. Scopigno. MagicSphere: an insight tool for 3D data visualization. *Computer Graphics Forum*, 13(3):317–328, 1994.

6. L. De Floriani, P. Magillo, and E. Puppo. Building and traversing a surface at variable resolution. In *Proceedings IEEE Visualization 97*, October 1997.

7. C. Elliot. The cahn-hilliard model for the kinetics of phase separation. *Num. Math.*, 1988.

8. T. Gerstner. Adaptive hierarchical methods for landscape representation and analysis. In S. Hergarten and H.-J. Neugebauer, editors, *Lecture Notes in Earth Sciences 78*. Springer, 1998.

9. M. H. Gross and R. G. Staadt. Fast multiresolution surface meshing. In *Proceedings Visualization*, pages 135–142, 1995.

10. R. Grosso, C. Luerig, and T. Ertl. The multilevel finite element method for adaptive mesh optimization and visualization of volume data. In *Proceedings Visualization*, 1997.

11. B. Hamann and J. Chen. Data point selection for piecewise trilinear approximation. *Computer Aided Geometric Design*, 11, 1994.

12. H. Hoppe. Progressive meshes. In *SIGGRAPH 96 Conference Proc.*, pages 99–108, 1996.

13. J. Maubach. Local bisection refinement for n-simplicial grids generated by reflection. *SIAM J. Sci. Comput.*, 16:210–227, 1995.

14. R. Klein, G. Liebich, and W. Straßer. Mesh reduction with error control. In *Proceedings Visualization*, 1996.

15. Y. Livnat, H. W. Shen, and C. R. Johnson. A near optimal isosurface extraction algorithm using the span space. *Transaction on Visualization and Computer Graphics*, 2(1):73–83, 1996.

16. W. F. Mitchell. A comparison of adaptive refinement techniques for elliptic problems. *ACM Trans. on Math. Software*, 15(4):326–347, 1989.

17. R. Neubauer, M. Ohlberger, M. Rumpf, and R. Schwörer. Efficient visualization of large scale data on hierarchical meshes. In W. Lefer and M. Grave, editors, *Visualization in Scientific Computing '97*. Springer, 1997.

18. M. Ohlberger and M. Rumpf. Adaptive projection methods in multiresolutional scientific visualization. *IEEE Transactions on Visualization and Computer Graphics, Vol 4 (4)*, 1998.

19. M. C. Rivara. Algorithms for refining triangular grids suitable for adaptive and multigrid techniques. *Internat. J. Numer. Methods Engrg.*, 20:745–756, 1984.

20. W. J. Schroeder, J. A. Zarge, and W. A. Lorensen. Decimation of triangle meshes. In *Computer Graphics (SIGGRAPH '92 Proceedings)*, volume 26, pages 65–70, 1992.

21. R. Shekhar, E. Fayyad, R. Yagel, and J. F. Cornhill. Octree-based decimation of marching cubes surfaces. In *Proceedings Visualization*. IEEE, 1996.

22. U. Weikard. Finite-Element-Methoden für die Cahn-Hilliard-Gleichung unter Einbeziehung elastischer Materialeigneschaften. Diplomarbeit, 1998.

23. J. P. Wilhelms and A. Van Gelder. Octrees for faster isosurface generation. In *Computer Graphics (San Diego Workshop on Volume Vis.)*, volume 24, 5, pages 57–62, 1990.

24. Y. Zhou, B. Chen, and A. Kaufman. Multiresolution tetrahedral framework for visualizing volume data. In *IEEE Visualization '97 Proceedings*. IEEE Press, 1997.

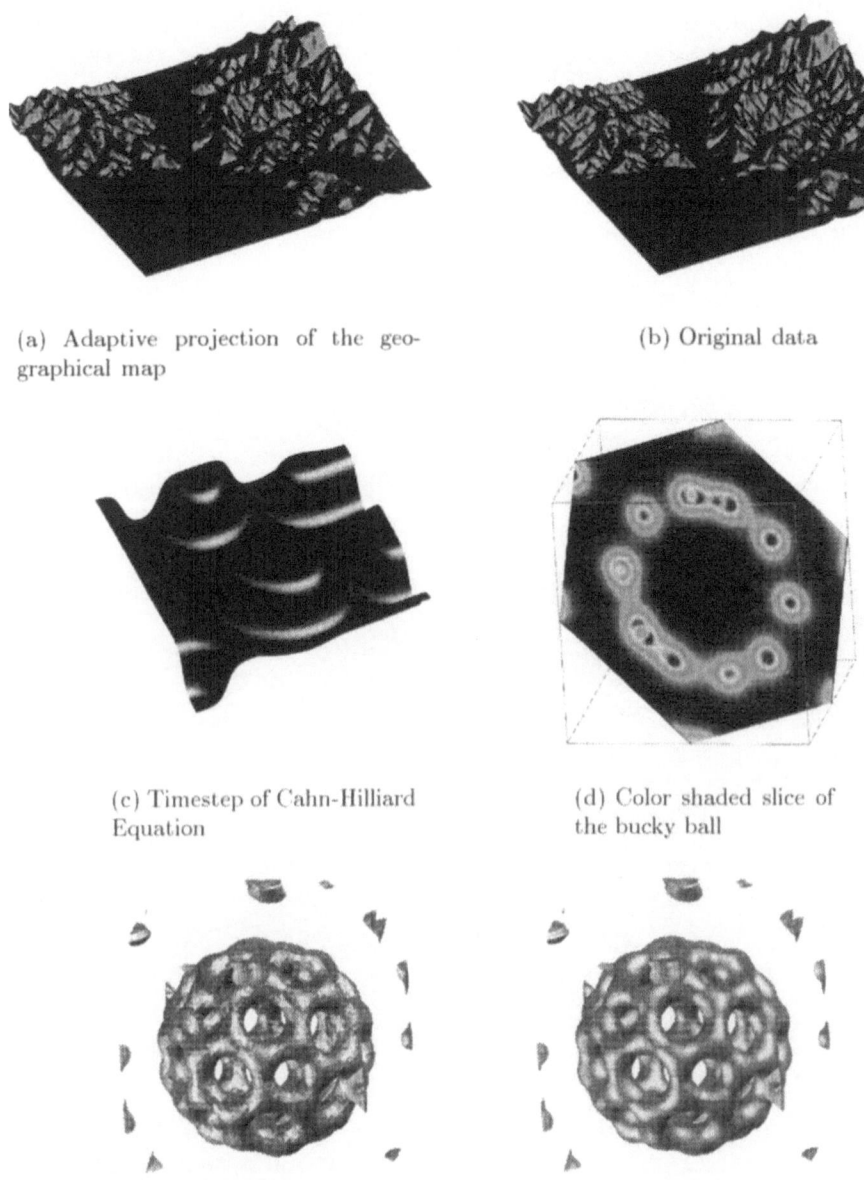

(a) Adaptive projection of the geographical map

(b) Original data

(c) Timestep of Cahn-Hilliard Equation

(d) Color shaded slice of the bucky ball

(e) Adaptive projection of the isosurface

(f) Original data

Fig. 4. Above the graph of a geographic height field, its adaptive projection and a timestep of the Cahn-Hilliard-Equation are shown. Of the bucky ball data set we show a color shaded diagonal slice, an adaptive projection and a full resolution isosurface.

Editors' Note: see Appendix, p. 329 for colored figure of this paper

Efficient Ray Intersection for Visualization and Navigation of Global Terrain using Spheroidal Height-Augmented Quadtrees

Zachary Wartell, William Ribarsky, and Larry Hodges

College of Computing
Georgia Institute of Technology, Atlanta GA 30332-0280, USA

Abstract. We present an algorithm for efficiently computing ray intersections with multi-resolution global terrain partitioned by spheroidal height-augmented quadtrees. While previous methods support terrain defined on a Cartesian coordinate system, our methods support terrain defined on a two-parameter ellipsoidal coordinate system. This curvilinear system is necessary for an accurate model of global terrain. Supporting multi-resolution terrain and quadtrees on this curvilinear coordinate system raises a surprising number of complications. We describe the complexities and present solutions. The final algorithm is suited for interactive terrain selection, collision detection and simple LOS (line-of-site) queries on global terrain.

1 Introduction

The increasing computation power, memory and rendering rates coupled with efficient data organization make it feasible to interactively visualize global 3D terrain with varying resolutions down to a centimeter. Interactively rendering these large-scale terrain databases places increasing demands on the software system. Real-time level-of-detail management, efficient spatial subdivision and the use of an accurate two-parameter ellipsoidal coordinate system are a must.

This paper describes the impact of this geodetic coordinate system on quadtree spatial subdivision with respect to computing ray-terrain intersections. We extend a well-known ray-casting method for height-augmented quadtrees defined on Cartesian coordinates. The extension also handles multi-resolution terrain.

2 Background

Our terrain visualization software is VGIS [8]. VGIS uses automatic, continuous level-of-detail management for geometry and imagery. The data is partitioned into 32 spheroidal quadrilaterals called zones. Each zone contains its own quadtree. Each quadtree supports terrain at resolutions varying from 8km down to a centimeter. Currently, individual VGIS applications contain datasets with a mixture of resolutions with the higher detail insets covering the more important regions. The current maximum resolution data set is at 1 ft resolution.

214

To accurately model this global terrain, VGIS uses a two-parameter ellipsoidal coordinate system commonly used in geodesy [12].

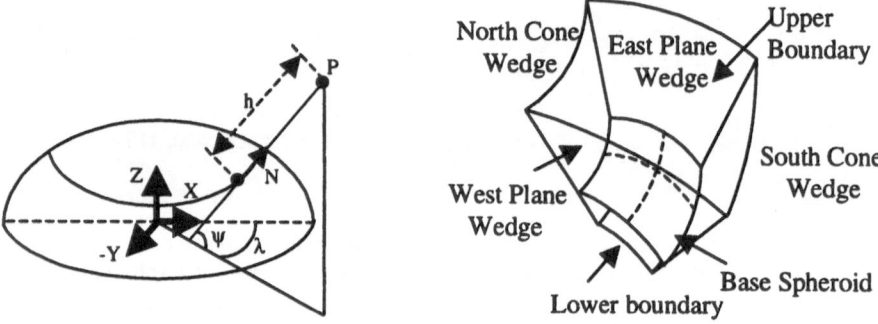

Fig. 1. Spheroidal Coordinate System   **Fig. 2.** Spheroidal Height-Quad

This two-parameter ellipsoidal coordinate system is based on a spheroid. A spheroid is subclass of ellipsoid created by rotating an ellipse about its major or minor axis. It is synonymous with "rotation ellipsoid" [4] ,"biaxial ellipsoid" [12] and "ellipsoid of revolution"[11]. Figure 1 illustrates this coordinate system. The two parameters are the spheroid's major semi-axis, a, along X and Y, and minor semi-axis, b, along Z. In this system longitude, λ, is equivalent to the longitude in polar coordinates; however, latitude, ψ, is the angle between the surface normal and the equatorial plane. Height, h, is measured parallel to the normal between the point in question, P, and the underlying surface point.

VGIS builds its terrain database and quadtree subdivision on this curvilinear coordinate system. In this quadtree the quads are bounded by meridians and parallels. This divides the spheroid into triangles at the poles and quadrilaterals elsewhere. (Note that since meridians are not geodesics, these quadrilaterals and triangles are not true spheroidal quadrilaterals and spheroidal triangles; however, for brevity we will ignore this distinction.) Finally each quad is augmented with a height attribute equal to the maximum spheroidal height of the contained data.

Using this terrain structure, we must provide an efficient method for finding arbitrary ray to terrain intersections. Such an algorithm serves as a basis for interactive terrain selection, collision detection and simple line-of-site queries.

While an efficient method for ray-casting through Cartesian coordinate height-augmented quadtrees is well-known [2], this method assumes that the bounding volumes are bounded by their Cartesian coordinates. Extending the algorithm to handle spheroidal based height-quadtrees for multi-resolution terrain poses a number of problems. We present our solutions in order of their generality with respect how terrain is modeled. First we address tracing through the spheroidal bounding volumes. The presented algorithm applies to terrain modeled either as voxels, triangles, or bilinear patches. Next we address tracing through individual terrain elements. Here our solution is specific to terrain modeled as a regular triangle mesh. Third we address complications added by triangle models using multi-resolution data sets. Finally, we discuss surface continuity issues that are specific to VGIS's continuity preservation methods.

3   Traversing Spheroidal Height-Quads

Cohen et al.'s [2] method for efficient ray-terrain intersection, is similar in spirit to Bresenham line drawing. It traces the XY projection of a ray through the XY footprints of a height-augmented quadtree based on Cartesian coordinates. Upon entering a height-quad the entering and exiting z-coordinate of the ray is compared to the height of the quad. If the ray intersects the quad, the algorithm steps into the child quad at the next resolution level. Otherwise, the algorithm steps into the next quad at the same resolution level. The algorithm is so efficient that it is targeted towards real-time rendering of terrain. Figure 3 (see Appendix) illustrates the high-level functionality of the algorithm. The figure is a side view with the ray in red and 3 levels of recursive height-quads. Blue volumes are intersected by the ray. Solid black volumes are not intersected, but the ray does enter their X-Y footprint. Dash black volumes are not examined by the algorithm at all. The red volume is the lowest level intersected volume. This figure illustrates the recursive nature of the bounding volumes and of the algorithm.

Ideally, a spheroidal extension would use incremental integer calculations similar to Cohen's midpoint method. Unfortunately, while the basic high-level algorithm still applies, the midpoint technique that works so beautifully in the Cartesian setting appears to have no similarly efficient analog in this spheroidal case.

An exact analog would require a spheroid to plane mapping in which the spheroidal projection of a ray in 3-space maps to a line and in which the spheroid's quads are mapped to a regular square grid. The only common sphere to plane mapping that maps parallels, meridians and projected rays onto lines is the gnomonic projection [9, 236]. The gnomonic mapping centrally projects the sphere through the sphere center onto a plane. The plane is placed tangent to the sphere at an arbitrary intersection point. Unfortunately, this mapping projects spherical quads onto planar rectangles with varying sizes. As we examine rectangles farther and farther away from the plane-to-sphere intersection point, the rectangles' areas grow towards infinity. The gnomonic map will not allow us to translate the Cartesian algorithm to the spheroidal case.

A partial analog to the Cartesian algorithm would require a spheroid to plane mapping in which quads map to a regular square grid and a projected ray maps to a curve. A cylindrical mapping can map quads onto a regular square grid. Unfortunately, this mapping maps the projection of a 3D ray to a complicated curve containing multiple embedded transcendental functions (equation 3-1 [14]). While efficient methods for discretizing lines [6], ellipses [6], and cubics [15] are known, a similarly efficient method of discretizing a curve of this complexity is not available.

Without an incremental integer approach for stepping through quads, we resort to floating-point computation of ray-quad boundaries. Unfortunately, there appears to be no closed form solution for solving t in terms of the latitude, ψ [14] for the cylindrical projection. This would be needed for computing the projected ray's quad intersections with closed form arithmetic. We therefore perform the ray-quad intersection tests in 3 dimensions where closed form solutions exist.

3.1 Bounding Surfaces

We begin by describing the bounding surfaces of a spheroidal height-quad. Generally these boundaries consist of 4 side boundaries formed by 2 plane wedges and 2 cone wedges, and of upper and lower boundaries formed by quadrilaterals on the normal expansion of the spheroid (Fig. 2). We now give the equations of these surfaces. They are derived in [14].

A longitude side boundary at longitude λ is a wedge of the plane:

$$\cos \lambda x + \sin \lambda y = 0 \tag{1}$$

A latitude side boundary at latitude ψ is a wedge of the cone:

$$x^2 + y^2 - \frac{z^2}{m^2} + \frac{2zk}{m^2} - \frac{k^2}{m^2} = 0$$

where

$$m = \tan \psi, k = Z - Xm,$$
$$X = \frac{a \cos \psi}{\sqrt{1 - e^2 \sin^2 \psi}}, Z = \frac{a(1 - e^2) \sin \psi}{\sqrt{1 - e^2 \sin^2 \psi}}, e^2 = \frac{a^2 - b^2}{a^2} (\text{see } [4])$$

\tag{2}

For the upper boundary surface of a quad with maximum height, h, the true boundary surface is not amenable to analytic intersection computations [14]. Therefore we use the approximation spheroid, B_h:

$$B_h = \begin{cases} major Axis = (a + h), minor Axis = \left(b + \frac{ha}{b}\right), \text{if } h \geq 0 \\ \\ major Axis = \left(a + \frac{hb}{a}\right), minor Axis = (b + h), \text{if } h \in \left(\frac{-b^2}{a}, 0\right) \end{cases} \tag{3}$$

The stipulation that $h \geq -b^2/a$ is necessary due to degenerate cases of the true boundary surface. In practice, surface terrain models well exceed this lower boundary because a and b are typically close while the minimum h is orders of magnitude greater than a or b. For example in the WGS-84 Earth datum, $-b^2/a$ is -6,335,439m while the minimum h is around -15,000m.

Finally, since we assume that traced rays begin outside the planet, it is sufficient to choose a single global lower bounding surface. We model the lower boundary simply as a sphere whose radius equals the distance from the spheroid center to the closest terrain vertex. B_h is inappropriate here since it lies outside the true boundary while a lower boundary approximation must lie inside the true boundary.

4 The Algorithm

While the high-level principles of the Cohen algorithm (Section 3), apply to the spheroidal case, the details differ. The spheroidal algorithm is divided into two procedures. A user called procedure performs setup and zone traversal and a recursive procedure then recursively traverses through each zone's quadtree. The user called procedure first clips the ray to the volume bounded by a global upper boundary and the global lower boundary. The global upper boundary is the upper bounding surface, B_h, with height equal to the maximum global height. As part of this clipping, we compute, t_global_exit, the ray parameter value of the ray's global exit point. Next, we determine which zone contains the ray origin. Starting with this zone, we step through successive zones until either an intersection occurs or the ray exits the global boundaries. Zone traversal is quite similar to quad child traversal which is discussed in detail below. For each zone we call the recursive procedure to recursively traverse the zone's quadtree.

The recursive procedure must first determine whether the ray, which is assumed to enter a quad's side bounds, truly intersects the quad volume. Since the upper boundary is curved, it is insufficient to check the height of the ray's entering and exiting intersections with the side boundaries. Instead, we compute the ray's parameter values, t_in and t_out, at these side intersections and we compute the ray's intersection parameters, t_0 and t_1, with the quad's upper boundary surface (Fig. 4, see Appendix). If and only if these two parameter intervals overlap, then the ray has entered the height-quad volume and we step through the quad's children.

If the quad volume is intersected, the algorithm must traverse the quad's children and recurse at each child. The first encountered child is determined from two factors. The first factor is which side boundary of the parent the ray entered. The second factor is in which half-space of one of the parent's internal partition surfaces the entrance point lies. In Figure 5 (see Appendix) these internal partitions are shown in blue. They partition the quad into four sections. The latitude partition surface is a cone wedge stretching east-west. The longitude partition is a plane wedge stretching north-south. Knowing which side boundary is intersected and which internal partition half-space contains the entrance point, we know which child quad to visit. For example, in Figure 5 the ray (red) enters the west side boundary and the entrance point (marked by a red X) is north of the internal latitude partition. Hence, the ray enters the north-west child. Four side boundaries and two internal partitions yield eight combinations. Each combination maps to one child. By determining which combination occurs, the algorithm determines which child to visit. Note an exception arises when the current quad contains the ray origin. In this case, we visit the child containing the ray origin. This child is determined by examining which half-spaces of both internal partitions contain the ray origin.

Having visited the first child, we must determine the other child quads intersected by the ray. Note that in the spheroidal case, a ray may intersect all four of a quad's children or may enter a quad twice. This can occur since a ray can have two intersections with a quad's latitude cone boundary. Given these complications we determine the next child quad by computing the ray's intersection with the current child's boundaries. Note these boundaries are subsets of the parent's boundaries and internal partitions. The child exit boundary is the child boundary whose ray intersection's ray parameter value is the smallest while still being greater than the child's entrance point's value. This exit point is illustrated by the second red X in Figure 5. So given the current child, we compute t_child_out, the ray parameter where the ray exits the child, along with $side\_out$, the boundary of the child at this exit. With knowledge of $side\_out$, we know what child is entered next. Child traversal terminates when either a child reports terrain intersection, all children are visited, or t_child_in of the current child is greater than t_global_exit, the ray intersection with global upper boundary.

5 Traversing Individual Terrain Elements

While the methods of the previous section apply to terrain regardless of the modeling method (voxel, bilinear patch, or triangles), the issues raised when traversing individual terrain elements are model dependent.

In ray-casting methods [2] the height-quad tree recurses down to the level of the smallest modeled terrain element. In regular triangle mesh methods, however, the height-quad tree typically does not recurse down to the level of the smallest modeled terrain element. Instead, a quad contains a fixed size matrix of triangles such as in Figure 6 (see Appendix). Within this block there is no further quadtree subdivision. This means that for triangle modeled terrain, once we trace a ray to a leaf quad, we must then separately trace the ray through that quad's block of triangles. Additionally, the modeling method affects the mathematical surface in between the sampled elevation points. If we render with ray-casting we might model the surface as set columnar voxels which project radially out from the zero-elevation surface. (Note on the spheroidal coordinate system, these voxels are not cubes as in traditional Cartesian based terrain.) Alternatively, we can define the surface to be a set of bi-linearly interpolated patches. This is the typical method of interpolating height fields in geodesy [3]. Unfortunately, while these are the most mathematically robust surface definitions, a practical polygon graphics pipeline based system must interpolate between sampled elevation points by treating these points as vertices of triangles.

Here we will focus on the triangle model. In order to minimize the number of triangles tested, we treat each triangle-pair as if it was contained in its own small height-quad and we then visit only those height-quads whose sides are intersected by the ray.

In Figure 7 (see Appendix), four triangle pairs are drawn in red on a part of a spheroidal quad in black. The blue arrows are extensions of the spheroid normal at the quad's terrain grid points. Triangle vertices are confined to these lines. Furthermore, the lines delineate plane wedges defining four-sided volumes (blue). Note, triangle edges are confined to these plane wedges. These four-sided volumes can serve a similar purpose to the high level height-quads. If the ray intersects the first triangle pair's volume, A, (in bold blue), we determine which of the 4 neighboring triangle-pairs to visit next by intersecting the ray with the volume's planar wedge sides. If the ray intersects the side shared by volume A and B, this immediately tells us to visit the triangle-pair volume B. Similarly if the ray next intersects the side shared by B and C, we step into volume C. At each volume we test for ray-triangle intersections with the triangles in that volume. We continue traversing and testing triangle pairs until either a triangle is intersected, the quad boundary is reached or t_volume_exit, the ray parameter at its exit from the current triangle-pair's volume, is greater than t_global_exit. Figure 6 (see Appendix) illustrates a typical pattern of examined triangle pairs in red.

Unfortunately the triangle model poses a theoretical problem that the other surface models do not have. Since the spheroidal height-quads are concave volumes, they will not contain all parts of the triangles whose vertices are contained in the volume and assigned to the quad. Specifically, the latitude conical boundaries do not contain all parts of the planar terrain triangles along this border. This problem is illustrated in Figure 8 (see Appendix). Figure 8 shows 3 terrain triangles in red at the corner of a quad whose east, north, south and lower boundaries are drawn in black. The upper triangle is assigned to the illustrated quad while the lower 2 triangles are assigned to the adjacent quad across the south border. The green highlighted portion of the lower 2 triangles is the portion of these lower triangles not contained in the adjacent quad.

It is important to note that the containment problem is fundamental to any recursive spheriodal partitioning. Using Bowring's theorem on normal sections [3], it is easy show that all spheroid partitionings, such as [1], [5], [7] and [10], have this problem [14].

The containment problem can potentially cause the ray intersection algorithm to miss an intersection. Referring to Figure 8, the ray could first pass over the adjacent southern quad without intersecting it and then enter the illustrated quad. If the ray is at a steep angle, it could then pierce the green area. Since the illustrated quad does not contain the triangles associated with this green area, the ray will exit the global lower boundary and the algorithm would falsely indicate no intersection occurred.

When using this algorithm for interactively pointing at and grabbing terrain, however, it has been our experience that such pathological cases never occur [13]. The reason is that each quad contains a relatively dense 128x128 triangle-pair block making the green area in Figure 8 extremely small. While the increasing curvature of the cone wedges at extreme latitude quads could exacerbate the containment problem, the increasing surface density of the triangles at these extreme latitudes counteracts this effect. This increase in surface density occurs because the quad surface area grows smaller at extreme latitudes.

6 Managing Multiresolution Aspects of Terrain

While covering the general traversal of the high-level spheroidal quads and the specific traversal of triangle-modeled terrain elements, we glossed over how a multi-resolution terrain model interacts with the ray casting algorithm. A typical multi-resolution model such as VGIS stores terrain data in 2^nx2^n blocks at resolutions at varying powers of 2. For rendering purposes, the system then goes to great lengths to ensure that the rendered terrain is a continuous surface. The algorithm uses a visual error metric to render the lowest-detail level necessary to maintain visual quality while preserving mesh continuity [8].

As previously mentioned, contrary to ray-casting models where the recursive subdivision of height-quads continues down to individual voxels, in a regular mesh model a leaf quad contains a N by N array of triangle pairs called a block. Equally important, the quadtree is not a full tree. Instead a branch is only as deep as necessary to reach the highest resolution block available on disk. Moreover, while the complete quadtree is always in main memory, the actual triangle data is dynamically paged into main memory as dictated by the rendering algorithm. In Figure 9a, a flattened and zoomed in view of a single zone is shown. The outlines of the sub-quads existing in the zone are also shown. High resolution data is only available for the north-eastern most quad. This is indicated by the presence of the higher resolution quads in this region. Figure 9b shows the corresponding quadtree data structure.

a)   b)

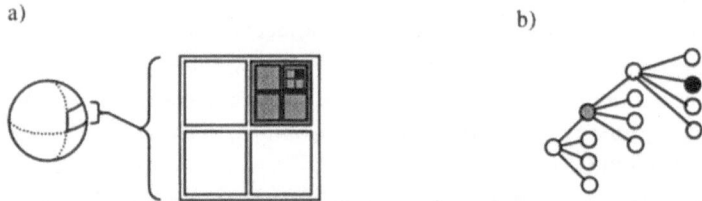

Fig. 9. Multi-resolution quadtree.

These complications lead to the following modification to the high-level quad traversal algorithm. We traverse the quad-tree as detailed in Section 4 until either the ray exits the tree or the ray enters a leaf quad. If the ray enters a leaf quad, we need to find the highest resolution in-memory block covering the quad. So if the entered quad's block is paged out, we must find the first quad ancestor whose block is paged in. A loop accomplishes this traversal. At each iteration we attempt to trace the ray through triangles of the ancestor quad. There are three possible results. Either the ray intersects a triangle, intersects no triangle, or the ancestor quad has no in-memory data available. The loop continues until reaching an ancestor with data available. For example, assume in Figure 9 we have recursed down to and intersected the north-eastern most quad. This quad is shown as a solid square in the geometric diagram (9a) and a solid circle in the tree diagram (9b). Let's assume data for this quad is paged out. Then the ancestor loop climbs up the tree to the first ancestor that contains terrain data. In Figure 9 we assume this is the 2nd level quad which is tinted light grey.

When an ancestor with data is found, we should only trace through the rectangular subset of its block which covers the original leaf quad. We compute the boundaries of this subset using an incremental integer approach [14] since floating point methods allowed rounding errors that occasionally yielded invalid array indexes.

Whether the algorithm should only address paged-in data is application dependent. Therefore we also have a parameter controlling how ray traversal handles paged-out data. This allows the programmer to balance the accuracy of the terrain data used against the performance penalty of paging. The additional parameter, min_level ("minimum level") indicates the minimum tree depth of a quadnode that may be used during triangle traversal. We modify the high-level quad traversal algorithm as follows. Recall that high-level traversal continues until a leaf quad is encountered. Now instead of just using terrain data from the leaf quad's first ancestor with paged-in data, we add the constraint that we must stop at the ancestor whose depth equals min_level. If we reach this ancestor without finding paged-in terrain data, we must wait and page in this ancestor's terrain. Note it is possible that the leaf quad depth is less than min_level. This means that terrain data at the desired resolution does not exist on disk. We must make do with the terrain data from this leaf quad and page it in as needed. With this modification, setting min_level to zero will use only the data in primary memory and will never wait for new data to be paged-in (unless even the lowest resolution data is absent). Setting min_level to the maximum possible depth will page-in whatever data is necessary to ensure that ray traversal uses the greatest resolution data on disk. Setting min_level to some other value allows the programmer to balance the accessed terrain's resolution with the algorithm's real-time performance.

7 VGIS Surface Continuity

Surface continuity issues add further complications. When two adjacent terrain blocks have different resolutions the edges of triangles along the shared border will not match. When rendering, VGIS uses a set of rules to discard certain vertices and generate a triangle mesh using this vertex subset. This resulting mesh has no cracks along block borders. How and/or when should we apply such rules to the terrain traversed by the ray intersection algorithm? The answer depends on the application of the intersection algorithm.

In our current VGIS virtual reality application, we use the ray-terrain intersection for terrain selection with a hand-held virtual laser pointer [13]. Whenever, the ray crosses block boundaries during triangle-pair traversal we compute temporary geometry for these polygons and test them for intersection with the ray. This method is simple and fast but it violates the spirit of the VGIS rendering algorithm. We continue to research how to modify VGIS's rendering rules for applications where the temporary geometry approach may be inappropriate.

8 Results and Conclusions

Figure 10 (see Appendix) illustrates our complete algorithm in operation. Here *min_level* is zero and we use the simplest continuity algorithm appropriate for fast interactive terrain selection. The application is running on a virtual workbench [13]. The red ray is a virtual laser pointer interactively manipulated by the user. In order to better distinguish the terrain from the quad outlines, the terrain has been altered to appear black-and-white. The yellow lines indicate the projection of the ray origin onto the spheroid and the point on the ray where it exits the global boundaries. The visited height-quads' upper boundaries are outlined in green, black, red and blue. Blue indicates the quad volume was intersected. Red indicates the quad was intersected and is a leaf. Green indicates the quad's polygon data was used for polygon traversal. Black indicates that while the quad side bounds were intersected the quad volume was not, i.e. the upper boundary was not pierced. The small streaks of green inside the red quads are the outlines of the triangles which were tested for intersection. In 10a, the planet is at a resolution such that the polygon data associated with the leaf quad (red) is not paged in. The algorithm visits ancestor quads until reaching the first quad (green) with polygon data covering the leaf quad. Figure 10b is similar to 10a, but it shows a further zoomed in view.

We are successfully using this algorithm for navigating global terrain on the virtual workbench [13]. The algorithm is fundamental to our navigation method. The user navigates with a virtual laser pointer used to grab the terrain when panning, rotating and zooming. Empirically the intersection algorithm has had no affect on framerate, as is desired for VR interaction. Asymptotically, the algorithm is equivalent to standard quadtree and octree traveral methods. It is linear in the number of pierced bounding elements.

To conclude we have described the impact of the geodetic coordinate system on quadtree spatial subdivision with respect to computing ray-terrain intersections. We presented a new set of efficient methods for tracing a ray over the terrain. These methods go beyond the work of Cohen, promoting a complete approach for global terrain in a multi-resolution spheroidal quadtree structure.

9 Future Work

There are several avenues of future work. First, the continuity issues have yet to be fully resolved for all uses of ray-terrain intersection. Next it is probably possible to switch from the spheroidal approach to the much simpler Cartesian approach when the algorithm reaches high detail quads. This is plausible because at some point the results of these two approaches will be the same due to the finite precision of computer arithmetic. Finally, the algorithm can be extended to manage spheroidal octrees for partitioning aerial information.

10 Acknowledgements

This work was performed in part under contracts N00014-97-1-0882 and N00014-97-1-0357 from the Office of Naval Research. Support was also provided under contract DAKF11-91-D-004-0034 from the U.S. Army Research Laboratory. We thank Frank Jiang for help in setting up the workbench environment.

References

1. Borgefors, Gunilla. A hierarchical 'square' tesselation of the sphere. Pattern Recognition Letters 13 (1992), pages 183-188.
2. Cohen, Daniel, and Amit Shaked. Photo-Realistic Imaging of Digital Terrains. Eurographics '93, Volume 12, (1993), No. 3. Pg 363-373.
3. Hooijberg, Maarten. Practical Geodesy Using Computers. Springer. 1997.
4. Dragomir, V., D.Ghiţău, M. Mihăilescu, M.Rotaru. Theory of the Earth's Shape. Elsevier Scientific Publishing Company. Amsterdam. 1982.
5. Fekete, György, Rendering and Managing Spherical Data with Sphere Quadtrees. Proceedings of the First IEEE Conference on Visualization. Visualization '90. 1990. Pp. 176-86.
6. Foley, James D., Andres Van Dam. Fundamentals of Computer Graphics. Addison-Wesley. Reading, Mass. 1990.
7. Hwang, Sam C., Hyun S. Yang. Efficient View Sphere Tessellation Method Based on Halfedge Data Structure and Quadtree. Computer & Graphics, Vol. 17, No. 5 (1993), pages 575-581.
8. Lindstrom, Peter, David Koller, William Ribarsky, Larry Hodges, Nick Faust, and Gregory Turner. Real-Time Continuous Level of Detail Rendering of Height Fields. Computer Graphics (SIGGRAPH 96), pp. 109-118.
9. Maling, D.H. Coordinate Systems and Map Projections. London: George Philip and Son Limited. 1973.
10. Otoo, Ekow J., Hogwen Zhu. Indexing of spherical surfaces using semi-quadcodes. Advances in Spatial Databases. Third International Symposium, SSD '93 Proceedings, pages.510-529.
11. Smith, James R. Introduction to Geodesy. John Wiley & Sons, Inc. 1997.
12. Vaníček, Petr, Edward Krakiwksy. Geodesy: The Concepts. North-Holland Publishing Company. Amsterdam. 1982.
13. Wartell, Zachary, William Ribarsky, Larry Hodges. Third-Person Navigation of Whole-Planet Terrain in a Head-tracked Stereoscopic Environment. (to appear) Proceedings of IEEE Virtual Reality 1999 (March 13-17 1999, Houston TX).
14. Wartell, Zachary, William Ribarsky, Larry Hodges. Efficient Ray Intersection for Global Terrain using Spheroidal Height-Augmented Quadtrees. GVU Tech Report 98-45.
15. Watson, Ben, Larry Hodges. Fast algorithms for rendering cubic surfaces. Proceedings Graphics Interface '92 (May 11-15 1992, Vancouver, BC), 19-28.

Editors' Note: see Appendix, p. 330 for colored figures of this paper

VISSION: An Object Oriented Dataflow System for Simulation and Visualization

Alexandru Telea, Jarke J. van Wijk

[1] Eindhoven University of Technology,
Den Dolech 2,Eindhoven 5600 MB, The Netherlands,
alext@win.tue.nl, http://www.win.tue.nl/math/an/alext
[2] vanwijk@win.tue.nl, http://www.win.tue.nl/cs/tt/vanwijk

Abstract. Scientific visualization and simulation specification and monitoring are sometimes addressed by object-oriented environments. Even though object orientation powerfully and elegantly models many application domains, integration of OO libraries in such systems remains a difficult task. The elegance and simplicity of object orientation is often lost in the integration phase, so combining OO and dataflow concepts is usually limited. We propose a system for visualization and simulation with a generic object-oriented way to simulation design, control and interactivity, which merges OO and dataflow modelling in a single abstraction. Advantages of the proposed system over similar tools are presented and illustrated by a comprehensive set of examples.

1 Introduction

Better insight in complex physical processes requires the combination of the visualization and interactivity (seen as the user ability to interrogate and modify the simulated universe). This has led to the advent of computational steering systems, which allow the user to change and monitor various parameters on-line and perform direct manipulation on the visualized data. To extend the user's freedom from process steering to interactive process design, the dataflow concept is often used: networks of computational modules exchanging data to perform the desired task are created by connecting module icons in a visual programming tool.

Object-oriented (OO) design is, on the other hand, the favourite technique for building extensible and reusable component libraries. Making such libraries available in a dataflow steering environment would give the end-users the conciseness, elegance and reusability of OO code, often appreciated only by code designers, but not used by the target system or lost at integration. Many environments offer steering, visualization, and code integration in various amounts, but no single one addresses these and the extra requirement of existing OO libraries integration in a unitary, easy to learn manner.

We addressed the above problem by designing VISSION, a general purpose environment for VIsualization and steering of SImulations with Objectual Networks. Dataflow modelling familiar to visualization scientists [4, 6] is completely merged with the OO modelling used by component designers [5, 2] in a single new abstraction. Independently developed OO code integration is thus almost transparent, especially since VISSION automatically constructs its GUIs from the given code, extending the approach presented in [7]. This paper presents VISSION from a user perspective, its object-oriented design being detailed in [9].

This paper is organized as follows: Section 2 presents the main requirements of generic simulation systems and the main limitations of existing systems. Section 3 shows how VISSION fulfills these requirements. Applications of VISSION are presented in Section 4. We conclude the paper presenting further research directions.

2 Background

In the most demanding scenario, an open environment targets three user categories: end-users (EU) steer a simulation via virtual cameras, direct manipulation, GUI widgets, or interpreted command languages. Application designers (AD) build applications for a wide range of EU domains and thus require simple to use, yet generic interactive tools to select and assemble domain-specific components [15]. Component developers (CD) build these components and require that existing code should be easily extensible and reusable as modular components, and that the target environment should not constrain their design. Often the same person goes through all three roles (e.g. a researcher who develops his own code as a CD, then builds experiments to test algorithms as an AD, and finally monitors and/or steers the final application as an EU). The cycle repeats, (EU insight triggers application design revisions, which may ask for new/specialized components), so the role transition should be transparent: CD's code should be immediately available to the AD, who should easily produce the EU's end-application.

There is hardly any visualization/simulation system which fulfills the above requirements union and offers a simple, yet generic solution for the role transition. Turnkey systems (Fig. 1) have custom tools and GUIs to excel in specific tasks, are easy to learn and use, but are by definition not extensible or customizable. OO libraries [2, 5, 3] are highly customizable and extensible, but require manual programming of data flows and GUIs. Dataflow systems [4] are

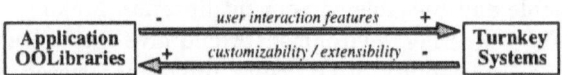

Fig. 1. A flexible system should combine the customizability/extensibility of OO libraries with the usability of the turnkey systems.

extensible and customizable, as simulations are interactively built by connecting user-written modules, but still have limitations. Few support both by-value and by-reference data transfer between modules (limitation *L1*), even fewer support user-defined types for the modules' inputs and outputs (limitation *L2*). Many such systems use different languages for module implementation, user interface, scripting, and dataflow control, making them hard to learn and use (limitation *L3*). As there is often no way to map constructs from an (OO) language to another, developers are forced to use the languages' common subset(limitation *L4*) [6,5,4], or to manually adapt their code (limitation *L5*). The set of system GUI widgets is usually not extensible to reflect directly e.g. user-defined types (limitation *L6*). *L5* and *L6* imply that GUI construction can not be automated (limitation *L7*). Few systems allow writing new modules by reusing and/or combining the existing ones and programming only the new features (no support for module inheritance, limitation *L8*). Some systems enhance monolithic simulations with dataflow-based tracking / steering features by manual insertion of data transfer and synchronization code [16,11,10,14]. These systems provide no inherent support for the CD, as they have no 'component' notion.

3   Overview of the System

The integration of dataflow/visual programming with component OO modelling comes naturally as the presented requirements are fulfilled complementarily by dataflow systems (interactivity,visual programming,GUI construction,steering) and OO application libraries (customizability, extensibility,high-level modelling), Some systems [4],[6] take this path, but none *combine dataflow modelling and OO modelling in a single abstraction*, so the listed problems are merely alleviated. VISSION completely merges OO and dataflow modelling in a new abstraction called a *metaclass*, used as its fundamental concept. The following shows how this addresses the outlined limitations and requirements.

3.1   The Metaclass Concept

From the OO modelling viewpoint, modules are implemented as C++ classes, organized by the CD as various application domain *libraries*. From the dataflow viewpoint, a module (called a *metaclass*) is an entity which enhances a C++ class with a *dataflow interface*, i.e. a set of typed input and output *ports* and an update procedure. The ports and update procedure are specified in terms of the C++ class's public methods and members: when a port is read/written, a C++ member's value is read/written or a method is called and the return value is used. Ports are typed by the C++ types of their underlying class members. Metaclasses are object-oriented entities, so they can inherit from each other, thus enabling the reuse of existing metaclasses to create new ones (addresses limitation *L8*). All information needed to 'promote' a C++ class to be directly loadable by VISSION resides in its metaclass. Our solution differs fundamentally from other systems asking the user to 'insert' system calls in his code to make it available to the framework [4],[10],[11] or to inherit from a system base class [2], [5] and hence addresses limitation *L5*.

Metaclasses:	C++ classes:

node **IVSoLight**
{ input:
   WRPort "intensity" (setIntensity,getIntensity)
               editor: Slider
   WRport "color"   (setColor,getColor)
   WRport "light on" (on)
}
	class **IVSoLight**
	{ public:
	   BOOL on;
	   void setIntensity(float);
	   float getIntensity();
	   void setColor(IVSbColor&);
	   IVSbColor getColor();
	};

node **IVSoDirectionalLight: IVSoLight**
{ input:
   WRPort "direction" (setDirection,getDirection)
}
	class **IVSoDirectionalLight: public IVSoLight**
	{ public:
	   void setDirection(IVSbVec3f&);
	   IVSbVec3f getDirection();
	};

Fig. 2. Example of C++ class hierarchy and corresponding metaclass hierarchy

Figure 2 exemplifies the above for two C++ classes and their metaclasses: the IVSoLight metaclass has three inputs for a light's color, intensity, and on/off value, implemented by the corresponding class's methods with similar names, and of types IVSbColor (a RGB triplet), float, and respectively BOOL. IVSoDirectionalLight extends IVSoLight with the light's direction, of type IVSbVec3f (a 3D vector). The user can easily specify other information in the metaclass, such as GUI and widget preferences, and help data. The appropriate widgets are automatically constructed based on the ports' types (3 float typeins for the vector and the RGB color, a toggle for the boolean, and a slider, as the preference specified, for the float). Separating this information from the C++ class lets us enhance existing classes with dataflow/GUI features non-intrusively. It has also let us develop a generic persistence scheme to save a simulation as a C++ source file. This addresses limitation $L4$, as no custom file format was needed (see [8] for a similar approach).

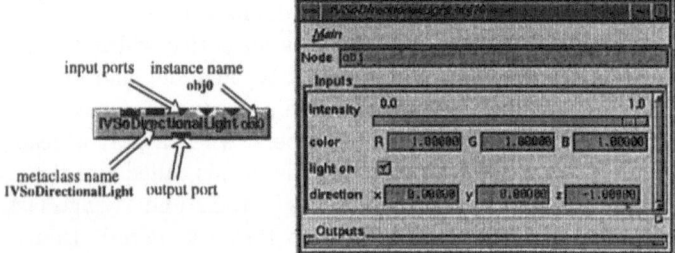

Fig. 3. Left: Visual representation of a metaclass. The various graphical signs used for the ports encode port C++ type, by value/by reference transfer, and other attributes. Right: automatically constructed GUI for the metaclass

3.2 Features

VISSION allows the user to load the desired metaclass libraries, browse a palette with the loaded metaclasses, create new nodes (i.e. instances of metaclasses), connect, clone, or delete existing nodes in a GUI similar to [4, 6] (Fig. 4). The fundamental differences between VISSION and similar systems appear as we look at the dataflow semantics.

The Dataflow Mechanism The dataflow mechanism is entirely based on the object-oriented typing offered by the C++ language, that is, data can be passed by value, by pointers or by reference, and can be of any type (addresses limitations *L1* and *L2*). If class types are used, constructors and destructors are properly invoked when data elements flow from the output to the input port. Secondly, connections between ports obey the full OO typing of by C++: a port of type *A* can be connected to a port of type *B* if the C++ type *A* conforms to the C++ type *B*. All C++ type conversions [1] are used: trivial conversion, subclass to baseclass, constructors, and conversion operators. The user interactively builds networks using the same types, rules and checking he would use in a C++ compiled program. We believe the above is a sound generalization of the dataflow typing used by other systems: The Oorange system, based on *Objective C*, offers by-reference but no by-value transfer. AVS/Express limits data types to its own *V language* which is far less powerful than C++ (e.g. it lacks constructors, destructors and multiple inheritance). Compiled libraries (e.g. vtk) are only statically extensible, as all types have to be known at compile time, which makes them unsuitable for a dynamic, interactive modelling environment.

To cope with complex networks, VISSION offers *node groups* containing subgraphs up to an arbitrary depth, which can be interactively constructed by adding nodes and ports to an empty group. This generalizes Oorange's nodes, AVS's macros, and Inventor's node kits. A less common feature is the support for networks having several *loops*, which allows a very natural way to describe iterative processes, or to implement direct manipulation as dataflows that go 'upstream' from the camera modules to other 'data processing' modules. We make no distinction between up and down stream (as compared to [4]), as the network traversal copes with any directed cyclic graph.

The GUI Interactors GUI interaction panels (shortly interactors) are provided to examine and modify the values of the nodes' ports, connect or disconnect ports, or perform actions on nodes. Interactors create the third object hierarchy in the system, isomorphic with the C++ class and metaclass hierarchies. The widgets of an interactor are based on the types of their ports: a *float* port can be edited by a slider, a *char** port by a textual type-in, a three-dimensional *VECTOR* port by a 3D widget manipulating a vector icon in 3-space, a *boolean* by a toggle button, etc (Fig. 5, 3). The set of GUI widgets for the basic types can be extended by the AD with widgets for user-defined types. This allowed us to provide GUI widgets for some types of specific libraries, such as 3D vectors, colors, rotation matrices, light values, etc. This addresses limitation *L6*.

230

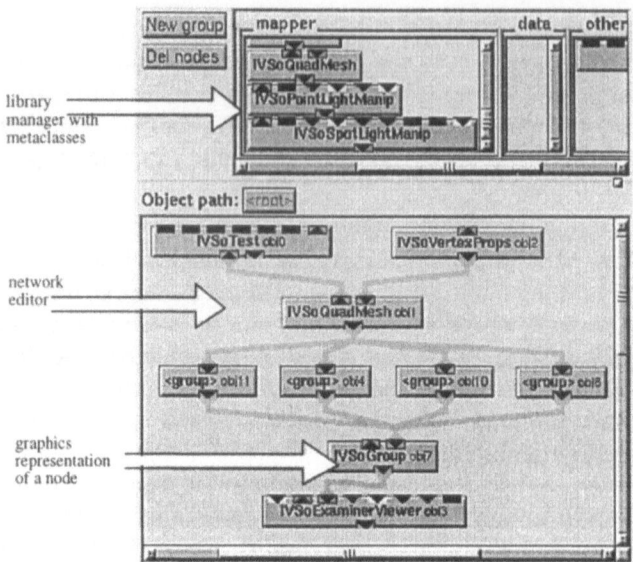

Fig. 4. The network editor and the dataflow graph. Nodes are created from metaclasses shown in the library manager's GUI.

VISSION automatically associates widgets with port types by picking out of the basic/custom widgets the one whose type best matches the port's type. The AD can thus customize the look of an application GUI, either by creating new GUI widgets or by associating the existing ones with other types (e.g. prefer a float type-in instead of a slider for a float port), and still have the interactors built automatically (this addresses limitation *L7*). Finally, the EU can instantly change a port's widget at run-time via a menu listing all widgets capable to edit that port. Dynamically associating OO widgets to OO types enables thus the creation of *user interface libraries* that can be transparently reused by VISSION to steer any network reading/writing compatible types.

The system offers also a GUI for direct inspection and modification of the C++ objects used by the metaclasses, similar to the visual class browsers of several OO compilers or debuggers. This allows CDs to directly test their C++ classes bypassing the metaclass abstraction level. Finally, the EU can type commands directly in C++ in a GUI window to be interpreted (Fig. 5), an interaction mode preferred by some users over the widget metaphor, or load and execute C++ source code. This allows the EU to write animations based on arbitrarily complex control sequences directly in C++ without having to learn a new animation language (see Fig 7c for an example of a finite element simulation based animation). Limitation *L4* is addressed, as C++ is VISSION's single language used for application class coding, dataflow typing, run-time command-line user interaction, and persistence.

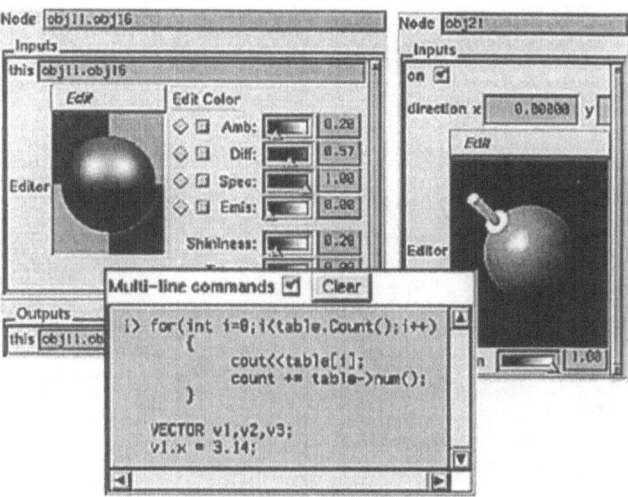

Fig. 5. GUI widgets for interaction with metaclasses

3.3 Implementation

VISSION consists of three main parts: the object manager, the dataflow manager, and the interaction manager (Fig. 6), based on two lower level components: the C++ interpreter and the library manager, communicating by sharing the dataflow graph. The key element (which enabled us to elegantly and easily re-move the limitations exhibited by similar systems) is a *C++ interpreter*. Port connections/disconnections, data transfer between ports, invocation of node up-date methods, GUI-based inspection and modification of ports, automatic GUI construction, and interpreted scripts are uniformly implemented as small C++ fragments dynamically sent to the interpreter. The interpreter cooperates with the *library manager* to dynamically load application libraries containing meta-class declarations and their compiled C++ classes, with the *object manager* to create and destroy the metaclasses, and with the *interaction manager* to build and control the GUIs. All application code is executed from compiled classes, leaving a very small C++ code amount to be interpreted. The performance loss as compared to a 100% compiled system was estimated to be lower than 2%, even for complex networks intensively accessing the interpreter.

4 Applications

The following presents some of the applications we have built with VISSION.

Scientific Visualization We have chosen the Visualization Toolkit (shortly vtk) [5], one of the most powerful freely available scientific visualization libraries,

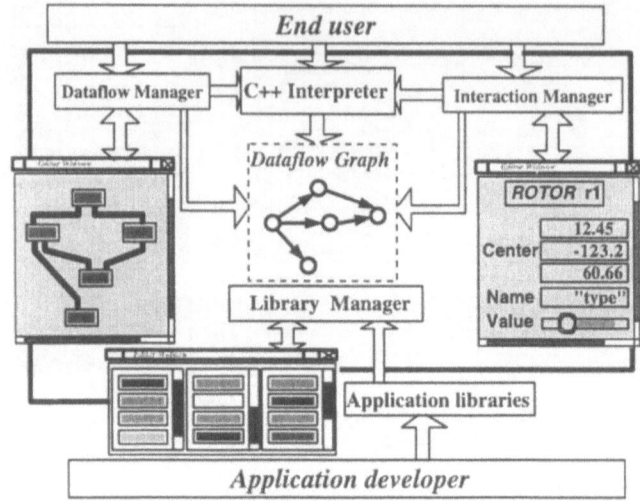

Fig. 6. Architecture of the simulation and visualization system

and integrated it into VISSION. Since vtk is a C++ library, its integration didn't pose any problems (keeping the constraint of not modifying its source code).

As a rendering back-end we used Open Inventor, which we also fully integrated. The EU can pick any vtk or Inventor class (of the total of approximately 250, respectively 70) in the visual browser, instantiate it, and connect it with other nodes, without knowing C++ or even knowing they are written in C++. We had to write a single 'adapter-like' class of around 120 C++ lines to connect all the Inventor rendering and direct manipulation facilities (superior to vtk's rendering classes, which we didn't use) to the vtk pipeline.

Scalar, vector, tensor, and medical visualizations were created with the vtk-Inventor metaclasses (Fig. 7 a,i,g, Fig 7 f, Fig 7 e, respectively Fig 7 j) with practically the same ease as if using AVS or other similar system. The integration required writing around 320 metaclasses, of an average length of 6 text lines, and absolutely no change to the two libraries (of which, Inventor was not even available as source code).

Global Illumination Radiosity simulation software often requires delicate tuning of many input parameters, and thus can not be used as black box pipelines. Testing new algorithms requires also the configurability of the radiosity pipeline. These options are however rarely available to non-programming experts in current radiosity software. We addressed this by including a progressive refinement radiosity system written in C/C++ by us before VISSION was conceived, into VISSION. The 3D world tessellation with vertex intensities output was easily made available for visualization in the Inventor library by the creation of an 'adapter' module. Users can now change all the 'hidden' parameters along the radiosity pipeline, easily insert new algorithms for e.g. sharp shadow detection [12] by subclassing, and visually monitor the process convergence (Fig. 7 d).

Finite Element Simulations Finite element (FE) applications mostly come as packages that limit the user's interaction input given as batch files, respectively an output visualized in a post simulation phase. We addressed these limitations by integrating our FE C++ library [7] in VISSION. Researchers can specify and solve FE problems interactively, experiment with different numerical techniques, and monitor error and convergence rates, without quitting the environment to redefine input files or recompile. Examples include 3D diffusion problems (Fig. 7 a), time-dependent free convection problems (Fig. 7 c), wave simulations (Fig. 7 h), or industrial steering turn-key software [13]. (Fig. 7 b). Visualization is performed again by the Inventor library.

5   Conclusion

We have presented VISSION, a general-purpose visualization and computational steering system built on an object-oriented foundation. VISSION is a generic environment the specification, monitoring, and steering of simulations which removes some limitations of similar systems by combining the powerful, yet so far independently used OO and dataflow modelling concepts.

We have enhanced the traditional dataflow mechanism used by simulation systems to an object-oriented one by introducing the metaclass concept, which extends C++ classes with dataflow semantics in a non intrusive manner. Adding application code to the system is greatly simplified as compared to similar systems. Application library design is clearly separated from the system-specific dataflow information held in the metaclasses. We have provided a mechanism for automatic GUI construction for application modules based on the OO metaclass semantics, and a way to add type-specific, user-defined widgets, based on OO typing.

Several applications illustrate the advantages of a fully object-oriented and single language architecture. Component designers have included libraries for scientific visualization and rendering (420 classes), radiosity (18 classes) and finite element analysis (25 classes) in the system in a short time (approximately 2 months, 5 days, 10 days respectively), while application designers and end users could effectively use the system in a matter of minutes.

Future work is aimed at the extension of VISSION's OO aspects with features such as class hierarchy browsing, automatic documentation, and a generalization of the dataflow model to include also *code flow*, that is to have modules synthesize, exchange, and execute C++ code fragments, creating multiple new possibilities for modelling simulations. Parallel work targets the inclusion of other application domains as numerical iterative solvers or computer vision interfaces and their coupling with the already available libraries.

References

1. B. STROUSTRUP, *The C++ Programming Manual*, Addison-Wesley,1993.
2. J. WERNECKE, *The Inventor Mentor: Programming Object-Oriented 3D Graphics with Open Inventor*, Addison-Wesley, 1993.
3. A. M. BRUASET, H. P. LANGTANGEN, *A Comprehensive Set of Tools for Solving Partial Differential Equations: Diffpack*, Numerical Methods and Software Tools in Industrial Mathematics, (M. DAEHLEN AND A.-TVEITO, eds.), 1996.
4. C. UPSON, T. FAULHABER, D. KAMINS, D. LAIDLAW, D. SCHLEGEL, J. VROOM, R. GURWITZ, AND A. VAN DAM, *The Application Visualization System: A Computational Environment for Scientific Visualization.*, IEEE Computer Graphics and Applications, July 1989, 30–42.
5. W. SCHROEDER, K. MARTIN, B. LORENSEN, *The Visualization Toolkit: An Object-Oriented Approach to 3D Graphics*, Prentice Hall, 1995
6. C. GUNN, A. ORTMANN, U. PINKALL, K. POLTHIER, U. SCHWARZ, *Oorange: A Virtual Laboratory for Experimental Mathematics*, Sonderforschungsbereich 288, Technical University Berlin. URL http://www-sfb288.math.tu-berlin.de/oorange/OorangeDoc.html
7. A.C. TELEA, C.W.A.M. VAN OVERVELD, *An Object-Oriented Interactive System for Scientific Simulations: Design and Applications*, in /textitMathematical Visualization, H.-C. Hege and K. Polthier (eds.), Springer Verlag 1998
8. B. MEYER, *Object-oriented software construction*, Prentice Hall, 1997
9. A. C. TELEA *Design of an Object-Oriented Computational Steering System*, in *Proceedings of the 8th ECOOP Workshop for PhD Students in Object-Oriented Systems*, ECOOP Brussels 1998, to be published
10. J. J. VAN WIJK AND R. VAN LIERE, *An environment for computational steering*, in G. M. Nielson, H. Mueller and H. Hagen, eds, *Scientific Visualization: Overviews, Methodologies and Techniques*, computer Society Press, 1997
11. S. RATHMAYER AND M. LENKE, *A tool for on-line visualization and interactive steering of parallel hpc applications*, in *Proceedings of the 11th International Parallel Processing Symposium*, IPPS 97, 1997
12. A.C. TELEA AND C. W. A. M. VAN OVERVELD, *The Close Objects Buffer: A Sharp Shadow Detection Technique for Radiosity Methods*, the *Journal of Graphics Tools*, Volume 2, No 2, 1997
13. M. J. NOOT, A. C. TELEA, J. K. M. JANSEN, R. M. M. MATTHEIJ, *Real Time Numerical Simulation and Visualization of Electrochemical Drilling*, in *Computing and Visualization in Science*, No 1, 1998
14. D. JABLONOWSKI, J. D. BRUNER, B. BLISS, AND R. B. HABER, *VASE: The visualization and application steering environment*, in *Proceedings of Supercomputing '93*, pages 560-569, 1993
15. W. RIBARSKY, B. BROWN, T. MYERSON, R. FELDMANN, S. SMITH, AND L. TREINISH, *Object-oriented, dataflow visualization systems - a paradigm shift?*, in *Scientific Visualization: Advances and Challenges*, Academic Press (1994), pp. 251-263.
16. S. G. PARKER, D. M. WEINSTEIN, C. R. JOHNSON, *The SCIRun computational steering software system*, in E. Arge, A. M. Bruaset, and H. P. Langtangen, editors, *Modern Software Tools for Scientific Computing*, pages 1-40, Birkhaeuser Verlag AG. Switzerland. 1997

Editors' Note: see Appendix, p. 331 for colored figure of this paper

Information Visualization and Systems

(Case Studies)

Application of Information Visualization to the Analysis of Software Release History

Harald Gall[1], Mehdi Jazayeri[1], Claudio Riva[2]

[1] Distributed Systems Group, Technical University of Vienna,
Argentinierstrasse 8/184-1, A-1040 Wien, Austria
{gall, jazayeri}@infosys.tuwien.ac.at
[2] Nokia Research Center, Software Technology Laboratory,
P.O. Box 45, FIN-00211 Helsinki, Finland
claudio.riva@research.nokia.com

Abstract. We present our experiences in applying information visualization techniques to the study of the evolution of a large telecommunication software system. We used the third dimension to portray the temporal evolution of the system and color to display software attributes. The visualization was surprisingly successful in uncovering interesting and useful patterns in the system's evolution. To do the visualization, we built a tool that combines off-the-shelf components: a database for storing software release data, VRML for displaying and navigating three-dimensional data, and a web browser for the user-interface. The tool is published on the web. The tool is capable of providing effective views of data that are always kept by software development organizations but are often ignored. Information visualization makes it possible to exploit such historical data about past projects to help in the planning stages of future software projects.

Keywords: software release history, third dimension, color, World Wide Web

1. Introduction

Large software systems undergo significant evolution throughout their lifetimes. Often, the structure of a system after a few releases is very different from the structure it had at its initial introduction. Such drastic changes in the structure are due to new functional requirements on the system after it has been released, improper initial design that does not accommodate enhancements easily, or inadequate performance that can only be ameliorated by major restructuring. The history of a software system's evolution contains information that can be used by managers and system developers in both understanding the reasons for the changes in the current system and in guiding the development of future systems. Unfortunately, such information is only passed on through informal anecdotes without any real understanding. Even though most companies collect lots of data about system releases, such information is rarely used in future planning. The main reasons for not using the information are the huge volume of data and the difficulty in drawing any coherent conclusions from the data quickly.

This paper reports our experience in trying to make sense of a database of release histories of a telecommunication system consisting of about 10 million lines of code (10 mloc). The company had developed a custom database to maintain data about system releases. The database provided basic tools that an experienced engineer could use to pose queries about the structure of the software. Our goal in the project was to find general ways to discover patterns in the evolution of the software. We identified some key metrics to measure and initially used simple two-dimensional graphs to plot them [8]. Already, these graphs showed significant trends in the system's evolution that were not known to the developers. We then applied three-dimensional graphics and color to visualize further information about the release histories. The techniques were useful in quickly summarizing the large volume of data in the release database [11]. We believe that the techniques are generally applicable to the analysis of software release histories and can be a valuable tool for software engineers and managers in the planning phases of system development and enhancement. Our general conclusion is that three-dimensional information visualization is readily applicable to the analysis of release histories and should become a standard tool in the software manager's toolbox. To our knowledge, no such techniques are currently used or are even being considered in software development organizations.

This work was supported by the European Commission within the ESPRIT Framework IV project ARES (Architectural Reasoning for Embedded Systems). The technique was developed in conjunction with an industrial case study as part of the project ARES.

2. Related Work

Information visualization is currently being explored to examine and comprehend masses of data regarding software systems. Most software visualization work has concentrated on code-level visualization or on performance visualization. Our focus is on structural and architectural level of large software systems. Several papers have addressed this area.

Eick et al. have developed a set of tools for visualizing several classes of data [5] [2] [7]. In particular, SeeSys [1] is a visualization system for displaying the statistics associated with code. Johnson and Schneiderman concentrate on visualizing hierarchical data using Treemaps [18] [12]. The approach uses the screen-filling method for displaying the attribute values of hierarchy components. The color of the region can provide an additional attribute.

Software visualization using three dimensional display is a current research area. Koike developed a 3-D framework to visualize the execution of two parallel/concurrent computer systems [14]. Koike et al. also proposed a 3-D framework for both version control and module management [13] [4]. Ware et al. investigate how to visualize object oriented software in three dimensions [19].

3-D graphics for displaying hierarchical structures have been development by Card, Mackinlay and Robertson [17]. Their work provides ConeTree which is a technique for displaying hierarchical objects in 3-D space. Improvements of Cone-Tree are described by Carrière and Kazman [3] and by Koike [15].

3. Visualizing the Software Release History

The case study's architecture is organized as a *layered system*. The structure is a tree hierarchy with four levels. The top level is the *system level*. It is based on the *sub-system level* (second level), *module level* (third level) and *program level* (fourth level). Each level consists of one or more elements. The elements in each level are named corresponding to the names of the levels: *subsystems, modules, programs*.

Table 1 shows an example of the software release history. Each row represents the version numbers of a *program* element. The first and second column (labeled Sub and Mod) contain respectively the *subsystem* and *module* to which the *program* belongs. The third column (labeled Prog) contains the name of the *program*. The columns labeled from 1 to 20 represent the twenty system releases. The version number of each element is the RSN of the release where it had the latest change. For example, if a *program* changes its implementation at release 1, 2 and 5 then its version numbers are the sequence < 1 2 2 2 5 5>.

The database is populated with 20 releases of the software product. Each release contains 8 *subsystems*, 47 to 50 *modules* and 1500 to 2300 *programs*.

Table 1. An example of the database in which the software release history is stored.

Sub	Mod	Prog	1	2	3	4	5	6	...	20
Sub1	Mod A	Prog A	1	2	2	2	5	5	...	18
Sub1	Mod A	Prog B	0	2	3	4	5	0	...	20

Following the success of the 2-dimensional graphs in the previous study [9], we decided to explore the application of three-dimensional visual representations for examining the software release history. The purpose of the representation is to present the data contained in the evolution history in a more comprehensible way.

The software release history consists of three entities: time, system structure and version numbers. These entities are visualized in one three-dimensional diagram. The coordinates are called *x, y, z*. The coordinate *z* stores the time information. This coordinate is expressed in release sequence number (RSN). The system structure is displayed by 2-D or 3-D graphs. The graph is spatially positioned along the coordinate *z* at the value of its own RSN. Version numbers are shown using colors. Colors are associated with version numbers through a color scale. The structure elements are painted according to their own version number. Fig. 9 provides an example of color scale. In this way, each color represents an attribute value.

Four 3-D graphical objects are used to visualize the abstract elements of the system structure: cubes, spheres, lines and bars. The composition of these graphical objects represents a graph and serves to visualize the system structure of one release. Cubes represent modules of the system structure (i.e. *subsystems*, *modules* and *programs*). The color of the cube denotes its version number. A sphere represents the topmost module of the system structure (i.e. *system*). The color denotes the release sequence number (RSN) of the system structure. A line represents the "contains" relationship between two modules. A percentage bar is a graphical object that offers a compact representation of a group of elements. It is composed of a set of colored blocks. Each block has two properties: relative size and color. The relative size is proportional to the percentage of modules that have the same version number. The color depends on the version number through the color scale. For comparative analysis, size is the most effective perceptual data-encoding variable [6].

Section 6.1 presents examples of three-dimensional visualization. Section 6.2 shows the use of 3-D diagrams for examining the historical evolution. Section 6.3 presents 2-D representations obtained projecting the three-dimensional graph.

3.1 3-D Visualization of structure

The case study has a tree hierarchical structure. A tree hierarchy can be visualized by 2-D and 3-D tree graphs. 3-D graphs are implemented using the Cone Tree technology [17]. 2-D graphs use the same Cone Tree technology in two dimensions. Fig. 7 shows the whole structure of one system release of the case study. The whole structure of the system is visualized in one view. The 3-D layout allows the viewer to navigate the system structure. Fig. 8 is a zoom on one *subsystem*.

3.2 3-D Visualization of historical evolution

The main purpose of our approach is to use 3-D diagrams to examine the historical evolution of the system structure. Three-dimensional diagrams help to visualize both historical and structural information. The third coordinate is expressed in RSN. For each value of RSN the associated system structure is visualized by one 2-D tree graph. Fig. 1 shows an example. The 3-D diagram represents 10 releases of one of the *subsystems*. For each release, the structure of one *subsystem* is visualized. The structure contains all its *programs* and *modules*. In Fig. 1 the *program* elements are visualized by percentage bars. In Fig. 2 the same *subsystem* is visualized without percentage bars. In this way, *program* elements are distinctly visible.

3.3 2-D Visualization

2-D visualizations can be obtained by projecting the 3-D diagram onto 2-D space. Fig. 2 shows how to make a simple projection by compacting and rotating the diagram. The picture obtained by the projection reveals a concise and informative representation of system evolution. Fig. 3, Fig. 4, Fig. 5 and Fig. 6 contain examples obtained in this way. Each picture visualizes the evolution of one *module* element through the percentage bars (RSN is vertically directed from up to down like in Fig. 6). Each row is associated with a system release. For each system release (from 1 to 20), the percentage bars are displayed. The percentage bars show with the same color the percentage of modules that have the same version number. This representation highlights the critical times—when major changes seem to have happened— in the evolution of the whole *module*.

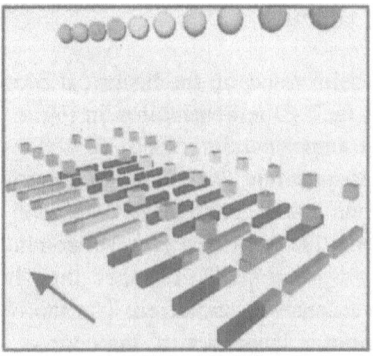

Fig. 1. Visualizing the history

Fig. 2. Rotating the 3-D graph

4. Visualizing to understand

We applied the graphical representation to the examination of the case study. This section presents the results and the advantages we achieved by visualizing the data contained in the database. A more detailed description of the results, of interest to software engineers, can be found in [11][16].

3-D Layout
Three coordinates allow to visualize both system structure and historical evolution of the system. The viewer can perceive both structural and historical information looking at one view. Most important, the viewer can view the changing system structure over time. This important view never comes out in software projects. The viewer can also navigate into the representation to find the best perspective for looking at the data. The advantages are summarized below:
− Visual perception of the structure: in understanding abstract information human minds create visual representations to simplify the process. Visualization can provide these representations for the viewer and relieves his/her mind. In this way imagination and creativity are free to address new ideas. Moreover, three-dimensional layout, rendering and shading is pleasant for humans because they can simulate the reality to which our mind is naturally accustomed.
− Visual retrieval of data: the visualization system we have developed allows the user to click on an element to retrieve its properties. For example, clicking on a *program* element the user can extract its name and version number. This is faster and more attractive that setting up a textual query and submitting it to the database. 3-D layout combined with this functionality allows navigating the system structure and quickly retrieving the information stored in the database.
− 3-D Navigation: the user can virtually navigate the 3-D graphical space for examining the system structure. Users can choose the best point of view to concen-

trate on specific data sets. New representations can be easily obtained by just rotating, zooming, projecting or moving the graph.

Observations on the historical evolution

The 2-D representations make evident some useful trends in the historical evolution of the modules. As and example, we examine the 2-D representation in Fig. 6. This view can be obtained as a projection of the 3-d representation of Fig. 2. It visualizes the evolution of *program* elements belonging to a *module*. At release 1 (first row) all programs have the same version number 1 (red color). At release 2 (second row) 96% of *programs* have version 2 (pink color) and the rest (4%) have version number 1 (red color). This means that a majority of *programs* (96%) changed their implementation and therefore the *module* has been extensively modified. The motivation for such a big modification may be found by direct inspection of the code or *module*'s documentation. Whatever the reason, such a view is a cause of concern and should trigger a closer examination.

Still in Fig. 6, at release 8 the *module* has its last major modification. In fact at release 8 (eighth row) the large dark green zone shows that many *programs* have changed their implementation. Then from release 8 until release 20 many of these *programs* maintain their version number: for each release from release 8 to 20 the green color zones are the biggest ones. Between release 8 and 20 a small fraction of *programs* change their version number: in Fig. 6 this is reported by the regions on the right colored with blue, purple and dark green. The *programs* change at releases 9, 10, 11, 12, 14, 15, 17, 19. It is clear that the *programs* have stabilized.

Changing rates

Changes of version number are visualized as changes in colors. This allows a qualitative and intuitive indication of the changing rates. High changing rates are identified by regions where colors change very quickly. Low changing rates are identified by regions with plain color. In Fig. 3, the first two *modules* have very high changing rates, almost all the *programs* change their version number every release. This anomalous behavior of the module was already detected with the statistical analysis [9]. In Fig. 4, the third *module* has very low changing rates, only at release 19 and 20 some *programs* change their version.

Growing rates

In visualizing the *program* elements, we decided to use a black color to represent those elements that have been removed by the module or that have not been implemented yet. The system has a common structure in all the releases, and the black colored elements are elements not present in a particular release. This visualizes the growing rate of the *module* in terms of amount of *program* elements. In fact, the size of the black region is associated with the amount of not-present *programs*. The modification in size of this region is an indication of the growing rate. If the black region diminishes with time, the *module* is adding *programs*. If the black region is not present, it means that the *module* has reached its maximum size. If the black region increases with time, the *module* is removing *programs*. In Fig. 3, the third *module* has high growing rates. In Fig. 5 the eighth *module* on the second row has been removed from the system.

Patterns

One advantage of visualization is that it enables and exploits the humans' pattern matching skills. This ability relies on the human visual system that is able to detect regions characterized by repetitive use of the same shape, the same color, the same filling or the same texture. Identification of patterns is needed to discover dependencies and relationships between system elements. An example can be identified in Fig. 5 where all the *modules* of one subsystem are visualized using the 2-D.

Considering Fig. 5, we can identify three *modules* that have the same color filling. They are the last *module* on the right of the first row, the fifth and the sixth *modules* on the second row. The common attribute is that their representations have a large region, filled with the same pink color. This region begins at release 2 and extends until release 20. This means that for each *module* many *programs* have the same version number 2 (pink color) and these *modules* don't change in any of the releases. This observation could lead to identify relationships or commonalties among the *programs* or could lead to identify more generally that the *programs* of different *modules* have the same behavior.

Comparisons

The visual representation allows the user to make comparisons between different data sets:

- Comparisons among modules: 2-D graphs and 3-D graphs allow to visualize the structure of the system. With this representation the viewer can navigate through the data to compare the modules of the system. Different views of the same configuration of modules can make such comparisons easier. Comparing colors the viewer automatically makes comparisons of values.
- Comparisons of different measures: a limitation of the case study is that it contains data only about version numbers. Therefore, it is not possible to investigate different measures. The idea is that by comparing the same modules visualized with different attributes, it should be possible to identify commonalties or differences.
- Comparisons of different releases: the historical (temporal) evolution of a *module* can be visualized in a compact form that shows the changes of its *programs*. Such view allows the comparison of the same *programs* in different releases and the historical evolution of different *modules*.

5. Visualization on the web

To support the 3-D representations reported in this work, we have developed a visualization tool called 3DSoftVis. The tool consists of three components: a database, a

3-D visualization engine and the graphical user interface. The database contains the data regarding the evolution of a software system. The 3-D visualization engine transforms the data extracted from the database into 3-D models. The user interface presents the graphs to the user through multiple windows and allows the viewer to customize the views interactively.

The tool has been developed for the World Wide Web platform using web components such as Java, VRML (Virtual Reality Modeling Language), JDBC and HTML. The platform-independence technologies which compose the system allow to publish it on the web [20], making it available to any user with access to the network. The major features of 3DSoftVis are summarized below. A more detailed description can be found in [10].

- World Wide Web Application: the web architecture makes the tool accessible from the network. The application and the database reside on a remote server and they can be easily upgraded by developers. On the client side 3DSoftVis requires a minimal configuration and relieves users from installation problems.
- Graphical engine based on VRML: VRML is an easy solution for developing 3-D graphical tools. We are not 3-D graphical experts and the choice of VRML simplified considerably the implementation of the graphical engine. VRML is also a standard web component that can be easily integrated in a web application.
- Collaborative environment: data and that application for their analysis are bound together on the server machine, so that the most updated database and the most recent release of the tool are immediately available. Groups of researchers can use 3DSoftVis by sharing the same working environment without the constraint of having to be located nearby or working on the same machine-platform.

6. Lessons Learned

Our experience with software visualization has been successful in providing insights into the evolution of the software system in the case study. Since the use of 3-dimensional and color visualization is rather rare in software engineering studies, here we summarize some of the more general lessons we have learned in the hope that they will be useful in future studies.

1. Information visualization is a powerful technique for examining huge data sets and for understanding abstract information. Visual representations aid humans comprehension because they change the process of understanding from being a cognitive task to being a perception task. Visual patterns are easy to detect and support discovery of hidden dependencies in the data. These advantages, which have been exploited in other application areas, are also available in the study of abstract entities such as software structure and qualities.

2. Three-dimensional visualizations allow the viewer to visually perceive the abstract information about the structure of a software system. Navigation makes it possible to change the viewpoint and quickly extract the data from the database. It could serve as a functional interface to a configuration management system.

3. Color and region filling are two effective techniques for presenting data and enable to quickly detect patterns. In our work, 2-D visualizations have been used to visually detect unstable modules and to discover unknown dependencies among system components. The fact that color was helpful in detecting patterns is rather surprising because of the lack of any natural relationship between software and color.

4. The visualization system has been developed with components that are available in the public domain. Without any particular expertise in graphics, we could build a 3-D graphical application in a reasonable time. The web platform allows publishing the tool on the web and simplifying the process of maintaining the database.

5. Evolution history of software systems contains valuable information for both software developers and managers. This information concerns the quality of the system. Visualization is the first technique we know of to make sense of the data that quality assurance groups collect (and usually stay dormant).

7. Summary and conclusions

In this paper, we have presented our experiences in applying information visualization techniques to the study of the evolution of a large telecommunication software system. It rather surprising that the more advanced computer science technologies are not usually applied within computer science itself. Although information visualization has been successfully applied in many application areas, its application to the study of software has been rare. It is certainly not part of any organized software process model. Our experience shows that the planning phases of software development can certainly benefit from the use of information visualization on software release data. Traditionally, data are kept and analyzed by quality assurance groups with hardly any feedback to, or influence on the work of development groups. Information visualization could provide an effective medium of communication and collaboration between the quality and development groups.

Also surprising was the fact that rather sophisticated graphics techniques are readily available and can be used by non-experts to build useful visualization tools. Such tools can then be published on the web for use by large communities of researchers and engineers.

We were surprised by how well color depicted the patterns in the system's evolution. Based on this experience, we believe that color has many potential applications in the display of different software attributes.

References

1. Baker M. J., S. G. Eick, Visualizing Software Systems, AT&T Bell Laboratories, 1994.
2. Ball T. A., Eick S. G., Software visualization in the large, IEEE Computer, April 1996, pp. 33-43.

3. Carrière J. and Kazman R., Interacting with Huge Hierarchies: Beyond Cone Trees, Proceedings of Information Visualization '95, Atlanta, Georgia, Oct, 1995, pp. 74-81.

4. Chu H. and Koike H., How does 3D Visualization Work in Software Engineering ?: Empirical Study of a 3D Version/Module Visualization System, International Conference Software Engineering 98 (ICSE 98), 1998.

5. Eick S. G., Fyock D. E., Visualizing corporate data, AT&T Technical Journal, January 1996, pp. 74-76.

6. Eick S. G., Engineering perceptually effective visualizations for abstract data, in Gregory M. Nielson, H. Mueller, and H. Hagen, editors, Scientific Visualization Overviews: Methodologies and Techniques, IEEE Computer Science Press, February 1997, pp. 191-210.

7. Eick S. G., Steffen J. L., Sumner E. E. Jr, Seesoft - A Tool For Visualizing Line Oriented Software Statistics, IEEE Transactions on Software Engineering, Vol. 18, No. 11, Nov 1992, pp. 957-968

8. Gall H., Hajek K., Jazayeri M., Detection of Logical Coupling Based on Product Release History, International Conference on Software maintenance (ICSM '98), Washington, DC, 1998.

9. Gall H., Jazayeri M., Klösch R., and Trausmuth G., Software evolution observations based on product release history, International Conference on Software maintenance (ICSM '97) (Bari, Italy), pages 160-6, IEEE Computer Society Press, September 1997.

10. Jazayeri M., Riva C., Experiences with developing Native World Wide Web Applications, http://www.infosys.tuwien.ac.at/~riva/Docs/bibl/jaz98a/full/, to submit.

11. Jazayeri M., Riva C., Gall H., Visualizing the Software Release Histories: the use of Color and Third Dimension, ACM Transactions on Software Engineering and Methodology, 1998, submitted, also in Technical University of Vienna, Distributed Systems Group, technical report TUV-1841-98-14.

12. Johnson B., Visualizing Hierarchical and Categorical Data, Ph.D. Thesis, Department of Computer Science, University of Maryland, 1993.

13. Koike H., Chu H.: VRCS: Integrating Version Control and Module Management using Interactive Three-Dimensional Graphics, Proceedings of 1997 IEEE Symposium on Visual Languages (VL'97), 1997, pp.170-175.

14. Koike H., The role of another spatial dimension in software visualization, ACM Transactions on Information Systems, 11(3), July 1993, pp. 266-286.

15. Koike H., Yoshihara H., Fractal Approaches for Visualizing Huge Hierarchies, Proceedings of the 1993 IEEE Symposium on Visual Languages (VL'93), 1993, pp.55-60.

16. Riva C., Visualizing Software Release Histories: The Use of Color and Third Dimension, Master's Thesis, Politecnico di Milano, Milan, Italy, June 1998, also in http://www.infosys.tuwien.ac.at/~riva/Docs/bibl/riva98/full/

17. Robertson G. G., Mackinlay J. M. and Card S. K., Cone Trees: Animated 3D Visualizations of Hierarchical Information, Proceedings of the ACM Conference on Human Factors in Computing Systems (CHI '91), ACM Press, 1991, pp. 189-194.

18. Shneiderman B., Tree Visualization with Treemaps: A 2D Space Filling Approach, ACM Transactions on Graphics, Vol. 11, No. 1, 1992, pp. 92-99.

19. Ware C., Hui D., Franck G., Visualizing Object Oriented Software in Three Dimensions, Conference Proceedings of CASCON' 93, Toronto, Canada, October, 1993, pp. 612-620.

20. 3DSoftVis Demo, Technical University of Vienna, Distributed Systems Group: http://www.infosys.tuwien.ac.at/~riva/vis

Editors' Note: see Appendix, p. 332 for colored figures of this paper

Internet-Based Front-End to Network Simulator

Taosong He

Bell Laboratories
taosong@research.bell-labs.com

Abstract. We present a Java-based interactive visual interface to network simulators. Having successfully incorporated network visualization technologies, our system provides an effective front-end interface supporting real time control and monitoring of the back-end simulator through the Internet or intranets. We extend a traditional spring embedding graph layout model, and propose a new hierarchical node placement algorithm for presenting the structure of a complex network.

1 Introduction

Network simulators are the tools that enable users to simulate the operations and behavior of a network. For network designers and managers, these tools provide a cost-effective way of evaluating a network under different scenarios. This project focuses on a proprietary Lucent Technologies network simulator for switching-based voice networks. By manipulating the network parameters such as routing, topology, and traffic through the simulator, users would be able replicate an existing network, simulate different kinds of nodes/links failures and observe their impact, analyze the effectiveness of certain network management control, and optimize their network designs.

Network simulation is a computational intensive process. For this and other proprietary and security reasons, our simulator is implemented with C and running on a back-end multi-processor server. To support the simulator, it is of crucial importance to develop an interactive visual front-end system. Typically, users of our system range from the network designers working on high-end graphics workstations to the marketing managers working on their laptop PCs. They would have to be able to access and control the remote simulator from their local machines in *real time* through an *intranet* or the *Internet*. To satisfy these requirements, we have developed a Java-based interactive visual interface as the front-end to network simulators. We have chosen Java to implement our system for its free portability, strong support of network programming, and object-oriented environment. The main contributions of our system include a generic Java-based framework for designing network visual interfaces, as presented in Section 2; a new graph layout algorithm for interactively presenting hierarchical structure of complex networks, as introduced in Section 3; and the discussion of some issues and lessons related to the development of such an Internet-based application, as presented in Section 4.

2   Java Framework for Network Interfaces

Developing a network visualization system usually requires programmers to start from scratch and deal with all the issues such as handling of network data, rendering, and interactions. To address this problem, we have developed a generic framework for network interface development[5]. In this paper, we extend the framework from C++ in [5] to Java and enhanced it with networking capability. The new framework provides the tool developers an OO and Internet-based programming environment for the easy implementation of application dependent network visual interfaces. The main idea is to abstract the most fundamental functionalities of all the network views, and encapsulate them into several independent "building blocks". Some of the major building blocks include:

Network Data This building block handles the acquisition, storage, and communication of network data. Traditionally, a network is defined as a set of nodes and links. To handle the more general cases, we adopt the tuple-based network data representations of [5] where a network is defined as a set S of $n - tuples$: $(a_0, a_1, ..., a_n), a_i \in S, n \geq 0$. Each data entry (a tuple) in a network is associated with static dynamic attributes such as trunk capacities or current traffic volume on a switch. This building block supports some basic data operations including entry create/delete, information retrieve/modification, and message receive/send from/to other building blocks. The new extensions to [5] include a JDBC-based class for querying the possibly remote databases storing network configuration information; and a communication component for propagating data update messages between remote clients. Compared to the more commonly used table-based representation of network data, our tuple-based representation supports a unified handling of heterogeneous network data. It is more also flexible for dealing with edit operations such as delete, add, and connect.

Interactive Operations This building block supports three classes of operations: *selections*, *display control*, and *data* operations. The purpose of *selection* is to identify a set of display entities for a future operation, such as copy or delete. A more detail discussion of different *selection* operations can be found in [7]. *display control* operations are used to control the rendering environment, including how to display each individual element on the screen. They can be classified into *visual* control (highlight, visibility, and glyph property change) and *coordinate* control (scale, move). *Data* operations are used to modify the underlined network data, which could in turn change the display of the network. They include interactive editing of network topology such as create, delete, copy, and paste; and modification of data attributes such as trunk capacity. Data operations in this buidling block invoke the correspondent functions in the *network data* class discussed above.

Most of the views in our system for the displaying of networks are based on the traditional node and link diagram [2]. A main problem of such a diagram is that the display can be cluttered when the number of nodes or especially links is large. To address this problem, our display control operations also support some standard clutter reduction methods including transformation, merging, fisheye[6], and interactive filtering (subnet, focus) [1]. We have also invented during this project a new clutter control operation: *aggregate*. The basic idea of "aggregate" is to recursively merge the network elements by their types and create "virtual" elements with user-defined aggregated attributes in a network. Our experiments have shown that as a generic network visualization tool, this operation is very effective for reducing the display complexity, while presenting a *skeleton* structure of a complex network.

View Linking At a top level, a network visualization system consists of a set of views presenting from different perspectives the status of a (dynamic) network. One of the main features of our system is that all the views are appropriately linked together through this building block, based on interactive operations such as selection, focus, and interactive editing. When an operation is performed in one view, it is passed to a view linking engine together with its parameters. Based on the type of the operation, the engine sends out the appropriate view link messages to all the corresponding views. In our framework, we have designed a three level view linking mechanism. At the *local* level, all the operations are restricted to the current display without propagating to other views. Most of the rendering control operations such as zooming abd grouping are by default local. The second view linking level is the *display* level where an operation propagates to all the other views that are displaying the same data. Selection and brushing fall into this level. An interesting feature of our framework is that linking between any two views at this level can be interactively turned off by users. The highest level of view linking is the *data* level. Similar to those at the *display* level, operations at this level affect all the related views. The difference is that a data level view link can not be turned off. Examples include editing operations such as create or delete.

Other important building blocks include *rendering* and *glyph design*. Each building block is implemented as a Java "interface". Since Java does not support multiple inheritance, our framework also includes an implementation of a *BaseView* class for gluing all the building blocks. To develop a new view, programmers could simply inherit from the BaseView, and overload some of its functions. All the views in our front-end system are implemented within this framework. One of the biggest advantages of the framework is that each view developed based on it automatically supports generic network data and a rich set of interactive operations. Our experiences have also demonstrated that by using this framework, tool development time can be dramatically reduced.

3 Network Layout

The back-end network simulator in this project is designed to simulate two-level switching-based voice networks. The higher level, or the backbone, of this kind of networks is usually a fully meshed network deployed with several types of switches such as service nodes and mobile service centers. The lower level consists of switches configured as local exchanges (LE). Each switch is located within a Local Calling Area (LCA), and trunks with different capacities are used to connect the switches. The main view of the front-end is a network editor; cf bottom left of Fig. 1. In this figure different kinds of switches are represented by different icons, and trunks are represented by blue lines. A user can modify a network with a variety of interactive operations as discussed in Section 2 through the element icon platter.

Network layout is essentially the most important task for network visualization. With the help of background maps, geographical-based layout is a straightforward yet powerful visual metaphor. Our system supports both polygon and image based maps. On the other hand, logic layout presenting hierarchical network structures or dynamic simulation patterns are usually more important for users. Particularly, our clients have identified two requirements for a logical view. First, all the switches within the same LCA need to be put into the same area on the screen (cluster problem). Second, within each LCA, switches with stronger task dependent relationships needs to be put closer to each other (standard node placement problem [8]). To satisfy these requirements, we have developed a new algorithm based on the classical spring embedding model (SEM) [4].

In an SEM, nodes are considered as mutually repulsive charges and edges (relationships) as springs attracting connected nodes. Starting from an initial layout, the nodes will move along the force directions until achieving minimal total energy. The force F is defined as the sum of repulsion and attraction forces:

$$F(n,m) = F_{att} + F_{rep} = \lambda_{att}\Delta(n,m)||\Delta(n,m)||^2 - \lambda_{rep}\frac{\Delta(n,m)}{||\Delta(n,m)||^2} \qquad (1)$$

where $\Delta(n,m)$ is the distance vector between two nodes n and m. An edge (n,m) is at equilibrium if $F(n,m) = 0$. Unfortunately, classical SEM approaches can not be directly applied here since the repulsion and attraction coefficients λ_{rep} and λ_{att} are usually assigned based on physical models. They do not generate results satisfying the node placement requirements [8].

Our contribution is to prove that an SEM can be extended to provide an effective node placement solution. Mathematically, assuming that the weight on (n,m) is w, we simply assign $\lambda_{att} = w^4$ and $\lambda_{rep} = 1$. Edge (n,m) will then be at balance when $||\Delta(n,m)|| = 1/w$, exactly as required. Wills[8] has pointed out that for a good node placement solution, higher weight edges have to affect the layout output more than those more irrelevant edges. To prove this, a small perturbation is added to the optimal solution, $||\Delta(n,m)|| = 1/w + \epsilon$, and resulting:

$$F_{att}(n,m) + F_{rep}(n,m) = \Delta(n,m)w^2[(1 + w\epsilon)^2 - \frac{1}{(1 + w\epsilon)^2}] \qquad (2)$$

As desired, Equation 2 illustrates the importance of stronger edges. Compared to NicheWork[8] and simulated annealing [3], our method is simpler, usually faster, and generates consistent results.

To solve the cluster problem, the classical SEM is extended so that a node is represented by a region instead of a point. $\Delta(n, m)$ is consequently defined as the shortest distance between region n and m. Starting from a random initial layout, Equation 1 guarantees that no regions intersect each other in the final layout. Our complete algorithm is therefore a two-step process. First, different LCA regions, represented by circles for simplicity, are placed using the extended SEM. To better use the screen estate, the area of each circle is proportional to the total number of switches in the correspondent LCA. The edge weights between different LCAs are task dependent. Second, switches within each LCA, represented by zero radius circle, are placed within the area.

Fig. 1 presents a logic layout of a network with 324 switches located within 12 LCAs. The bottom left displays the overall structure of the network. Inside each LCA, switches are placed by assigning higher weights to relations between an LE and its *toll home*. That is, the long distance traffic through an LE, represented by a smaller icon, has to be first routed through a "toll home" backbone switch before reaching other LCAs. The tree structured homing information is clearly presented in the overall view. In the bottom right of Fig. 1, we have focused on a specific region, and demonstrate the effectiveness of our algorithm by verifying that the outlier LE does not have toll home information. Our clients have reported that this is due to database incompleteness.

4   Implementation and Discussion

Our system supports real time visualization through a variety of monitoring and analytical tools such as histogram and table view. Fig. 2 presents a snapshot during a simulation run. In the main window, colors of switches and trunks are encoded to represent the current states of equipments, where red indicates emergencies, green for normal, and other colors for in-between situations. The dynamic strip charts present real time statistic information on certain parameters interactively demanded by users through the front-end. An animation control tool with full play-back functionality allows users to interactively examine simulation results at any time from the start of a simulation to the current time.

There are several important issues related to the development of such a real time Internet-based application:

Synchronization Two kinds of information generated by the back-end simulator need to be displayed in the front-end: statistics information interactively required by users such as number of calls on certain switches or trunks; and "alarm" information such as node/link failure. Since alarm messages require immediate attention, an independent thread with infinite loop is applied to continuously update the network views. To synchronize the statistical information, a "simulation clock" message is sent from the simulator every 10 secs to update the display.

Bandwidth To simulate a large network such as that displayed in Fig. 1, the back-end simulator could generate huge amount of log data. Fortunately, not all of them need to be sent to the front-end in real time through the socket connection. For example, statistical information are generated every 10 secs and only the demanded parameters are sent back. In the mean time, limited by the screen size, a user can not simultaneously monitor too many strip charts. There could be a high number of alarm messages, especially when some key nodes and links fail. However, the length of each message is very short. Our experiences have shown that we can run the system through a 28.8Kb modem across the states with US, and through an intranet across the continents. In general, we suggest that current network bandwidth is enough for many non-graphica; real time applications.

Speed Not surprisingly, our system bottle-neck is the interactive rendering of large networks, especially those with large number of trunks. Generally, the performance of Java is not as good as native C++, and sometimes causes delay on low-end laptops. We expect this problem to be greatly alleviated with the advance of graphics card and Java implementations.

Security Our system can be run both as an application and as an applet through Netscape. We have applied the netscape.security package and digital ID to guarantee the security through browser.

5   Conclusions

In this paper we present an Internet-based visual interface to a switching-based voice network simulator. Implemented with a generic Java framework for developing network visualization package, our system allows users to interactively create and modify a network, remotely control the network simulation, and visualize simulation results with a variety of tools in real time. We have also proposed an innovative hierarchical network layout algorithm.

We are currently investigating connecting the front-end system to other kinds of network simulation and management systems. We have also finished a prototype of incorporating 3D views through VRML and Java3D into our system.

References

[1] R. Becker, S. Eick, and A. Wilks. Visulizing network data. *IEEE Transcations on Visualization and Computer Graphics*, 1(1):16–28, March 1995.

[2] J. Bertin. *Graphics and Graphics Information Processing*. Walter de Gruter & Co., Berlin, 1981.

[3] R. Davidson and D. Harel. Drawing graphs nicely using simulated annealing. *ACM Transactions On Graphics*, 15(4):301–331, 1996.

[4] P. Eades. Heuristic for graph drawing. *Congressus Numerantium*, pages 149–160, 1984.

[5] Taosong He and Stephen G. Eick. Constructing interactive network visual interfaces. *Bell Labs Technical Journal*, 3(2):47–57, April 1998.

[6] M. Sarkar and M. Brown. Graphics fisheye view. *CACM*, 37(12):73–84, 1994.

[7] G. Wills. Selections: 524288 ways to say "this is interesting". *Information Visualization'95*, pages 54–60, October 1996.

[8] G. Wills. Nicheworks - interactive visualization of very large graphs. *Graphics Drawing'97*, 1997.

Editors' Note: see Appendix, p. 333 for colored figures of this paper

Visualization of Grinding Processes

Markus Fiege, Gerik Scheuermann, Michael Münchhofen, and Hans Hagen

Institute for Computer Graphics and CAGD
Department of Computer Science
University of Kaiserslautern, Postfach 3049, D-67653 Kaiserslautern, Germany
E-mail: {scheuer,mmuench,hagen}@informatik.uni-kl.de,
markus_fiege@yahoo.de
WWW home page: http://davinci.informatik.uni-kl.de/~m_fiege

Abstract. In grinding technology, the application of superabrasives and increasing demands for higher productivity and higher quality require an appropriate selection of optimum set-up parameters. An efficient way to determine and test these parameters is modeling and simulating the grinding process. A visualization of the results can support the choice of the parameters and increase the knowledge of the complex grinding process.
This paper describes a web-based visualization tool on the basis of a kinematic simulation. The tool allows the visualization of the surface of an already ground workpiece as well as the changing shape of the workpiece during the grinding process. Two methods for the visualization of the grinding-objects are implemented. One method describes the scene with the Virtual Reality Modeling Language, the other one uses a renderer to create the images.

Keywords: Internet, visualization, VRML, grinding

1 Introduction

The application of superabrasives and increasing demands for higher productivity and higher quality requires the optimum selection of set-up parameters for high performance grinding processes, in order to use the complete potential of this manufacturing process. One possible way of reducing the time consuming and costly grinding tests to determine the set-up parameters, is modeling and simulating the grinding process. The Institute of Manufacturing Engineering and Production Management of the University of Kaiserslautern has developed such a kinematic simulation [1],[7],[8].

A visualization of the simulation gives the possibility to view parts of the grinding process that are not or hardly ascertainable in reality. For example, the engagement of grinding wheel and workpiece in the contact area can be viewed. This is covered in the real process and therefore not visible. The knowledge of the grinding process can be increased by such detail-examinations. Thereby the set-up parameters and the choice of tools can be optimized in order to use the full potential of the grinding process.

The presented paper describes a web-based visualization tool [3], based on the data of a kinematic simulation. The visualization includes two different aspects of the grinding process. On one hand, an already ground workpiece can be visualized, called in the following *static grinding-object*. This gives the user the possibility to view the surface of the workpiece in detail. On the other hand, the workpiece can be shown during the grinding process (called *dynamic grinding-object*) so the user is able to view the changing shape of the workpiece.

Two methods for the visualization of the grinding-objects are implemented. The first one describes the scene with the Virtual Reality Modeling Language (VRML). This gives the facility of an interactive exploration of the workpiece in a VRML-viewer. The second method uses a renderer to create images. This has the advantage of a more realistic view of the scene and the possibility to use a higher level of detail.

2 Data formats and format-transformations

From the raw data, that means the output data of the simulation, to the final images, several data transformations have to be done. Figure 1 gives an overview about the structure of these transformations.

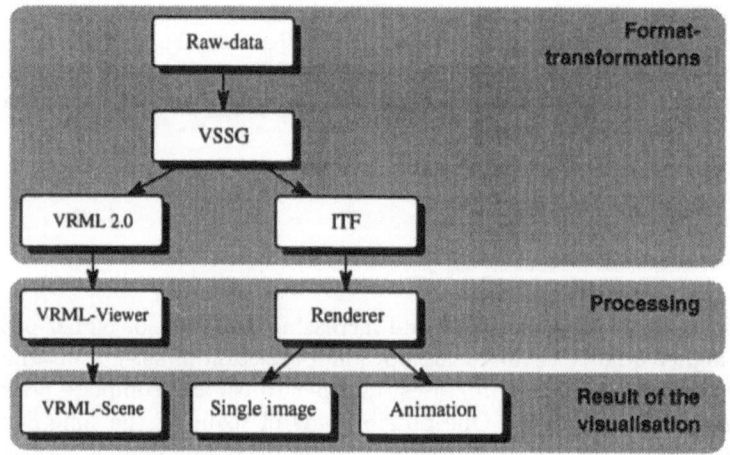

Fig. 1. The format transformations for the visualization

2.1 Raw data

The raw data is given from the simulation in an ASCII-file that describes the ground surface of the workpiece. The file consists of a header, describing the extensions of the workpiece, and a body, that defines the surface. It is defined by a height field over a rectangular grid (see Fig. 2). Therefore just one coordinate of each data point must be stored, the other ones could be calculated by the index and the distance of the grid.

For the grinding-objects the following coordinate-system is used. The x-axis is pointing in grinding direction. The y-axis defines the height of the workpiece, the depth is described by the z-axis. The data points of one line in grinding direction, that means with the same z-value, are defining a *grinding-track*.

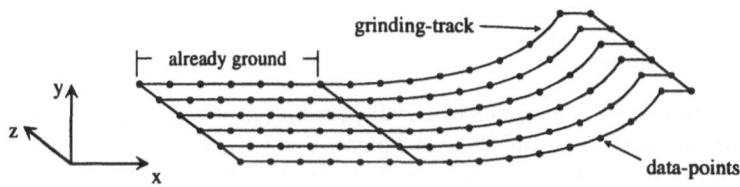

Fig. 2. Height field over a rectangular grid with coordinate system

To describe a static grinding-object all information is given by such a raw data file. The surface of a dynamic grinding-object changes with the time. For such objects, the shape of the workpiece is described at different times. Here the surface at any time is determined by a linear interpolation.

2.2 The Viper Symbolic Scene Graph (VSSG) / VIPER-System

The VSSG is a dynamic application independent data structure, developed in the VIPER-Project [2],[5]. All data is treated as *data objects* (so-called VSSG-Nodes), i.e. instances of generalized data types – *classes* – that may have attributes, which are either elementary or (reference to) other data objects. Classes are organized in an inheritance hierarchy. Type-specific behavior is added through modular actions (methods) which may be attached to data types individually for each application. The VSSG is chosen as the basic data structure for the visualization.

2.3 VRML and ITF

VRML [4] is a language to describe virtual environments, which recently has become the standard for interactive 3D-graphic in the Internet. One reason is the independence of hardware. With a Plug-In, for example the CosmoPlayer from Silicon Graphics, a common web-browser can handle VRML-scenes.

The Indexed Triangle Format (ITF) is a data structure to describe surfaces with triangles. The ITF is the input format for a renderer, developed at our institute.

2.4 The transformation out of the VSSG

The process of the transformation from VSSG into VRML or ITF is the same. The data processing is done by new actions, built in the VIPER-system. For every grinding-object the following steps are done:

1. Define the materials
2. Generate the geometry:
   - Reduce the data points
   - Define the data points
   - Create a triangulation of the data points
   - Define the faces by using the triangulation
3. Define the illumination
4. Define the viewpoints

A default material is assigned to all grinding-objects, for that a metal according to the usual material of a workpiece is chosen. To create an illumination corresponding to daylight, a distance light with a high part of ambient light is used to illuminate the scene. This should give the user an impression of the ground workpiece as realistic as possible.

For examining the workpiece, different viewpoints are defined. Some viewpoints are giving an overview of the grinding-object, others are focusing special details like the grinding edge. As special feature for the static grinding-objects an animated flight of the camera is implemented.

3 Data processing

3.1 Data reduction

A raw data set of a grinding-object is huge, it contains millions of data points. Especially the data files describing dynamic grinding-objects can reach a size of several hundred Mbytes. To reduce the amount of data points to a sufficient size, a data reduction is necessary. Two requirements are made:

- The data must only be reduced in grinding direction.
- The conversion into the VSSG should be made without any reduction.

The first consideration is made because the topological characteristic of the workpiece should be kept for every grit of the grinding-wheel. Each grinding-track represents an engagement of one or more grits. A reduction in direction of the z-axis would cause a reduction over the grits. In addition the height differences in grinding direction are very small in comparison with the changes across this direction – a reduction in both directions would not gain much.

The second requirement follows the idea, to create the VSSG as a 1:1 copy of the raw data. Only the data processing after the VSSG should contain a data reduction, which can be individually optimized according to the further application. In practice this proceeding isn't always optimal. Especially with the data sets of a dynamic grinding-object, a small reduction is sometimes useful. The following two algorithms for data reduction are implemented for the different steps in the data processing.

Data independent reduction. The data independent reduction is very easy and fast. Simply whole "rows" are removed out of the height grid, i.e. only every second (or third, fourth, ...) data point of a grinding-track is used. Despite its simplicity and missing consideration of the data, this proceeding turns out to be useful. The changes in height of the data points in grinding direction are very small and regular, so the result does not differ much from the result of a data dependent reduction. A reduction up to $\frac{1}{10}$ or $\frac{1}{20}$ of the original data shows no considerable difference between the data independent and data dependent algorithms. It is a further advantage of this method, that after the reduction all data points are still ordered above a regular rectangular grid.

For these reasons, the data independent reduction is used for the (possible necessary) reduction from the raw data into the VSSG. In comparison with the whole data reduction just a small reduction is required, a reduction up to the factor ten should be enough.

Data dependent reduction. The data dependent reduction is more costly than the method described above, because the values of the data points are taken in consideration. Instead it allows very strong reductions, because only the points with the least influence on the shape are removed. This algorithm for data reduction is used for the conversion from the VSSG into the ITF or VRML.

The data points of a grinding-track can be seen as an open polygon, whose vertices are represented by the data points. The idea of this method consists of creating a line from the beginning of the grinding-track as long as possible, without exceeding a maximum predefined error between the points and the line (see Fig. 3). From the final data point, a new line is restarted. Because in grinding direction the difference in height of the data points is very small and often some points are located on a line, this procedure removes many data points.

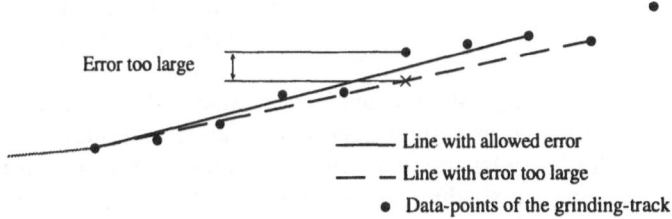

Error too large

——— Line with allowed error
— — Line with error too large
• Data-points of the grinding-track

Fig. 3. The data dependent reduction with approximated calculation of the distance

The calculation of the distance between the data points and the line is the most costly part, because every point between beginning and end of the line must be checked. Therefore, two different algorithms are implemented where the user can choose the one he prefers according to his purpose.

Calculation of distances. The first algorithm approximates the exact distance by calculating the distance between the data point and the point on the line, that has the same x-coordinate. In addition, data point and line are on the same grinding-track, so they have the same z-coordinate. Therefore the distance is the difference between the y-coordinate of both points (see Fig. 3). So the algorithm just determines both y-coordinates, followed by calculating the difference – simple and fast. It overestimates the error, so no necessary points are removed.

The exact algorithm calculates the distance between data point and line with Hesse's Normal Form. Because of the exact distances the optimum reduction result is reached for a given error with this reduction algorithm. Despite the bigger costs for the calculation, for very strong reductions the exact calculation of the distance is often useful.

Reduction for dynamic grinding-objects. In the dynamic case each time-slice has a separate height field with the same grid structure – thus any data point has different heights for the different times. The reduction should keep this property. This means that all data points should exist in all time-slices, so only the height is allowed to change. A time-dependent grid would increase the computing time without improving the visual appearance.

The data independent reduction can be used for dynamic grinding-objects without any changes. However, the data dependent reduction must be used in a special way to keep the requirement. Each time-slice is checked parallel. For that, a line is built with the "common" procedure from the "same" data point of each time-slice. As soon as the first line reaches its full length, its final data point is kept in all time-slices as necessary. From this data point the algorithm is restarted.

3.2 Creating a triangulation

The surface of a workpiece in VRML as well as in the ITF format is composed of triangles. To extract the triangles from the data points, a triangulation is used. The triangulation is not created over the whole grinding-object. A VRML-viewer and the renderer are doing a shading over a face. Such a shading across the whole surface is not desired. In grinding direction a shading is desirable, because the edges between the triangles are smoothed according to the real workpiece. In direction of the z-axis a shading would smooth the engagements of the single grits. Certainly it is decisive to view the structure of the single grinding-tracks. Therefore the surface is divided into several stripes, with the area between two neighboring grinding-tracks building a stripe. In this way, the shading is done only in grinding direction. Now for each stripe a triangulation is created.

4 Conclusion and Future Work

Our tool allows the visualization of an already ground workpiece as well as the examination of a workpiece during the grinding process. The two implemented ways of visualization are by VRML (see Fig. 4) or by images, created by a renderer (see Fig. 5). The used data formats and the data transformations are explained. To carry out the transformations, two algorithms for data reduction are presented. For the data dependent reduction, two implemented strategies for calculating the distance are described.

One target during the development of the visualization tool was to use the World-Wide-Web for the visualization. So the programs are created with the ability to create a web-based client/server architecture.

In the future, some additional features can be integrated in the visualization system. Besides the geometry, a number of cutting parameters can be determined as the grinding forces, the grinding power or the heat flux in the contact area. To increase the knowledge of the grinding process, these quantities can be shown by special visualization techniques. A further improvement could be the creation of a material database.

Acknowledgments

The simulation data is courtesy of the Institute of Manufacturing Engineering and Production Management. We like to thank for providing the data. Especially, we want to thank Udo Zitt, Oliver Braun and Günther Warnecke for helpful discussions, their time and inspiration.

References

1. Braun, O.: Analyse und Modellierung der Abtrags- und Verschleißmechanismen beim Hochleistungsschleifen. Master-Thesis, Institute of Manufacturing Engineering and Production Management, University of Kaiserslautern (1998)
2. Disch, A.; Jacob, Ph.; Münchhofen, M.: VIPER – A 3D-Modeling- & Visualization-Toolkit for web-based Applications. WWW7 Conference, Brisbane, Australia (1998)
3. Fiege, M.: Visualisierung von Schleifprozessen. Master-Thesis, Institute for Computer Graphics and CAGD, University of Kaiserslautern (1998)
4. Kloss, J. H.; Rockwell, R.; Szabó, K.; Duchrow, M.: VRML 97: Der internationale Standard für interaktive 3-D Welten im World-Wide-Web. Screen-Multimedia Edition, Addison-Wesley (1998)
5. Seck, A.: VSSG nach VRML 2.0 Konvertierung. Projektarbeit, Institute for Computer Graphics and CAGD, University of Kaiserslautern (1997)
6. The Virtual Reality Modeling Language
   http://vrml.sgi.com/basics/index.html
7. Warnecke, G.; Zitt, U.: Kinematic Simulation for Analyzing and Predicting High-Performance Grinding Processes. Annals of the CIRP, volume 47/1/1998 (1998) 265–270
8. Zitt, U.: Modellierung der Schleifscheibentopographie von CBN-Schleifscheiben zur Simulation und Analyse des Flachschleifprozesses im Hinblick auf prozeßstabilisierende Korrektiven. Report to the research project Wa 501/19-1, Institute of Manufacturing Engineering and Production Management, University of Kaiserslautern (1996)

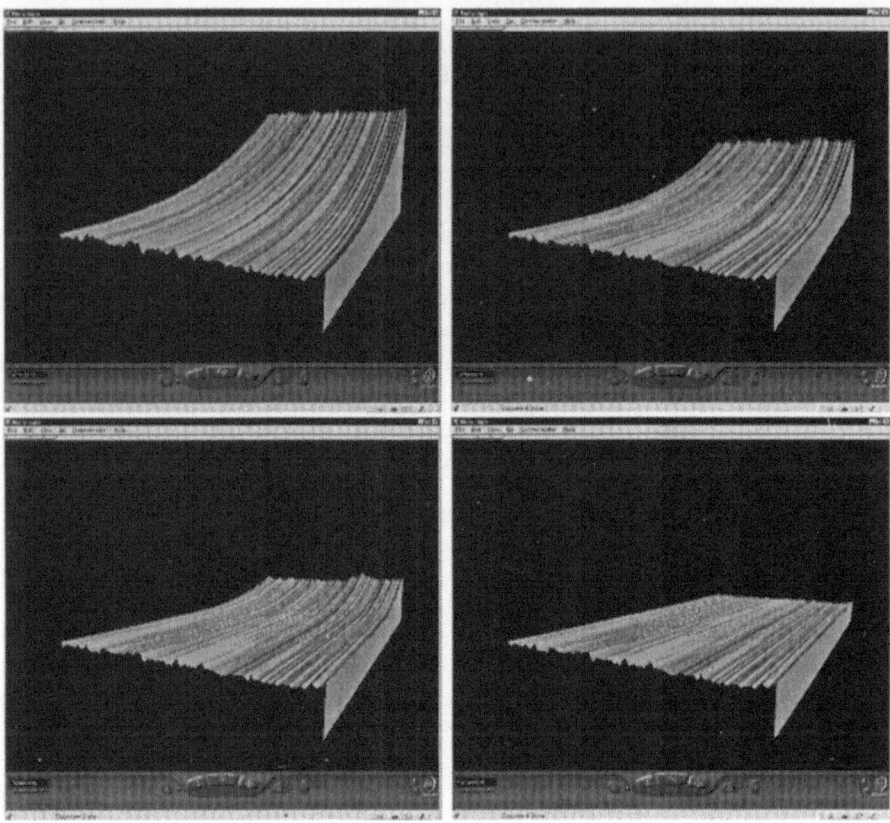

Fig. 4. A VRML-scene with a dynamic grinding-object

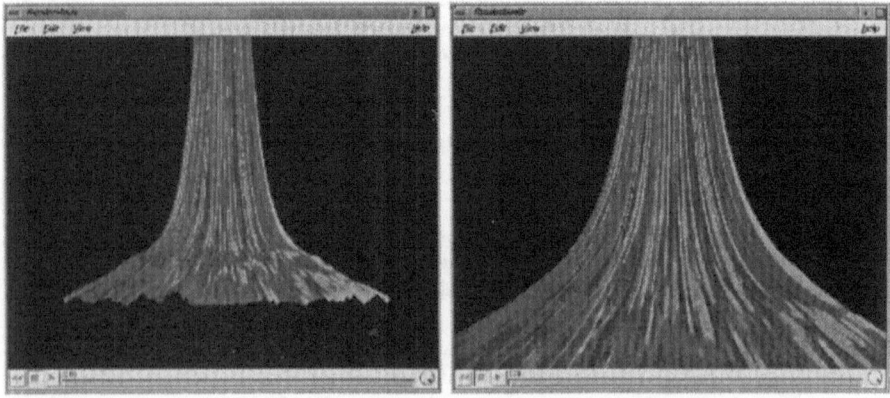

Fig. 5. A scene with a static grinding-object calculated by the renderer

Where Weather Meets the Eye –
A Case Study on a Wide Range of
Meteorological Visualisations for Diverse
Audiences

H. Haase[1], M. Bock[1], E. Hergenröther[1], C. Knöpfle[1], H.-J. Koppert[2],
F. Schröder[1], A. Trembilski[1], and J. Weidenhausen[1]

[1] Fraunhofer Institute for Computer Graphics (IGD),
Rundeturmstraße 6, 64283 Darmstadt, Germany
http://www.igd.fhg.de/www/igd-a4
[2] German Meteorological Office (DWD),
Kaiserleistraße 42, 63067 Offenbach, Germany
http://www.dwd.de

Abstract. Sophisticated visualisation enables experts as well as lay persons to extract knowledge from complex data. This is particularly true for visualising the massive amounts of data involved in meteorological observations and simulations. These are of interest to scientists, to forecasters, and to the general public. The paper presents and discusses a range of solutions for meteorlogical visualisation. Topics covered include systems for the production of TV weather forecasts, for the analysis of simulation output by experts, for personalised weather information in the Web, and for meteorological visualisation using Virtual Studio and Augmented Reality technology.

1 Introduction

The visualisation of meteorological data [3] is the last element in the pipeline of meteorological data processing. Only sophisticated visualisation enables the forecaster, the researcher, or the general public to understand complex circumstances [6]. However, users are requesting tailored visualisation tools [2]. While researchers need interactive scientific visualisation, forecasters at the media department are requesting sophisticated TV presentation systems. Private and business customers must be supplied with easy to understand, sometimes even entertaining, and probably personalised presentations of the weather data.

Fraunhofer IGD and the German Meteorological Office (DWD) jointly collaborate in weather visualisation since early 1992 [4]. The TriVis system for professional TV production is in daily operation at several TV stations since 1993. It offers animated 2D weather maps or 3D scenes. In order to fully exploit the results from the numerical weather prediction (NWP) forecasts and observations, a 3D system called VISUAL has been developed. It is installed in the operational environment of DWD's central forecasting office. VISUAL is a scientific visualisation system which is used by both forecasters and researches to "understand data". In order to supply the general public with tailored information, IGD has developed a system called Weather on Demand (WxoD) which is on-line in the WWW since 1998. The WxoD software is based on Java, JavaScript, VRML, and CGI-Scripts. Upon request it creates images and animations for customer definable locations, areas and time intervals. A user-database can be used to personalise the system. Virtual studio technology helps to achieve appealing, cost effective, improved TV weather products. Augmented Reality can also lead to additional insight and ease of use of weather visualisation.

2   Professional TV production

Modern numerical weather prediction models offer rich information full of dynamics that is of high interest to the television spectators. In the early 90s when the traditional weather production was mostly done by hand at the TV station with the help of weather faxes, it could not convey this growing information content anymore. Direct and highly automated access to the full model output was necessary. This directly led to the development of the TriVis system which can produce animated 2D weather maps (fig. 1) or 3D scenes (fig. 2, 3). TriVis can show clouds with snow precipitation animated over time with the snow accumulating on the ground and melting away during the following hours in a single smooth forecast video (fig. 2).

TriVis stands today as a fully or semi-automatic production environment for broadcast-ready weather forecast videos and is well suited for the generation of on-line weather products or cost-effective weather show production In addition, it offers the possibility for the interactive manipulation of all visualisation parameters and attributes at all times during the production process.

The very flexible import functionality of all design relevant objects like fonts, maps, or weather symbols allows an individual design for every television station. When desired, complex visualisation functions and effects algorithms generate attractive and sometimes even spectacular computer graphics.

The two-dimensional presentation offers mainly clear and easy-to-perceive weather maps, while the three-dimensional modules bring a high realism together with necessary abstraction. TriVis smoothly integrates into existing production environments at television stations or weather services.

TriVis can process many meteorological data types. Satellite images, radar data, any scalar data (e. g., temperatures, precipitation amounts, snow heights, UVB, or wind speeds), and multivariate cloud information can be imported from observations or numerical weather prediction models. It also offers a smooth integration of micro animations (e. g., falling raindrops) according to simulated data (e. g., precipitation type and strength) plus the possibility to display animated weather symbols, overlay texts, contour lines, and fronts.

During the complete production process, TriVis offers all standard visualisation techniques plus spectacular 3D effects. The lay audience is considered in each visualisation step. One special feature is certainly that clouds are visualised with advanced fractal functions to use familiar naturalistic cloud objects for easier perception by the visual simulation of clouds [5].

Currently, TriVis itself is installed and running daily at meteorological offices in two countries and at about a dozen TV stations. In addition, America Online gets its daily weather from the TriVis system and many sites on the Internet feature TriVis images or movies with current weather information.

3   Interactive visualisation for experts

Since the mid-90s, new generations of NWP models have made effective interactive 3D visualisation eminently important. Local effects and also larger simulated complex weather structures can be understood much faster and better if all spatial dimensions can be directly controlled during the visualisation process.

VISUAL allows the highly interactive and combined visualisation of direct model output, historic or experimental model run data, observation values, and arbitrary curvilinear volume data. In addition to this support of data analysis, the observation of a numerical model's behaviour or the comparison of predicted and measured values are possible to gain additional insight. All these methods can be applied to different simulation models and arbitrary data sets at the same time (see fig. 4 and 6).

One main feature are the accurate computations on the original model's data grid (fig. 5) for the generation and presentation of all visualisation objects. They are generated in the precise coordinate system of the numerical weather prediction model (computational space) before being displayed. Within VISUAL, time is considered as an equally important fourth dimension of the data and treated by analogy with the transformation of all spatial information.

In order to visualise new parameters, a user with programming skills can write a plug-in to create, e.g., the Richardson number, the relative humidity, the cloud cover, the pseudo-potential temperature, or the potential vorticity on the fly.

VISUAL is installed at the central forecasting office of the DWD in Offenbach. VISUAL is considered as a powerful 3D visualisation tool to investigate NWP data. In research it is already accepted as an effective tool to visualise the results of model development, especially when doing non-hydrostatic modelling. Interpolation is avoided as much as possible, and so the visualisation is done on the model's grid, and the result is transformed to the current map projection for display. High-quality rendering is one of VISUAL's strong features. Therefore, a sophisticated contouring package has been implemented (fig. 7) that offers both hardware colour shading and the possibility to do solid contours with splines. Cutting planes can be dumped in various pixel and vector formats.

In contrast to the mentioned NWP models, the new Global Model of DWD is best visualised on a sphere. Flexible, interactive visualisation of this data can be done with the ISVAS visualisation system (see fig. 10).

264

4   Personalised Web access to weather data

While TriVis fulfills the high demands of daily TV broadcast services and VI-SUAL offers first-class interactive visualisation for researchers, the special needs of individual on-demand weather visualisation for lay persons are met by the WxoD system. Due to the broad availability of powerful Internet technology, new ways of interactive personalised weather information can be realized to the benefit of both private and business customers. For a general discussion of possible distribution schemes of Web-based visualisation see [11], one example of such a service is described in [9].

In a joint project with the German and Swedish meteorological offices (DWD and SMHI), IGD has realized Web based meteorological products. Here, distributed visualisation is used to deliver a variety of meteorological information to many recipients with various client platforms across the Internet. This requires a range of different solutions utilizing HTML, VRML, CGI, JavaScript and Java technology.

A number of weather products have been realized, including CityWeather and Meteogram. In CityWeather (fig. 8), the user can select visualisation parameters to see the predicted weather situation at a particular city in 3D. The request is sent to a graphics server executing the TriVis software which produces the appropriate, individual image according to this request, which then is sent back to the customer. Thus, high quality images can be tailor-made disregarding the client hardware. For the Meteogram (fig. 9), on the other hand, raw data is sent to the client and visualised locally using Java. This gives the user full control over the visualisation process, resulting in high quality 2D-graphs. The different Web-based visualisation solutions which were realized for Weather on Demand implement a range of distribution schemes for different tasks: low or medium bandwidth connection, computation on server or on client.

Further discussions, including personalisation by means of user databases for WxoD and push technology, can be found in [1].

5   3D Weather in the Virtual Studio and in outdoor scenes with Augmented Reality

For the IFA'97 exhibition in Berlin, a first broadcast of three-dimensional weather forecast scenes from a virtual studio was prepared and implemented. DWD's data, the design and broadcast experience of the Hessischer Rundfunk TV station, and the weather visualisation expertise of Fraunhofer IGD made a perfect match for this pioneer project. The TV presentation of weather forecasts from within a virtual studio offers many important advantages. Apart from the significantly lower production costs, there are now many new possible ways and forms of presenting the weather. The weather of tomorrow can be happening today inside or around the studio. At the exhibition, we were able to fully integrate weather in a virtual studio in a world premiere. During the exhibition, weather forecasts produced at the IFA with sophisticated technology were broadcast all over Germany in the ARD TV channel daily (fig. 11). The basis of this successful work was the TriVis system with its extensive 3D weather capabilities.

Augmented Reality (AR) will become an alternative to the conventional techniques of desktop and VR environments for applications where scientific visualisation is needed in a real environment, of course the reuse of conventional, approved scientific visualisation methods in Augmented Reality systems is a must, but first they have to be evaluated and possibly adapted for the new, special demands of AR. Still, only a few AR-based scientific visualisation systems have been realized so far [7][8].

In the field of AR for visualisation, we currently concentrate on two main aspects [10]. Firstly, since AR requires the visualization software to run on *portable* computers, it is necessary to minimize the system's resource consumption. Secondly, we believe that the visualisation techniques will have to be adapted for the usage in AR, e.g., in order not to cover all of the real environment with augmented visualization objects.

But how can meteorological visualisation benefit from AR? The latest numerical weather simulation model developed by DWD is called "Lokalmodell". It scales the horizontal simulation mesh size down to 2.5 km. This makes it possible to visualise the forecast data directly corresponding to the real terrain where the specific weather situation is going to happen. We chose the Feldberg mountain close to Frankfurt to show the next day's horizontal wind direction. We used local terrain data for computing the occlusion of the environment with the augmented objects. The example in fig. 12 suggests that in an outdoor Augmented Reality application the user should consider to use a constant icon size, since a constant size can be one of few remaining depth cues to perceive his real distance to the icon position. Still it can be an advantage to additionally use an artificial depth cue, as demonstrated in the same figure, where a vertical red line cutting the terrain shows the true spatial position of two arrows. If we compare this image to one without vertical red line, we realize the need of "second-order visualisation objects" in Augmented Reality, which do not visualise any value themselves but only help the observer understand the meaning of the actual visualisation. This is similar to auxiliary lines etc. in classical desktop visualisation.

AR is a new, very promising technology for a new class of applications of scientific visualisation. In some cases the known Scientific Visualisation techniques have to be modified and adapted to the special demands of AR.

6 Conclusion

With appropriate tools to process and visualise large volume multiple data sets, it is possible to exploit the inherent information. There are many areas, like aviation meteorology and local scale problems, where animated 3D visualisation is indispensable to understand the nature of the underlying processes. VISUAL will be further developed into an application that allows to effectively support the forecaster during his daily work. Concurrently, consumers of weather information demand better products with appealing graphics that convey more meteorological information in an intuitively understandable way. This must be possible via broadcast channels and increasing use of virtual studio technology or through interactive on-demand applications with individual and tailored forecasts based on WxoD technology. Another emerging technology, namely Augmented Reality, will also be of great benefit in a number of applications.

Next generation numerical weather prediction models simulate very small-scale physics on fine grid meshes that result in extremely large data sets. As the TriVis system demonstrates, simple cloud animations that contain mainly the knowledge about 3D humidity, temperature, and wind fields can compress quite diverse data sets in an intuitively understandable way. Therefore, the future of visualisation lies not only in the depiction of large amounts of 3D, 4D, or 5D data. It rather requires also an intelligent and condensed postprocessing for exploration and presentation. We are working on doing just that for a continuing development of visualisation software where weather meets the eye.

References

1. Bock, M., Haase, H.: Easi2Vis: User-Centered Visualization Services for the World Wide Web. *to be published* (1999)
2. Haase, H.: Mirror, Mirror on the wall, who has the best visualization of all? – A reference model for visualization quality. In: Bartz, D. (ed): Visualization in Scientific Computing '98, Springer-Verlag (1998) 1–13
3. Hibbard, W., Santek, D.: Visualizing Large Data Sets in the Earth Sciences. IEEE Computer (1989)
4. Koppert, H.-J., Schröder, F., Hergenröther, E., Lux, M., Trembilski, A.: 3D Visualization in daily operation at the DWD. Proc. Sixth ECMWF Workshop on Meteorological Operational Systems, Redding, U.K. (1998)
5. Sakas, G., Schröder. F., Koppert, H.J.: Pseudo-Satellitefilm – Using Fractal Clouds to Enhance Animated Weather Forecasting. In: Hubbold, R.J., Juan, R. (eds): Proc. Eurographics '93, Computer Graphics Forum, NCC Blackwell Publishers, Vol. 12, 3 (1993) C329–C338
6. Schröder, F.: Audience Dependence of Meteorological Data Visualization. In: Grinstein, G., Levkowitz, H. (eds): Perceptual Issues in Visualization, Springer-Verlag (1995) 157–165
7. State, A., Livingston, M., Garett W., Hirota, G., Whitton, M., Pisano, E., Fuchs, H.: Technologies for Augmented Reality Systems: Realizing Ultrasound-Guided Needle-Biopsies. Proceedings of the ACM SIGGRAPH 1996, (1996) 439–446
8. Szalavári, Z., Schmalstieg, D., Fuhrmann, A., Gervautz, M.: 'Studierstube': An Environment for Collaboration in Augmented Reality. Virtual Reality, Springer Verlag, 3 (1998) 37–48
9. Trapp, J.C., Pagendarm, H.-G.: A Prototype for a WWW-based Visualization Service. In: Lefer, D., Grave, M. (eds): Visualization in Scientific Computing, Springer Verlag (1997) 21–30
10. Trembilski, A., Weidenhausen, J.: A Room with a View – Scientific Visualization of Spatial Data in Augmented Reality. *to be published* (1999)
11. Wood, J., Brodlie, K., Wright, H.: Visualization Over The World Wide Web And Its Application To Environmental Data. Proc. IEEE Visualization '96 81–86

Editors' Note: see Appendix, p. 334 for colored figures of this paper

Volume, Medical, and Molecular Visualization

(Case Studies)

Parallel Ray Casting of Visible Human on Distributed Memory Architectures

Chandrajit Bajaj[1], Insung Ihm[2], Gee-bum Koo[2], and Sanghun Park[2]

[1] Dept. of Computer Sci., Univ. of Texas at Austin, U.S.A.
[2] Dept. of Computer Sci., Sogang Univ., Seoul, Korea

Abstract. This paper proposes a new parallel ray-casting scheme for very large volume data on distributed-memory architectures. Our method, based on data compression, attempts to enhance the speedup of parallel rendering by quickly reconstructing data from local memory rather than expensively fetching them from remote memory spaces. Furthermore, it takes the advantages of both object-order and image-order traversal algorithms: It exploits object-space and image-space coherence, respectively, by traversing a min-max octree block-wise and using a run-time quadtree which is maintained dynamically against pixels' opacity values. Our compression-based parallel volume rendering scheme minimizes communications between processing elements during rendering, hence is also very appropriate for more practical distributed systems, such as clusters of PCs and/or workstations, in which data communications between processors are regarded as quite costly. We report experimental results on a Cray T3E for the Visible Man dataset.

1 Introduction

A few years ago, the National Library of Medicine (NLM) of the U.S.A. created huge volume datasets made of computer tomography (CT), magnetic resonance imaging (MRI), and color cryosection images of male and female human cadavers in an effort to offer a complete digital atlas of the human body [12]. The "Visible Man" data consists of axial scans of the entire body taken 1 mm intervals at a resolution of 512×512, in which the whole data set has over 1870 cross-sections. The "Visible Woman" data is made of cross-sectional images taken at one-third the interval of the male. The data sets amount to 15 Gbytes and 40 Gbytes, respectively.

Visualizing such very large volume data requires a great deal of computing time and memory space. In particular, ray-casting such volume data is one of the most compute- and memory-intensive tasks for volume rendering, while the ray-casting algorithm produces the highest quality of rendered images. The motivation for this work is to develop an effective parallel ray-casting scheme for visualization of very large volume data on distributed systems. In this article, we are particularly concerned with parallel ray-casting of the Visible Human datasets, on a Cray T3E, a distributed-memory parallel computer (For the previous works on parallel rendering of the Visible Human, such as MPIRE, refer to

[12].). Our new method tries to achieve high performance by minimizing communications between processing elements during rendering through compression, hence is also very appropriate for more practical distributed systems, such as clusters of PCs and/or workstations, in which data communications between processors are regarded as quite costly.

Our parallel ray-casting scheme is different from the previous approaches in that it is based on a compression method that is well-suited for developing interactive applications. In [3, 4, 13], Ihm et al. developed a new compression method, based on 3D wavelets, that provides very fast random access ability to compressed volume data. Most parallel rendering algorithms for very large volumes partition the data into subblocks that can fit into local memory of processing elements, and distribute them over the local memory spaces in the system. During rendering, load balancing is usually done dynamically for efficiency, and this often causes data redistribution between processing elements. The data redistribution, or remote memory fetch, when implemented carelessly, is one of the most serious factors that deteriorate the speedup of parallel volume rendering, especially when the data is very large [11].

In our implementation, the whole CT dataset of the Visible Man is compressed, and is replicated at each processing element. Since the entire dataset that is necessary for generating image segments, is available at the local memory, no data communication is needed between processors for data redistribution. As briefly explained in Section 2, the compression method we use, guarantees very quick random access which is faster than remote data fetch, hence produces a better speedup than the previous methods based on data redistribution.

2 Wavelet-Based 3D Compression of Volume Data

Our parallel ray-casting algorithm requires a volume data compression scheme which has the following properties: High compression ratio, minimal distortion in the reconstructed images, and fast random access ability. First, very large volume dataset (say, several hundred mega bytes to a few giga bytes) should be compressed into smaller sizes (say, 64 to 128 mega bytes) that can fit into local memory spaces. Second, the contents of the images should be retained as best as possible after reconstruction. Lastly, when an individual voxel in the compressed data is accessed in a random fashion, the data item should be reconstructed quickly during run-time.

Ihm et al. [3, 4] have compromised between these factors, and developed a wavelet-based 3D compression scheme for interactive visualization of very large volume data, and the timing performance has been enhanced further in [13]. Table 1 and 2 brief the performances of the compression method described in [13]. The experiments are performed on an SGI machine with 195MHz R10000 CPU using the fresh CT dataset of the Visible Man that amounts 720Mbytes ($512 \times 512 \times 1440 \times 2$bytes). We tested with the four ratios of wavelet coefficients that are used after wavelet transforms. Table 1 shows the compression ratios and quality of the reconstructed images. When more than 7% of wavelet coefficients are used, the ray-cast images are virtually identical with those generated from the uncompressed dataset.

Table 2 indicates how fast the compression scheme reconstructs an individual voxel from compressed data. Two situations were considered to evaluate reconstruction overheads: First, the timings (in seconds) for Pure Random access were taken by repeatedly fetching voxel values one million times with randomly generated indices (i, j, k) from uncompressed and compressed data, respectively. The test results indicate that fetching voxel values from compressed data is about 1.20 to 1.38 times slower. The timing differences can be ignored in many compute-intensive applications such as volume rendering, which usually take more than a hundred seconds. Secondly, the timings for Cell-Wise access were taken when voxels are grouped into $4 \times 4 \times 4$ subblocks, called *cells*, and are reconstructed cell-by-cell. Cell-wise reconstruction is more efficient for the applications, such as volume rendering, where data are accessed with some regular pattern. The results show the timings taken for accessing, cell-wise, all cells in the dataset (All), and only cells classified as skin (Skin). Notice that the access speed is faster when the voxels are accessed from compressed data. This is because most of the null detail coefficients are not even traversed in reconstruction.

Table 1. Experimental Results on Compression Quality

		Desired Ratio of the Wavelet Coef's Used			
		3%	5%	7%	10%
Compression Performance	Compressed Data Size (MB)	24.59	35.06	45.43	60.50
	Compression Ratio	29.27	20.54	15.58	11.90
Errors in Voxel Values	SNR (dB)	22.64	25.94	28.57	31.99
	PSNR (dB)	44.49	47.79	50.41	53.84

Table 2. Experimental Results on Voxel Reconstruction Time

		Uncompressed	Desired Ratio of the Wavelet Coef's Used			
			3%	5%	7%	10%
Pure Random		2.78	3.33	3.49	3.62	3.85
Cell-Wise	All	18.88	3.88	4.88	5.99	7.52
	Skin	6.50	2.28	2.91	3.53	4.36

3 Compression-Based Parallel Ray-Casting

3.1 Image-Order and Object-Order Volume Rendering

Volume ray-casting is an *image-order* volume rendering algorithm that is most popularly used since it produces the best rendering quality [8]. Various optimization methods have been proposed for ray-casting. The early ray termination technique allows to stop sampling along the rays as soon as accumulated opacities reach a pre-specified threshold value [9]. Hierarchical data structures such as octrees and pyramids, and k-D trees, have been applied to exploit object-space coherence inherent in volume data [9, 2, 15]. The data access pattern during parallel ray-casting is very irregular, and that makes such data structures less natural than the *object-order* algorithms.

In the object-order algorithms such as splatting, on the other hand, the data are traversed in a regular manner, hence, the data coherence can be exploited very easily [16]. While the object-order algorithm is more amenable to parallelization, it has the disadvantage that it is difficult to apply the optimization techniques for image-order algorithms, such as early ray termination. As described below, our *object-order* ray-casting exploits the advantages from both algorithms.

3.2   Our Parallel Ray-Casting Scheme

In our parallel rendering scheme, image screen is divided into regular spaced pixel tiles of small sizes, and these form a pool of tasks. During run-time, processors are assigned tiles from the pool of tasks waiting to be executed. The processors perform ray-casting repeatedly on tiles until the task pool becomes empty. Load balancing is carried out dynamically during rendering.

As mentioned previously, the entire CT data of the Visible Man are replicated at each local memory space in a compressed form. For compression, the volume data are partitioned into $16 \times 16 \times 16$ subblocks, called *unit block*, and unit blocks become the basic unit for encoding and decoding. Each compressed unit block is associated with the min-max values of voxels within it, which prevents from reconstructing voxels of no interest. The processor computes image segments corresponding to tiles using ray-casting in the object-order fashion. When a tile is assigned, its view volume is computed, and the unit blocks that intersect with the view volume are listed, on demand, in the front-to-back order by traversing an octree with a proper height, constructed from min-max unit blocks. Each unit block is, then, projected into the tile on the image plane, and the rays for the pixels within the projected area are advanced simultaneously through the unit block by accumulating colors and opacities.

The block-by-block access to the volume during ray-casting is more amenable to exploiting the data coherence in object space than the ray-by-ray access. In most previous methods, non-local voxels are fetched from remote memory on demand, and are usually cashed locally. When the cache is not large enough, the system starts to thrash. Since it results in a poor speedup, it is important to utilize voxels maximally once fetched. In [6], Law et al. also ray-cast volumes block-wise, and showed the object-order traversal reduces the costs for data redistribution. In our method, the entire volume data is available at local memory, but only at the cost of reconstruction. Although our compression method reconstructs voxels very fast, it is also important to minimize the number of reconstruction operations for the same voxels. Contrary to ray-by-ray traversal, block-wise traversal guarantees each voxel in the view volume is reconstructed at most once.

A problem with ray-casting based on the object-order traversal, is that it is no longer natural to apply the early ray termination technique which allows significant savings in computation. In our implementation, we solved this problem by using an image-space quadtree that exploits the image-space coherence as in [7]. To efficiently determine the visibility of regions in a tile, we construct a quadtree in image space as follows. Each leaf node of the quadtree, corresponding to each pixel in the tile, has value 1 when the pixel has been made opaque enough, that is, the opacity has reached a threshold value, and 0 otherwise. The four adjacent values at each level are then combined into one value at the next, coarser level by adding them. When a non-terminal node in the tree has value 4, it indicates that the corresponding region in the image plane is opaque, and one is recursively added to its parent node. Otherwise, its region is not opaque, and more ray sampling operations are necessary.

When a unit block is projected into the tile, the opaque regions are quickly rejected by recursively traversing the quadtree against the projected area, and the resampling operations are performed only on the transparent rays. When the entire tile gets opaque, that is the quadtree's root have value 4, traversing the list of unit blocks stops. In this way, we can simulate early ray termination in the object-order traversal algorithm.

3.3 Experimental Results

We implemented our parallel scheme on a Cray T3E-900 with 136 processors. The Cray T3E processing element (PE) includes a 450 MHz Alpha processor and 128 Mbytes local memory, and is connected by a high-bandwidth, low-latency bidirectional 3-D torus system interconnect network. In implementing our method, we used the Cray Shared Memory Access Library (SHMEM) which provides faster interprocessor communication than MPI and PVM do.

For a performance test, we have generated a $512 \times 512 \times 1440$ volume data set from the original fresh CT data of the Visible Man, which takes up 720 Mbytes (Note that some portion of slices in the Legs section of the fresh CT are missing.). For rendering, it was compressed into a dataset of size 45.43 Mbytes, using the 3D wavelet compression scheme. This example data uses 7% of wavelet coefficients, and the rendered image quality is visually identical to that from the uncompressed data.

Timings were taken in seconds for generating 512×1024 images (Skin) using 16×16 and 32×32 tiles (Figure 1). Figure 2 shows the performance results that compare very favorably with existing results for direct volume rendering, say, [1, 5, 6, 10]. These timings do not include data replication and image display. When 32×32 tiles were used, it took 357 seconds per frame on one processor, and 4.9 seconds on 96 processors. We observe that higher than 80% efficiency is achieved for up to 80 processors, which surpasses most of the recently reported parallel implementations for direct volume rendering.

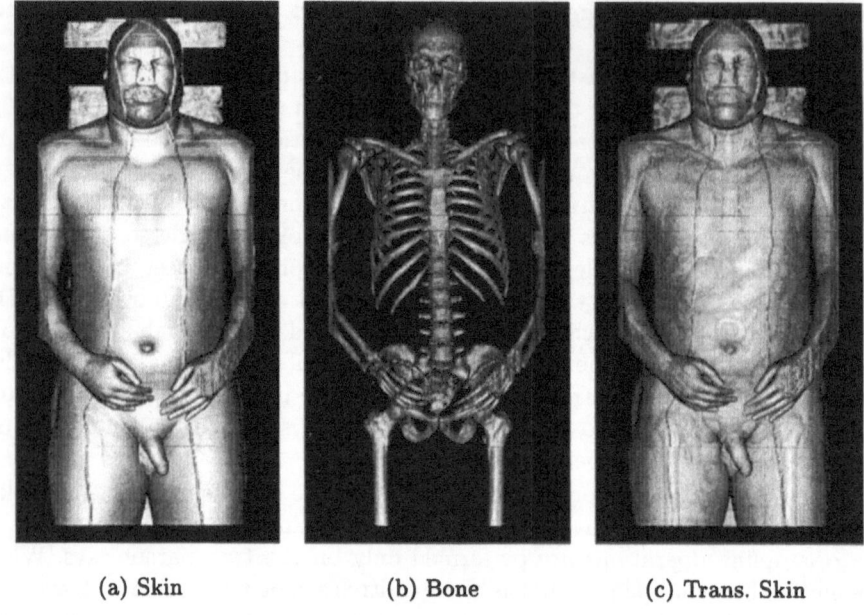

(a) Skin   (b) Bone   (c) Trans. Skin

Fig. 1. Parallel Ray-Cast Visible Man

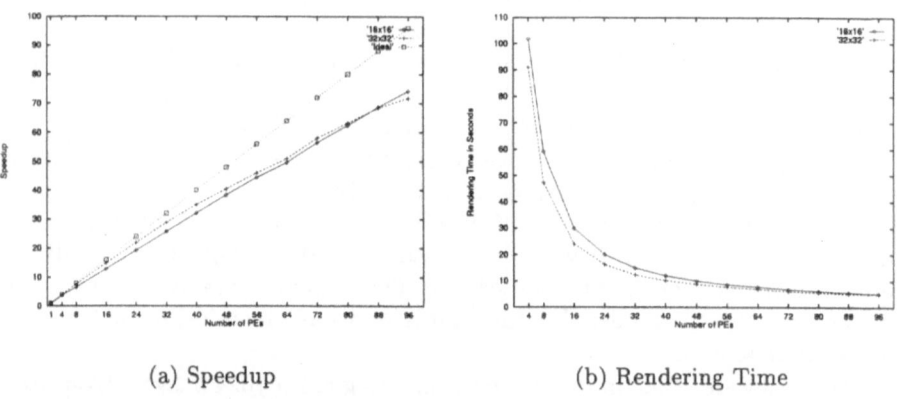

(a) Speedup   (b) Rendering Time

Fig. 2. Speedup and Rendering Time

The primary reason for getting the good speedup is that our compression-based parallel ray-casting scheme minimizes the data communication overheads during rendering. Only communication for task assignment and image segment collection is necessary, and breakdown of execution time for processors shows that the time taken for communication is very small compared to the time taken for rendering computation. For instance, when 64 processors are used, the average ratio of communication time to rendering time is less than 0.0001, which is negligible. This property becomes crucial when our method is implemented on such other platforms as PC/workstation clusters with slower Ethernet links, in which data communication is usually very expensive.

Notice that the performance depends on the tile sizes: As more processors join in computation, smaller tiles achieve better dynamic load balancing. However, the larger number of tasks increases overheads of managing tasks. At some point, there is no net gain from having the tiles smaller. The extra overheads such as task assignment and partial image collection, are very small in our scheme, and we observe that smaller task sizes, such as, 16×16 or 32×32 tiles, produce a bet-

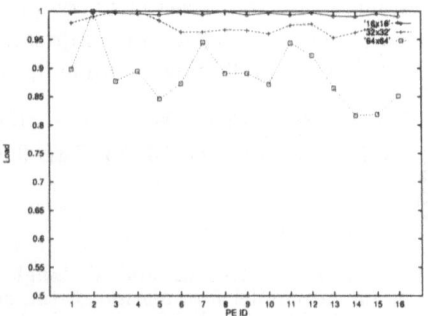

Fig. 3. Load Balancing

ter performance. Figure 3 shows how evenly the tasks are distributed among processors when 16 processors are used. Currently, we are modifying the task assignment strategy so that the task sizes vary dynamically according to various computing parameters.

4 Conclusions and Future Work

In this paper, we proposed a new compression-based parallel ray-casting scheme aimed at the distributed-memory architecture, and showed that it can be used very effectively for very large volumes. Our current result may not be the fastest in terms of frame rates. For example, interactive rendering of iso-surfaces of Visible Woman was achieved on an SGI Reality Monster, which is a scalable shared-memory multiprocessor [14]. Our parallel scheme targets visualization of very large volume data on distributed systems, and tries to achieve high performance by minimizing communications between processing elements during rendering through compression. Our scheme is more practical than the previous works in that it is also very appropriate for distributed systems with low bandwidth links, as well as distributed-memory multiprocessors, such as easily available clusters of PCs and workstations, in which communications between processors are regarded as quite costly.

We have been optimizing our parallel volume render to enhance its performance. Currently, a $16 \times 16 \times 16$ unit block is the basic unit for compression and decompression. Sometimes, for example, when the normal and gradient at a voxel is to be approximated using central differences, unnecessary voxels may have to be decoded. We are testing with $4 \times 4 \times 4$ cells as the basic unit, and a preliminary experiment with the optimized renderer shows almost two times faster rendering. There are very few works to compare with our results on parallel direct volume rendering of Visible Human. We are trying to achieve one frame per second on 96 processors for the Skin classification (Figure 1 (a)). Considering that most portion of the 512×1024 image is opaque, and our ray-caster is a true volume renderer, the goal appears to be high enough.

Our scheme also needs to be modified for huge data sets that do not fit into local memory spaces even after compression. We believe that a rendering scheme based on data partition and compression, will also perform very well, and make it possible to handle huge volume data more effectively.

Acknowledgements We wish to thank the ETRI supercomputing center in Korea for access to the CRAY T3E-900.

References

[1] M. Amin, A. Grama, and V. Singh. Fast volume rendering using an efficient, scalable parallel formulation of the shear-warp algorithm. In *Proceedings of the 1995 Parellel Rendering Symposium*, pages 7–14, Atlanta, October 1995.

[2] D. Cohen and Z. Sheffer. Proximity clouds - an acceleration technique for 3D grid traversal. *The Visual Computer*, 11:27–38, 1994.

[3] I. Ihm and S. Park. Wavelet-based 3D compression scheme for very large volume data. In *Proceedings of Graphics Interface '98*, pages 107–116, Vancouver, Canada, June 1998.

[4] I. Ihm and S. Park. Wavelet-based 3D compression scheme for interactive visualization of very large volume data. *Computer Graphics Forum*, 1999. *To appear*.

[5] P. Lacroute. Real-time volume rendering on shared memory multiprocessors using the shear warp factorization. In *Proceedings of the 1995 Parellel Rendering Symposium*, pages 15–22, Atlanta, October 1995.

[6] A. Law and R. Yagel. Multi-frame threshless ray casting with advancing ray-front. In *Proceedings of Graphics Interface '96*, pages 70–77, Tronto, Canada, May 1996.

[7] R. Lee and I. Ihm. On enhancing the speed of splatting using both object- and image-space coherence. Submitted for publication, 1998.

[8] M. Levoy. Display of surface from volume data. *IEEE Computer Graphics and Applications*, 8(3):29–37, 1988.

[9] M. Levoy. Efficient ray tracing of volume data. *ACM Transactions on Graphics*, 9(3):245–261, July 1990.

[10] P. Li, S. Whitman, R. Mendoza, and J. Tsiao. ParVox – a parallel splatting volume rendering system for distributed visualization. In *Proceedings of the 1997 Symposium on Parallel Rendering*, pages 7–14, Phoenix, U.S.A., October 1997.

[11] U. Neumann. Communication costs for parallel volume-rendering algorithms. *IEEE Computer Graphics and Applications*, 14(4):49–58, July 1994.

[12] NLM. *http : //www.nlm.nih.gov/research/visible/visible_human.html*, 1998.

[13] S. Park, G. Koo, and I. Ihm. Wavelet-based 3D compression schemes for the Visible Human dataset and thier applications. In *CD-ROM Proceedings of Visible Human Project Conference '98*, Maryland, USA, October 1998.

[14] S. Parker, P. Shirley, Y. Livnat, C. Hansen, and P. Sloan. Interactive ray tracing for isosurface rendering. In *Proceedings of IEEE Visualization '98*. IEEE, 1998.

[15] K. Subramaanian and D. Fussell. Applying space subdivision techniques to volume rendering. In *Proceedings of Visualization '90*, pages 150–159, 1990.

[16] L. Westover. Footprint evaluation for volume rendering. *Computer Graphics*, 24(4):367–376, 1990.

Editors' Note: see Appendix, p. 336 for colored figure of this paper

Exploring Instationary Fluid Flows
by Interactive Volume Movies

Thomas Glau

DaimlerChrysler AG, Research and Technology, Virtual Reality Competence Center,
Wilhelm-Runge-Strasse 11, D-89013 Ulm, Germany
Thomas.Glau@DaimlerChrysler.com

Abstract. Volume rendering offers the unique ability to represent inner object data and to realize enclosed structures "at first glance". Unlike software-based methods, the use of more and more available special-purpose hardware allows volume rendering at interactive frame rates - a crucial criterion for acceptance in industrial applications, e.g. CFD analysis. Careful optimizations and the exclusive use of hardware-accelerated data manipulation facilities even enable volume rendered movies supporting real time interactivity. This article presents the most important features and implementation issues of an *OpenInventor*-based stereoscopic, VR-featured volume rendering system for instationary datasets.

1 Introduction

In 3D fluid flow analysis simulations are usually performed by means of the finite element approach using locally adapted, unstructured meshes. As a result, scalar and vector quantities for a large number of cells (varying from 100.000 to >5.000.000) are generated and need to be made accessible to human perception.

Direct volume rendering proved to be an incomparable tool for getting a deeper insight into complex datasets. Because of its ability to show the dataset as a whole, flow structures (*features*) often can be recognized "at first glance". Nevertheless, volume rendering still suffers from computational expense. Hence, it can be hardly found in widespread industrial applications and it is almost exclusively used for stationary data analysis.

This article shows how to implement a volume renderer on a high end graphics workstation, fast enough for playing volume movies with full real time interactivity. Interactivity includes performing geometric transformations as well as feature extraction and data manipulation. By integrating the volume renderer into an existing virtual reality platform, hybrid rendering and stereoscopic viewing are provided as well as several virtual reality features, e.g. tracking.

Note that simulation and visualization are parts of an iterative procedure: parameters are modified until certain optimization criteria are fulfilled. Therefore, fast execution of each process involved in this loop is not only highly desirable, but rather absolutely essential. Considering this practice we can assume that in an early optimization phase interactivity is the major requirement for visualization. However, approaching to the final step fidelity turns to be the more important criterion. The user would neither accept a maximum quality rendering at low frame rates nor a high performance system with too much data loss. Before talking about movies we have to discuss some common details concerning fast volume rendering and feature extraction techniques.

2 Design issues

Even there are many visualization algorithms for direct volume rendering of unstructured grids, real-time interactivity is not achieved [6][16]. To realize a *today*-usable system meeting the requirements discussed above, the use of hardware-accelerated 3D textures seems to be the most promising way. In fact, the availability of this hardware increases permanently while its costs decrease rapidly. Meanwhile, 3D texture hardware is offered by a few vendors producing PC-based graphics systems. The OpenGL standard API enables flexible integration of volume objects into existing surface rendering systems. As three-dimensional texturing must be supported by all OpenGL 1.2 implementations, it actually becomes a standard feature.

The corresponding rendering approach, *slicing*, expects scalar volume data represented as a 3D texture. First, polygonal planes are rendered parallel to the viewport and stacked along the viewing direction. By mapping the 3D texture while rendering the polygons, the volume data are sampled along the planes according to a hardware-accelerated interpolation scheme, usually trilinear interpolation. The textured polygons are blended together in back-to-front order resulting in the final pixel image [4][14]. Here, the physical model for light transport is simplified according to the used blending operator [8][10].

Naturally, 3D textures are limited to Cartesian-grid data. Hence, we have to convert the simulation data from the original unstructured FE-grid by a preprocessing step, expecting some data loss (see chapter 3). After resampling, the voxel representation offers significant advantages for following postprocessing steps. Because it can be regarded as the three-dimensional analogue to the pixel model, a huge number of well-working image processing algorithms can be extended for use in three dimensions - often in a straightforward manner. This includes volume enhancement as well as segmentation and pattern recognition procedures [9].

The maximum resolution of the grid depends on the available texture memory and on further restrictions concerning the maximum texture size of the underlying hardware. Using a SGI *ONYX InfiniteReality* graphics workstation with 16 MB texture memory and four RM6 raster managers enables simultaneous visualization up to four 128^3 x 16 Bit Luminance-Alpha textures without swapping [1][11]. The luminance component is used for scalar value representation while the alpha component contains an optional classification tag. Displaying a dataset in a translucent view is the most important application for volume rendering. But often, the user gets confused with this because of missing depth cues and shape hints within the gel-like volume. Stereoscopic viewing in perspective projection avoids this weakness in a convincing fashion. So we emphasize that stereoscopic viewing is really an essential key feature for serious data analysis using volume rendering.[2][4][5]

3 Data generation

To generate Cartesian-grid data from the unstructured finite element grid input, a 3D *scan-conversion* has to be performed. Here, each FE-cell (hexahedron) is rasterized individually by slicing it into a set of polygons so that a standard 2D rasterization algorithm can be applied. After the voxel set occupied by the cell is determined, trilinear interpolation is used to obtain voxel data from the vertex data of the cells. Linear interpolation schemes are usually preferred because they are fast, first-order continuous and only require data at the vertices of the bounding cell [7]. Compared to image-order resampling techniques this object-order method avoids exhaustive cell searching for point location within the unstructured grid, but the higher the cell-to-voxel ratio becomes, the more the conversion quality reduces. For higher resolution, the conversion procedure can be spatially limited to a user-defined subvolume. To speed up the calculation, the algorithm was parallelized by scheduling chunks of cells to be scan-converted to all available CPUs. The volumetric dataset shown in Fig. 3 consists of 190.080 cells and needs 12.5 secs to be scan-converted into 2.097.152 voxels on a 4x*R10000* 194 MHZ SGI *ONYX*-system.

4 Minimizing the Rendering Costs

While rendering, performing the hardware-accelerated texture interpolation is the most expensive task. To overcome this bottleneck, the polygon stack is usually clipped to the limits of the texture volume instead of drawing simple rectangles (see Fig. 1). Considering the fact that all slicing planes are held parallel to the screen, the clipping algorithm is less complex compared to the general case and can be implemented quite easily with a few lines of code shown below: First, the object transform is applied to both the 3D texture and to each vertex of the bounding box. Then, the parametric representation and some auxiliary parameters of all edges are determined:

```
foreach edge
  if (dz != 0)
      h = dx / dz
      k = dy / dz
      a = p0.y - k * p0.z
      b = p0.x - h * p0.z
      zmin = MIN(p0.z, p1.z)
      zmax = MAX(p0.z, p1.z)
      is_parallel = False
   else
     is_parallel = True
 end
```

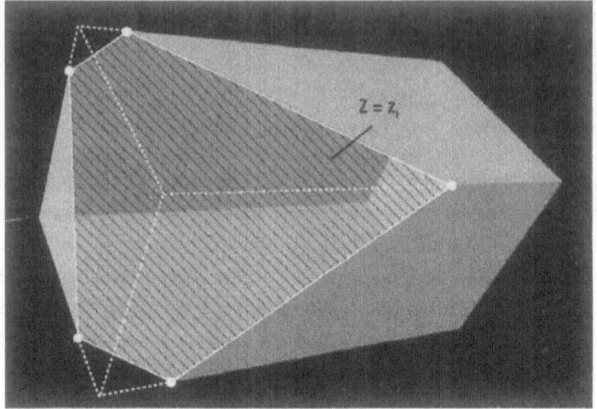

Fig. 1. Slicing a 3D texture (shown as bounding box)

where *p0* and *p1* are the end points of the edge and *dx* ... *dz* denote the differences between them with respect to the appropriate direction. With these parameters, the intersection points p_i of the edges with each plane of the polygon stack can be easily calculated, provided that the line is not parallel to the xy-plane and the intersection really exists. Finally, to render the polygon properly the convex hull of the intersection points has to be determined using a robust algorithm taken from [17].

```
foreach plane
  z += plane_distance
  foreach edge
    if (!edge.is_parallel && (z >= edge.zmin) && (z <= edge.zmax))
      p[nvertices].x   = z * edge.h + edge.b
      p[nvertices++].y = z * edge.k + edge.a
  end
  convex_hull(p, nvertices)
  render_polygon(p, z, nvertices)
end
```

The calculation procedure for a polygon stack composed of 128 slices takes typically 1 msec average calculating time on a *R10000* 194 MHZ CPU and improves rendering speed significantly when applied to non-uniformly scaled textures. Thus, frame rates of >30 Hz in a 1000x750 window are usually achieved. Using a software-clipping algorithm produces no load for the geometry engine and avoids wasting of hardware-supported clipping planes which are mostly limited to a small number even on high end graphics systems.

5   Feature Extraction

Usually, the user wishes to have a volume composed of *features* rather than visualizing the raw data. Preserving real time interactivity, threshold segmentation is performed by manipulating the hardware-accelerated RGBA-Texture-Look-Up-Table (LUT) which maps each luminance component to a RGB pseudocolor and each alpha component to another alpha value, respectively. To gain more flexibility, a virtual LUT (stored in main memory) is introduced for each single feature. The RGB LUT values are derived from an HSV colorbar where H originally moves from 0 ... 1 while S and V are set to unity to get a color spectrum. The H mapping can be modified by scaling and shifting and is therefore of first order (see Fig. 2). Unlike color mapping, alpha mapping is of zero order, i. e. constant over the whole luminance band delimited by the *value/width*-pair in Fig. 2. Finally, all tables masking and coloring a single feature (scalar band) are merged to yield the resulting transfer function.

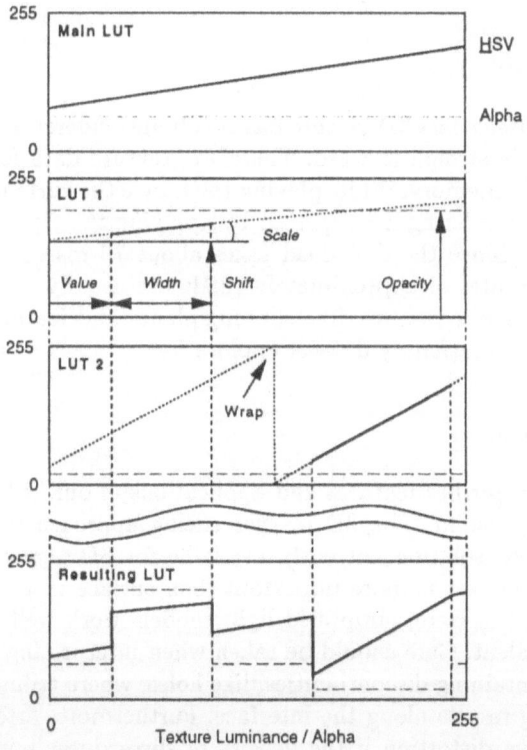

Fig. 2. LUT Merging. Both scalar band size and coloring of each feature can be adjusted very intuitively by five parameters shown in the second diagram.

Sometimes it is desirable to obtain iso-surfaces but to avoid the expensive extraction of a triangular mesh. We can achieve a surface-like impression by choosing a relatively small scalar band around the iso-value which is mapped to a full opaque, bright color (e.g. yellow). After removing dispensable pieces of the projection slices by an alpha-test, the polygon stack is depth shaded with decreasing intensity according to a square function. In addition, specular lighting improves the spatial impression when moving the object, although the surface normals of the polygons are considered for lighting calculation rather than the normals of the real iso-surface.

A more conventional way for volume exploration by clipping planes is also provided. They are realized by spacemouse-driven rectangles sampling the texture along the plane using the OpenGL texture coordinate generating function. Alternatively, OpenGL clipping planes can be applied to cut the polygon stack. Since texture sampling works with arbitrary geometries, data projection on more sophisticated surfaces (e.g. human model) is possible. More comprehensive techniques for hardware-assisted clipping and shading can be found in [15]. Furthermore, a dynamic refinement functionality is provided to improve the rendering quality of motionless volume objects by increasing the number of slicing polygons.

6 Volume Movies

Having implemented a fast 3D texture based volume renderer, generating volume movies is relatively straightforward. Therefore, texture data for each time step are kept into main memory. While playing the movie the current data in texture memory used for rendering are replaced by applying the OpenGL *glTexSubImage3D* extension. Since the download takes about 70 msecs on our hardware, volume exchange rates of approximately 10 Hz are usually achieved. Note that all capabilities described above, like clipping planes and feature extraction, are now available for instationary dataset exploration.

7 Conclusions

This case study describes features and applications of our of hardware-assisted hybrid rendering system. The 3D texture slicing approach is regarded as the most advantageous solution currently available for 3D *signal* analysis, where insight into the dataset is more important than surface inspection as required for volume graphics. Here, simplified light models work well because there is no natural equivalent. Care should be taken when interpreting rendered images from datasets containing discontinuities, like holes, where trilinear interpolation yields misleading results along the interface. Furthermore, artefacts may occur due to perspective distortion if the density of slices doesn't increase with distance from the viewer. A number of remarkable properties makes 3D texture slicing appealing for industrial use: High performance, good-quality rendering, real time surface visualization, immediate low-level feature extraction, arbitrary scan geometries, inherent stereo viewing, standard graphics API, compatibility with existing surface rendering systems and finally - volume movies. For CFD visualization, this technique requires an additional resampling step but offers real 3D data analysis and the application of a wide range of postprocessing algorithms already known from the 2D domain. Today, 3D texture hardware is fairly limited concerning resolution and data depth. Therefore, it is mainly used either to get a coarse overview over the entire dataset or to get a closer look at a relatively small subvolume. But considering the fact that 3D texture mapping becomes a standard feature, advanced hardware support is expected to be available soon.

References

1. K. Akeley, *Reality Engine Graphics*, ACM Computer Graphics, Proc. SIGGRAPH '93, pp. 109-116, July 1993
2. D. S. Ebert, R. Yagel, J. Scott, Y. Kurzion, *Volume Rendering Methods for Computational Fluid Dynamics Visualization*, Visualization '94, Washington, DC, 1994, pp. 232-239
3. T. Elvins, *A Survey of Algorithms for Volume Visualization*, Computer Graphics, Vol. 26, No. 3, 1992
4. R. Fraser, *Interactive Volume Rendering Using Advanced Graphics Architectures*, Silicon Graphics, Inc., Technical Documentation
5. van Gelder, Kim, *Direct Volume Rendering with Shading via Three-Dimensional Textures*, Proc. Symp. on Volume Rendering, San Francisco, CA, ACM 1996, pp. 23-29
6. A. Kaufman, *Volume Visualization : Principles and Advances*, SIGGRAPH '98 Course Notes, Orlando, Florida, 1998
7. D. Kenwright, *Visualization Algorithms for Gigabyte Datasets*, SIGGRAPH '97 Course Notes, Los Angeles, CA, 1997, pp. 4-1 - 4-31
8. W. Krueger, *The Application of Transport Theory to the Visualization of 3-D Scalar Data Fields*, IEEE Visualization '90, pp. 273-280, 1990
9. G. Lohmann, *Volumetric Image Analysis*, Wiley-Teubner, 1998
10. N. Max, *Optical Models for Direct Volume Rendering*, IEEE Trans. on Visualization and Computer Graphics, 1, **2** (1995), pp. 99-108, 1995
11. J. Montrym, D. Baum, D.Dignam, C. Migdal, *Infinite Reality : A Real-Time Graphics System*, Computer Graphics, Proc. SIGGRAPH '97, pp. 293-303, 1997
12. C. Stein, B. Becker, N. Max, *Sorting and hardware assisted rendering for volume visualization*, ACM Symposium on Volume Visualization '94, pp. 83-90, 1994
13. M. Teschner, Ch. Henn, *Texture Mapping in Technical, Scientific and Engineering Visualization*, Silicon Graphics, Inc., Technical Documentation
14. R. Westermann, Th. Ertl, *Efficiently Using Graphics Hardware in Volume Rendering Applications*, Proc. SIGGRAPH '98, Orlando, Florida, 1998
15. R. Yagel, D. M. Reed, A. Law, P.-W. Shih, N. Shareef, *Hardware Assisted Volume Rendering of Unstructured Grids by Incremental Slicing*, ACM Symposium on Volume Visualization '96, pp. 55-63, 1996
16. http://cm.bell-labs.com/who/clarkson/

The volume renderer introduced here is part of the virtual reality platform DBView, developed at the DaimlerChrysler Virtual Reality Competence Center. Simulations were performed by DaimlerChrysler Research and Technology, Dep. FT 1/AK.

Editors' Note: see Appendix, p. 337 for colored figures of this paper

Analysis and Visualization of the Brain Shift Phenomenon in Neurosurgery

C. Lürig[1], P. Hastreiter[1], C. Nimsky[2], and T. Ertl[1]

[1] University of Erlangen-Nürnberg, Computer Graphics Group,
Am Weichselgarten 9, 91058 Erlangen, Germany
{cpluerig,hastreit,ertl}@informatik.uni-erlangen.de
[2] University of Erlangen-Nürnberg, Department of Neurosurgery,
Schwabachanlage 6, 91054 Erlangen, Germany
nimsky@neurochir.med.uni-erlangen.de

Abstract. In this paper we present a method for analyzing the brain shift. The brain shift is a brain deformation phenomenon, that occurs during surgical operations on the opened head. This deformation makes navigation within the brain very difficult for the surgeon, as preoperative magnetic resonance images invalidate very quickly after the beginning of the operation. Up to now not enough is known about this deformation phenomenon in order to come up with solutions for corrective action. The aim of the tool which is presented here is to prepare ground for a better understanding by visualizing the deformation between two 3D brain data sets, where one has been taken preoperatively and the second one during the operation after the brain shift has occured. We propose a new method for the modeling of the deformation by means of efficient distance determination of two deformable surface approximations. Color coding and semi-transparent overlay of the surfaces provides qualitative and quantitative information about the brain shift. The provided insight may lead to a prediction method in future.

1 Introduction

The comprehensive diagnosis of diseases and lesions is considerably assisted by different tomographic imaging modalities like CT *(Computed Tomography)* and MRI *(Magnetic Resonance Imaging)*, since they provide various information and improve the spatial understanding of anatomical structures. Integrating such data into an operation with a target in the center of the brain makes the access easier and tremendously reduces the risk of hitting critical structures. Due to the rigid behavior of the skull it is possible to define a reliable transformation between the image data and the head of the patient in the beginning of an operation using a neuro-navigation system. Thereby, showing the position of an instrument in relation to the image data, it is intended to predict structures which are approached. However, depending on the drainage of cerebrospinal fluid and the movement or removal of tissue, the initial shape of the brain changes and leads to the brain shift phenomenon which results in great inaccuracies during the ongoing course of the operation [4]. Therefore, it is important to understand the correlation of all effects and to correct the shift of the brain. Data sets showing the head of the patient before and after the shift of the brain represent an important prerequisite of the analysis. Currently, they are mainly obtained with magnetic resonance scanners [3] or to a more limited range with ultrasound devices [2].

In order to gain some qualitative and possibly quantitative insight into the brain shift phenomenon we have implemented a tool, that is capable of visualizing the difference of a brain in a pre- and intra-operative stage. The main problem that is encountered in this assignment, is the fact, that the brains to be analyzed are not necessarily registered. The registration and the brain shift problem are coupled as it is not possible to compute an appropriate registration function, without knowing which parts of the data set are influenced by the brain-shift.

The analysis and visualization of this phenomenon is surface based. The surfaces are extracted from the two datasets separately, using the deformable surface approach presented by Lürig et al. [5]. This technique approximates boundary structures in the data set. In contrast to the original work on deformable surfaces of Terzopoulos [7] the presented approach does not require a certain surface topology. The two separately generated surfaces are then fused into a single visualization. First the two surfaces are registered in this fusing approach. This is done in two steps. The first step is a manual rigid registration. The second step the final registration is done by a variation of the ICP (iterated closest point) algorithm of Besl et al. [1]. Special considerations will be made to account for the brain shift distortion in the solution presented here, as this phenomenon may severly falsify the registration. The final visualization is based on the minimal distance of a point to the other surface. This distance is used for coloring and absolute measuring of the brain surfaces by assigning color values to the vertices of the surfaces.

The next section will give a detailed explanation of some of the details of this method. Section 3 finally gives some visualization examples including some interpretations and statements on the medical relevance of this method.

2   Visualization of the Brain Shift

The previous chapter has introduced the overall work-flow to generate the brain shift visualization. The main stages of this process, which are surface generation, registration, distance computing and visualization are explained in further detail in this chapter.

Surface generation is done by an iterative process which requires a coarse initial surface. The initial surface is modeled using a tool, that is based on Delaunay tetrahedrization. First a few surface vertices are specified in a slicing view. Then these vertices are connected using this tetrahedrization method. Non-convex parts of the object to be segmented may be modeled by interactive deletion of some of the tetrahedra. The resulting surface is the boundary surface of the tetrahedral complex which is then fed into the deformable surface module.

The deformable surface technique refines this initial triangular mesh to generate a refined hull, that describes the boundary of the brain. The transformation of the surface is formulated as a minimization problem, that reduces the curvature of the surface and its size and tries to push the surface into regions of high gradient magnitude. If the volume to be analyzed is described by a function $f : R^3 \to R$, and the surface to be generated by $\mathbf{v} : \Omega \subset R^2 \to R^3$, then the surface $\mathbf{v}$ has to minimize the following functional [7]

$$\int_\Omega \sum_{i=1}^{3} \left(\tau (\nabla \mathbf{v}^{(i)})^2 + (1 - \tau)((\Delta \mathbf{v}^{(i)})^2 - 2H(\mathbf{v}^{(i)})) \right) + P(\mathbf{v}) \, dA \to 0, \quad (1)$$

where $\mathbf{v}^{(i)}$ denotes the i-th component of $\mathbf{v}$, $H(\mathbf{v}^{(i)})$ the determinant of the Hessian Matrix of $\mathbf{v}^{(i)}$. The external energy term P is defined as

$$P(\mathbf{v}) = -(w_{edge} \|\nabla(G_\sigma * f(\mathbf{v}))\| + w_{image} f(\mathbf{v})) , \quad (2)$$

where G_σ denotes a Gaussian kernel with variance σ. This functional is minimized using a multi-level finite difference solver.

This procedure is applied to both data sets generating an approximating surface of the pre-operative and the intra-operative brain respectively. These two surfaces have then to be compared. In order to perform a visualization with measuring of distances, the two surfaces have to be registered in advance. This is done using a variation of the ICP algorithm [1]. To compute the transformation, which registers data set A with data set B, for every vertex of the data set A the nearest vertex in the data set B with respect to the Euclidean norm has to be found. The ICP algorithm minimizes the procrustean distance. The procrustean distance between two surfaces is the average value of the distances between the mentioned pairs of vertices. Using the estimated correspondences a least squares fit is applied, that computes the affine transformation between the two point sets. The fit is performed using a singular value decomposition from the Numerical Recipies [6].

The problem however is, that the least squares fit procedure has to account for the brain shift in order to avoid the compensation of this phenomenon. Our approach avoids this problem by excluding the vertices from the registration function, that are influenced by the brain shift. The user has the possibility to define these vertices by interactively specifying a maximum distance for the vertices to be considered. The problem however is, that this method only works, if the brain distortion of the surface is of larger magnitude than the registration error, that has to be compensated. As this is not the case in most situations, an additional manual rigid preregistration has been implemented.

In order to visualize the brain shift, we would like to annotate every vertex of one brain surface with the distance it has moved away from the other surface. The reasonable assumption, that is made here, is that every point on one of the brain surfaces corresponds to its nearest point on the other surface. If x is a point on one surface and Y is the other surface to which we have to compute the distance, we have to determine $d(x, Y) = \min_{y \in Y} ||x - y||_2$. In order to calculate this distance we would have to compute the smallest distance of the regarded vertex to each of the triangles, which is quite expensive. As the surface under discussion is also just an approximation with limited accuracy and relatively small triangles, only its vertices are considered to compute the distance. In this way the relationships between the vertices, which have been computed for the ICP registration, can be used directly to estimate the distance to every vertex.

The resulting values are used to color-code the registered surfaces. Either the whole distance value spectrum is mapped to different colors ranging from blue to red, or a threshold is visualized using two colors. It appeared to be useful to draw one brain surface colorized and the second one in a transparent way. This technique provides a good visual impression of the brain shift.

3   Results

We have applied the brain shift analysis tool to two data sets acquired pre- and intra-operatively from a Magnetom Open in the neurosurgical operating room of the University clinic Erlangen. For illustration purposes a volume rendering of the first brain is shown in a pre-operative stage in image 1(a) and in an intra-operative stage in image 1(b).

The first brain is shown in image 1(c) and 1(d). The two generated surfaces consist of 4061 and 5344 triangles. The time needed to compute the correspondences and to perform one step in the ICP algorithm lasted 7 seconds on a R10000 175MHz O2. The second brain, which has undergone a lateral approach is displayed in images 1(e) and 1(f). These two surfaces consist of 28928 and 30080 triangles. The needed computational time was three minutes in this case. Both brains are visualized using a color spectrum and a threshold visualization each.

The method of surface-based visualization of brain shift gives a quick overview and impression of the localization and extent of brain shift. In case of surgery with a slightly elevated head in the supine position as is shown in images 1(c) and 1(d). In this position the patient is lying on its back side. It is used for lesions in the frontal and fronto-parietal region, the main brain shift, amounting up to 1.5 to 2 cm, is localized in the frontal lobes, so that by drainage of cerebrospinal fluid and tumor removal the frontal lobe just follows gravity and moves to a great extent. In the case of a lateral approach the brain moves accordingly to the midline, also following gravity (see Fig. 1(e)). Additionally this method enables evaluation of the extent of brain shift at the borders, especially at the deepest point of a tumor, a crucial landmark during surgery, to know whether to continue with resection, or to encounter the risk of damaging vital brain areas. To allow this the tumor has to be segmented in the preoperative images and then these have to be compared to the pictures where the main tumor volume is already removed (see Fig 1(f)).

This visualization method emphasizes the problem of brain shift during neurosurgical procedures, which may result in great inaccuracies of neuro-navigation in the ongoing course of an operation due to leakage of cerebrospinal fluid, tumor removal and the usage of brain spatulas. The need for a correction of brain shift to allow further accurate neuronavigation, can be satisfied by the large scale approach of performing intraoperative magnetic resonance imaging which allows an update of neuro-navigation, thus compensating brain shift. It could also be achieved by predicting the brain deformation by some simple intraoperative distance measurements with the help of a navigation system which in combination of the knowledge of the extent and behavior of brain shift allows a recalculation of neuro-navigation which takes this deformation into account.

4 Conclusion and Future Work

After a brief discussion of the brain shift phenomenon in general, we have introduced a tool, that is capable of visualizing this phenomenon by comparing two extracted surfaces of different 3D images. These visualizations allow for first quantitative statements.

The next steps in visualization and evaluation of the extent of brain shift will be a combination of the demonstrated method with methods based on volumetric approach, to allow further descriptions of the brain shift phenomenon in the future, which will may be then enable us to perform a recalculation of neuro-navigation, thus compensating for the brain shift.

References

1. P.J. Besl and N.D. MacKay. A Method for Registration of 3D Shapes. *IEEE Transactions on Pattern Analysis and Machine Intelligence*, pages 239–256, 1992.
2. R. Bucholz, D. Yeh, J. Trobaugh, L. McDurmont, C. Sturm, C. Baumann, J. Hendersonand A. Levy, and P. Kessman. The Correction of Stereotactic Inaccuracy Caused by Brain Shift Using an Intraoperative ultrasound device. In *Proc. CVRMed-MRCAS*, Lecture Notes in Computer Science, pages 459–466. Springer, 1997.
3. R. Fahlbusch, C. Nimsky, O. Ganslandt, R. Steinmeier, M. Buchfelder, and W. Huk. The erlangen concept of image guided surgery. In H. Lemke, M. Vannier, K. Inamura, and A. Farman, editors, *CAR '98*, pages 583–588. Amsterdam, Elsevier Science B.V, 1998.
4. D. Hill, C. Maurer, M. Wang, R. Maciunas, J. Barwise, and J. Fitzpatrick. Estimation of Intraoperative Brain Surface Movement. In *Proc. CVRMed-MRCAS*, Lecture Notes in Computer Science, pages 449–458. Springer, 1997.
5. C. Lürig, L. Kobbelt, and T. Ertl. Deformable Surfaces for Feature Based Indirect Volume Rendering. In Nicholas M. Patrikalakis Franz-Erich Wolter, editor, *Proceedings Computer Graphics International 1998*, pages 752–760, 1998.
6. William H. Press, Saul A. Teukolsky, William T. Vetterling, and Flannery Brian P. *Numerical Recipies in C*. Cambridge University Press, 1982.
7. D. Terzopoulos. Regularization of Inverse Visual Problems Involving Discontinuities. *IEEE Transactions on Pattern Analysis and Machine Intelligence*, pages 413–424, 1986.

Editors' Note: see Appendix, p. 338 for colored figure of this paper

Advances in Quality Control of Intraoperative Radiotherapy

Stefan Walter[1], Gerd Straßmann[2], and Marco Schmitt[3]

[1] Fraunhofer Institute for Computer Graphics, Darmstadt, Germany
[2] Städtische Kliniken Offenbach, Strahlenklinik, Offenbach, Germany
[3] MedCom GmbH, Darmstadt, Germany

Abstract. Intraoperative radiotherapy is the kind of radiotherapy where the remains of a surgically not completely removed tumour are irradiated at the open situ of the patient. The current main drawback of this radiotherapy is the insufficient documentation of the applied radiation and the lack of a possibility for an individual treatment planning. This work presents a system that is a common development of Fraunhofer IGD, Städtische Klinik Offenbach and MedCom GmbH which offers a possibility for supervision of the placement of the irradiation flabs through interactive navigation in CT data acquired from the patient, the creation of a documentation of the applied isodose as well as the possibility for an individual treatment planning. ...

1 Introduction

Intraoperative radiotherapy (IORT) is a radiotherapy that is applied during the operation at the open situ of the patient after the surgical removal of a tumour with the objective to irradiate remains of the tumour that could not be removed surgically. Such an irradiation can be performed in a procedure where an Iridium radiation source is placed over the remains of the tumour with the help of an carpet like "flab" (see Appendix). Such a flab consists of a number of rubber pellets connected to each other (Freiburger Flab, s. articles in [8],[7]). The flab also contains small plastic pipes, so called "applicators", through which the radiation source is dragged or shifted. The accurate and effective placement of the flab inside the body of the patient for the irradiation highly depends on the experience and knowledge of the radiotherapist. So far neither an individual therapy planning could be performed nor a documentation of the applied isodose in the CT scans could be made because the position of the flab in its relation to a CT data set of the patient in which a radiotherapy planning can be performed is not known. Objective of this common project of the Fraunhofer Institute for Computer Graphics, the Städtische Kinik Offenbach and MedCom GmbH was the development of a system to overcome this rather inaccurate, person and experience depend technique by computer support. With a digital acquisition and visualisation of the geometry of the points of irradiation an accurate documentation of isodose distribution and in a further step a radiotherapy treatment planning in this kind of brachytherapy is made possible.

The original working steps (s. fig. 1) of the IORT procedure are:

– Acquisition of CT data: A CT data set of the patient tumour region is acquired, this data is solely used for surgery planning.
– Tumour Operation: The patient is operated, the tumour is removed surgically as far as this is possible.
– Placement of the flab: The flab is placed at the previous tumour position over the remains of the tumour according to the experience of the surgeon.
– Irradiation: The radiation sources are drawn by the irradiation device through the flab pipes with stop positions and irradiation times according to the experience of the radiotherapist.

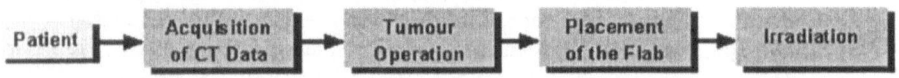

Fig. 1. Original IORT working steps

2 Implementation

The system (s. fig 2) is based on a PC workstation, with Microsoft NT operating system and an additional electromagnetic spatial tracking system (6 degrees of freedom, Polhemus or Ascension pcBird). Such a tracking system consists of a transmitter that builds up and controls an electromagnetic field and a receiver that detects its current spatial position and orientation within the electromagnetic field. The data with the geometrical information of the receiver is transferred to the workstation for further processing. The system is furthermore equipped with a foot pedal for controlling the recording of geometrical points with the tracking system.

The main part of the software is a system for visualisation of medical volume data, such as CT, MRI or three-dimensional ultrasound data, that has been developed over several years in the Fraunhofer Institute for Computer Graphics (s. [4],[5],[6]). In our case the data is acquired by a Siemens Somatom CT and transferred directly to the workstation via the hospital network in DICOM 3 format. With the visualisation software the data can be displayed as usual CT scans, oblique oriented slices in the volume data or as volume rendering. Additionally in the partner hospital the Nucletron Plato system (version BPS 2.4) is used for radiation treatment planning and visualisation of an isodose distribution in the CT data.

The procedure of registration of patient geometry and the CT data set of the patient assumes that the changes of the geometry of the body of the patient caused by the operation can be neglected. This is due to prior evaluations of the partner hospital not generally the case but can be assumed for the sacral region of the human body where the massive hip bones keep their previous geometry also during operation. Therefore this application focuses in the first step to intraoperative radiotherapy in the sacral region of the body.

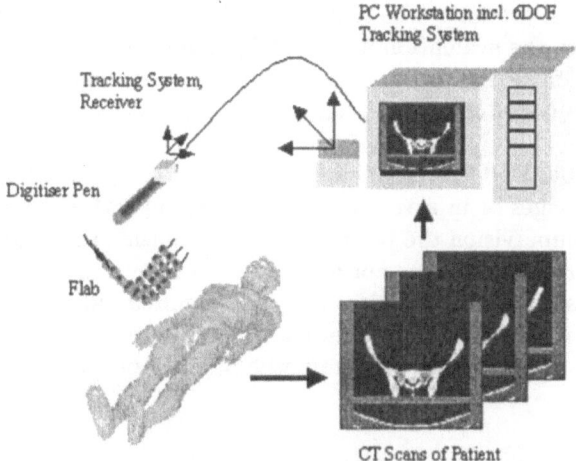

Fig. 2. System Overview

With this system the acquisition of the flab geometry can be performed with the following additional working steps (s. fig. 1,3) :

- Acquisition of CT data: A CT data set of the tumour region of the patient is acquired. This data can be used for operation planning as before, additionally it is used for the advanced IORT procedure in the steps described below. The CT scans must be acquired in the highest possible spatial resolution to guarantee the highest possible accuracy for the placement of the CT landmarks (see below). In our case the distance between two subsequent CT scans is 1mm, the spatial resolution within one scan is 0.625 mm.
- CT landmarks: As a preparation step landmarks (at least four) are manually selected in the CT data at typical anatomic positions, e.g. anatomical points at the Promontory or the Os Sacrum.
- Patient landmarks: The spatial coordinates inside the body of the patient of the anatomical landmarks defined in the CT data are acquired with the tracking system.
- Registration patient - CT: With these pairs of landmarks a transformation from the spatial position of the tracking system to the CT data can be calculated by the solution of the according (over estimated) equation system. Both kinds of data - a spatial coordinate and a position in a slice of the CT data - can be merged together afterwards.
- Navigation in CT data: When pointing at a specific anatomic position in the body of the patient with the digitiser pen mounted on the receiver of the tracking system the visualisation software displays the according position in the CT data. With this navigation tool a the physician can evaluate if a possible position for the placement of the flab is suitable for the irradiation of the remains of the tumour.

- Placement of the flab: The flab is placed at the previous tumour position according to the evaluation in the navigation step.
- Digitalisation of flab geometry: The flab is now digitised by recording a necessary number of spatial positions of pellets of the flab with the digitiser pen.
- Visualisation: The flab can now be displayed embedded in the CT data, as CT scan images or in a volume rendering to supervise the chosen position. With this supervision the position of the flab inside the body of the patient can be corrected if it does not guarantee a sufficient irradiation.
- Documentation / therapy planning: The acquired flab geometry can be exported to the Nucletron irradiation planning system where it is used for documentation and individual radiotherapy planning.

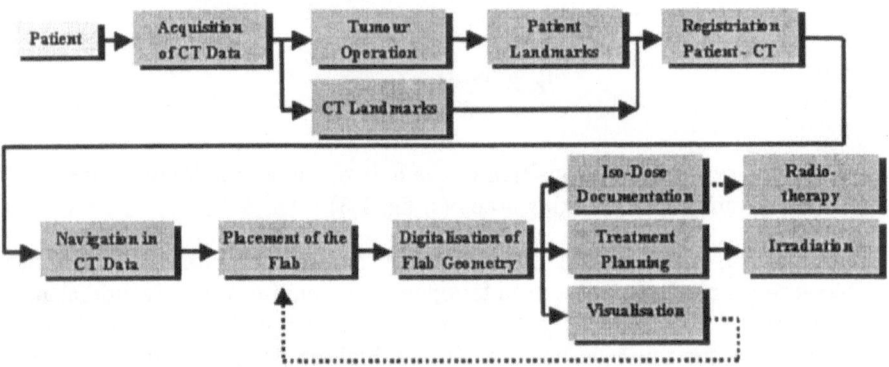

Fig. 3. Advanced IORT working steps

3 Results

3.1 Accuracy

The most problematic issue concerning the accuracy of the system is the electromagnetic tracking system. The overall accuracy of the system mainly depends on the distortion free recording of spatial points - both landmarks at the patient' s body and pellet positions - with the tracking system. All electromagnetic tracking systems have the common weakness that they easily can be disturbed by other electromagnetic fields or the influence of metal in the measurement range. The environment in a operation room contains lots of disturbing metals e.g. the surgery table, metal clamps, scalpels, etc. This influence of metal can be almost completely compensated by a careful definition and evaluation of the surgical working environment and selection of objects made of alternative materials like plastic.

An infrared tracking system cannot be used in this application because it would need a direct line of sight from the digitiser pen to a detection camera which cannot be guaranteed in the IORT scenario, for example when recording the applicator pellets inside the body of the patient.

The accuracy of the system has been validated with a special phantom (s. fig 4). This Phantom consists of a U - shaped plastic frame containing a small flab with two applicators mounted at a fixed position inside the frame. The phantom contains small marker holes that are additionally filled with a contrast media for easier detection of these markers in the CT data when the CT landmarks are placed. With this markers an optimal registration of the CT data with the phantom geometry can be performed. A CT data set of this phantom is acquired where the rubber pellets of the flab are clear to see in the CT data. The centre of each pellet is marked manually in the CT slices. After the registration of the CT data with the phantom has been performed and the flab pellets geometry has been recorded these marked pellet positions positions in the CT data can be compared with the recorded positions and the error can be calculated. We found that our system works under clinical conditions in the surgery room with an overall accuracy of 3 mm.

This procedure takes the advantage that the registration can be performed optimal because of the marker holes in the phantom. A "real life" registration must be performed with bone structures where the exact landmark position depends on the interpretation of the physician. The magnitude of this error has been evaluated in the same way as with the phantom described above but with a human hip bone. The location error found in this validation was below 2 mm.

It must be emphasised that it was not possible so far to obtain a documentation of what tissue of the patient is irradiated with a concrete placement of the flab. Also it could not be said if the remains of the tumour have been irradiated sufficiently. This application now enables the radiotherapist to document a specific constellation of patient, tumour remains and flab. Furthermore the acquired geometrical data can be used to calculate the stop positions for the Iridium radiation source to optimise the individual radiation dose.

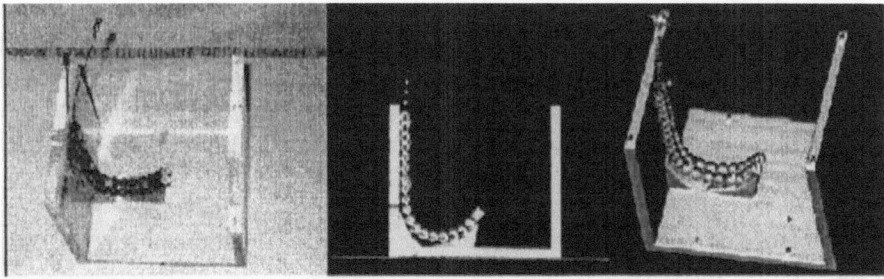

Fig. 4. Phantom for the validation of system accuracy: photo, CT scan and surface Volume Rendering

3.2 Visualisation

The visualisation software of the system is based on a direct volume rendering software which uses the ray casting algorithm. This method calculates the resulting image directly from the three-dimensional scalar data omitting an intermediate representation such as polygons (s. [1],[2],[3]). The available modes for the volume rendering are:

- Maximum/minimum intensity projection
- X-ray simulation
- Surface reconstruction (semitransparent cloud or gradient shaded)

3.3 Composition of Pellet Geometry and Rendered Images

The common visualisation of the acquired geometric coordinates of the applicator pellets and the volume rendering of the CT data is achieved with a bump mapping method. A pellet is hereby modelled by a two-dimensional array containing the gradients of the surface of a hemisphere. The dimensions of this quadratic bump map array represent the size of a pellet in the resulting image and depend for the concrete case on the current parameters of the viewing transformation of the volume rendering system.

The acquired real world coordinates of the pellet centre is first transformed into the volume coordinate system and then with the viewing transformation of the volume rendering system into the resulting image. A bump map with the proper dimensions according to the viewing parameters is vitually placed at this position in the resulting image and used to modify pixel values according to the gradient information in the bump map. The Z-position of an entry in the bump map is calculated accordingly and compared with the depth information in a surface Z-buffer calculated by the volume rendering system. If the calculated Z-coordinate of a bump map entry indicates that it is nearer to the spectator than the surface calculated by the volume renderer the new value of the according pixel is calculated. By combining the original pixel value with a gradient shaded illumination, in which the gradient of the according bump map entry and a virtual light source are combined a semitransparent sphere is placed over the surface rendering. Figure 5 shows images from two different views and according close-up views. The consideration of the Z-coordinate is not necessary for the transparent visualisation modes (e.g. maximum intensity projection) of the volume rendering system.

The system offers several modes for interactive visualisation during the intraoperative navigation. All modes support the display of the flab and the position of the digitiser pen.

1. CT Scans: Medical users of the software are currently used two work with 2D CT scan images and therefore the system supports a view of the original CT scans with an integrated view of the pellets of the flab intersected by this scan (see Appendix).

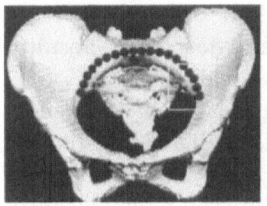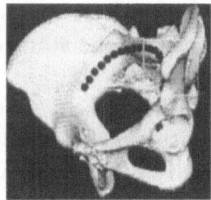

Fig. 5. Composition of pellet geometry and volume rendering with bump mapping: total view, close view, total view from a viewpoint where pellets are covered by parts of the hip bone, close view of the covered pellets.

2. 3 Orthogonal Cuts: The CT volume can be displayed as a wire frame cube with three intersecting orthogonal planes that can be chosen freely by the user. The cutting point of all three plans is coupled to the current position of the digitiser pen for navigation (see Appendix).
3. 3D Volume Rendering: In this mode 3D reconstruction of the volume with overlaid pellets can be displayed. The pellets are embedded in the volume rendering resulting image via a semitransparent bump mapping technique regarding the depth position of the surface that has been calculated by the volume renderer (see Appendix).

The first two modes allow a interactive working with the system which means the resulting images are calculated in less then 0.1 second on a double Pentium II 200Mhz system. The third visualisation mode described above is based on volume rendering and therefore the interaction rate is less good than for the first two modes. The volume rendering calculation time mainly also depends on the size of the CT data set and the size of the resulting output image. Typical data sets in the practical use of this application have a size from 20 up to 50 MB.

Table 1. Rendering times of CT data set with size 29MB, two sizes for the resulting image on a double Pentium II 200MHz system

Volume rendering mode	av. time (200x200)	av. time (400x400)
Maximum intensity projection	1.5 s	4.3s
Surface (gradient shading)	1.2 s	3.5s

4 Conclusions & Further Work

The introduced system offers a methods for optimised placement of the flab and an isodose documentation of applied radiation first as a quality control, second as an input to a long term radiation therapy. Furthermore the introduced working scheme offers an individual treatment planning for this kind of brachytherapy. The system is currently in the stage of a first clinical evaluation in the hospital in Offenbach, Germany.

One drawback of the current system is that all the positions of the pellets must be recorded with tracking system which means that depending on the concrete used flab up to fifty positions must be recorded. This high number and therefore the time consumption of this time critical procedure can be significantly reduced by introducing a mathematical description of a flab template (e.g. a spline interpolation) that is formed by the geometrical positions of some of the pellets. Another important issue is to improve the user friendliness of the system by introducing a voice control of the user interface.

Acknowledgement We herewith want to thank all the person that contributed during the development of the system with their opinions and their experience especially the medical supervisors in Städtische Kliniken Offenbach as there are Prof. Dr. Nier, Chirugische Klinik , Prof. Dr. Dr. Zambouglu and OA Dr. Kolotas.

References

1. R.Ohbuchi, D.Chen, H.Fuchs: Incremental Volume Reconstruction and Rendering for 3D Ultrasound Imaging, SPIE Visualization in Biomedical Computing, pp. 312-323, 1992.
2. D.H.Pretorius, T.R.Nelson: Three-dimensional Ultrasound, Ultrasound Obstet. Gynecol., Vol.5,pp. 219-221, April 1995.
3. M. Levoy: Display of Surfaces from Volume Data, IEEE Computer Graphics and Applications, Vol. 8,pp. 29-37, May 1988
4. G.Sakas, S.Walter: Extracting Surfaces from Fuzzy 3D Ultrasonic Data, ACM Computer Graphics, SIGGRAPH '95, Los Angeles, USA, pp. 6-11, August 1995.
5. G.Sakas, L.A.Schreyer, M.Grimm: Case Study: Visualization of 3D Ultrasonic Data, IEEE, Visualization '94, Washington D.C., USA, pp. 369-373, Oktober 1994.
6. G.Sakas, S. Walter, W. Hiltmann, A.Wischnik: Foetal Visualization Using 3D Ultrasonic Data, Proceedings CAR '95, Berlin/Germany June 1995
7. I.-K. K. Kolkman Deurloo, A.G. Visser, M. H. M. Idzes, P. C. Levendag, Reconstruction accuracy of dedicated localiser for filmless planning in intra-operative brachytherapy, Radiotherapy & Oncology 44 (1997) 73-81
8. J.M. Vaeth (Ed.), Intraoperative Radiation Therapy in the Treatment of Cancer, Front Ther. Oncol., Basel, Karger, 1997, Vol 31

Editors' Note: see Appendix, p. 339 for colored figures of this paper

Visualization of Molecules with Positional Uncertainty

Penny Rheingans[1] and Shrikant Joshi[2]

[1] Department of Computer Science and Electrical Engineering
University of Maryland, Baltimore County
Baltimore. MD 21250, USA
rheingan@cs.umbc.edu
http://www.cs.umbc.edu/ rheingan/index.html
[2] Celcore Inc.
3800 Forest Hill Irene Rd.
Memphis. TN 38125, USA
sjoshi@celcore.com

Abstract. Designing new and better chemotherapeutic compounds requires an understanding of the mechanism by which the drugs exert their biological effects. This involves consideration of the geometry of the active site, determination of the geometry of the drug, and analysis of the fit between them. This problem of drug-substrate fit, often called the docking problem, can be greatly influenced by uncertainty in the position of drug side chains. Traditional molecular graphics techniques fail to capture the distribution of likely atom positions. This paper describes a range of techniques for showing atom positions as probability distributions that more completely describe parameters which determine fit.
...

1 Introduction

The field of data visualization offers graphical solutions to interpreting large amount of data. Molecular graphics is one such area where a few good images can greatly enhance insight into the problem. Designing new and better chemotherapeutic compounds (drugs) requires an understanding of the mechanisms by which the drugs exert their biological effects. The structure of the drug molecule and that of the substrate molecule influence the binding property and the effectiveness of the drug docking process.

The docking problem involves determining the position and orientation of the drug molecule with respect to the substrate such that the energy of interaction of the two is minimized. The interactions are very specific in nature. Even a slight change in the side chain of a drug molecule can inhibit or enhance the interaction. Unfortunately, side chain position is not always well determined. The exact locations of atoms in a side chain can be subject to a substantial amount of uncertainty. The ability to visualize not only the best estimate of atom locations but also the range of likely atom locations would be a valuable

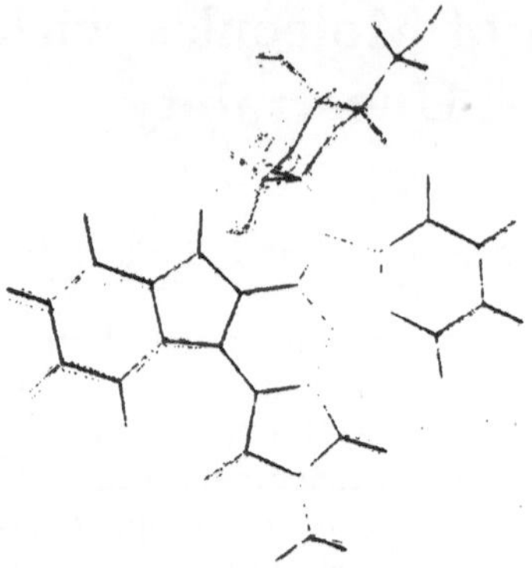

Fig. 1. Line drawing representation of eight family members.

tool in the study of structure and interactions between drug and substrate in the docking problem.

1.1 Drug Design

The drug docking problem is crucial in enzyme catalysis, antigen-antibody interactions, drug design, and advanced materials development. The particular driving focus addressed here is the mechanism by which anti-tumor drugs exert their biological effects [11]. Synthetic olegonucleotides are used as model DNA systems to study the drug binding phenomenon. High field NMR is used to examine each solution (drug and substrate mixture) at the molecular level. Initial geometries of molecules are constructed and these geometries are optimized using energy minimization calculations.

The drug molecule under study has a side chain which orients differently for each complex. Knowledge of the physical extent of the uncertainty of side chain position can be of great help in modifying, re-designing, and understanding the interaction of this drug with the substrate. Different versions of the drug molecule with identical atoms and connectivity but potentially different atom positions form a family of molecules. Each family member shares much of the molecule structure, differing mainly in side chain position and orientation. Previously a family of molecules was shown in wire-frame form in one image, with each bond shown by a line. Figure 1 shows an example of this technique. In sections where the members of the family have identical structure, multiple lines lie in the same place forming a heavy line. In sections where family members differ, a fan of options shows for the side chain. This representation is fairly successful at showing which sections have positional certainty and which do not, but is less successful at showing structural details. In sections with high positional uncertainty the display gets too cluttered to convey structure clearly. Even in

the more stable sections the 3D structure can be difficult to see due to the lack
of depth cues.

2   Related Work

Common discrete and static techniques used to display molecular structure include wire frame depictions of molecule bonds, ball and stick representations
showing both atoms and bonds, ribbons to trace out the backbones of complex
molecules, and space-filling models showing each atom's van der Waals radius [4].
Molecular dynamics are commonly displayed using animations. In situations of
limited dynamics, such as in crystal structures, line drawings showing atoms as
thermal-motion probability ellipsoids have been used [2]. A few researchers have
created more continuous representations of molecules using volume techniques
to represent electron density [3]. One focus of the research described was to
combine discrete and continuous techniques into a single unified view.

The representation of data with uncertainty is an active area of research in
the field of Geographic Information Systems (GIS), using such approaches as
the use of color, transparency, line width, blur, haze, interactive probing, and
animation [[1], [5], [10], [12]]. Within the field of visualization, researchers have
also recognized the importance of uncertainty data to accurate visualization and
have been active in developing new representation techniques. Their techniques
have included iterated function system fractal interpolation, fat surfaces, and
sonification of uncertainty [14], [6], [7]].

3   Techniques for Showing Positional Uncertainty

Both drug and substrate are sufficiently complex that judging the quality of the
fit is difficult. In order to more clearly illustrate the potential fit between a drug
molecule and the target site, a visualization should show the range of possible
drug configurations.

3.1   Discrete Representations

A common molecular graphics technique which addresses the problem of insufficient shape cues in the wire frame form is the ball and stick representation. In
this representation, atoms are represented as spheres of appropriate color and
size, while bonds are represented as shaded rods or tubes. Normally the ball
size is a function of the van der Waals radius for the atom and color is chosen
to conform to the standard molecular graphics conventions. Since solid objects
are used to represent the atoms and bonds, one gets better depth cues from the
rendered image. The solid objects and the accompanying shading of the atoms
and bonds create an object boundary for the atoms. These occlusion boundaries
convey a strong relative depth information about the superimposed surfaces.

Unfortunately, simply putting ball and stick representations of all family
members in the same image produces a very cluttered image. See Figure 2a.

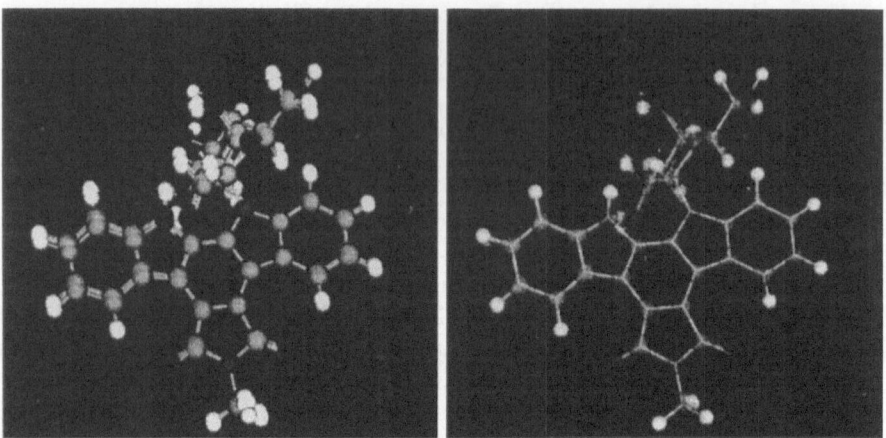

Fig. 2. Eight family members shown, a) each with full opacity, b) each with an opacity of 0.125. Carbon atoms are grey, nitrogen red, oxygen blue, and hydrogen white.

Worse yet, the least certain parts of the molecule produce the most solidly packed areas of the images. A slight modification of the ball and stick technique would assign to each family member a fraction of available energy corresponding to its likelihood. See Figure 2b.

In addition to making the image less crowded, this technique is effective in separating the image into two regions. The opaque part represents the more static portion of the drug molecule, whereas the more transparent part shows portion of the molecule that is dynamic. This gives sections of the molecule a visual weight corresponding to the likelihood that the section is actually at the displayed location. It also gives a sense of motion blur to this part of the image.

3.2 Likelihood Volume Representation

In the previous techniques each member of the molecule family is displayed as a separate and discrete possibility. Such representations could imply that these are the only configurations possible. It would be more accurate to consider molecule family members as samples of a presumably continuous possibility space. In such a space, the position of each atom is a probability distribution. In order to create this sort of continuous visualization, the first step is to create the likelihood volume that describes the probability of an atom being at each location. The original data set is discrete and unstructured; the likelihood volume is continuous and structured.

This transformation from an unstructured data set to a structured one can be accomplished using a variation of splatting. Westover introduced splatting as an object-order method for directly rendering structured volumes [13]. Using this method, 2D footprints of voxels of a 3D volume are composited onto a 2D image. Vtk includes a generalization of Westover's splatting technique which

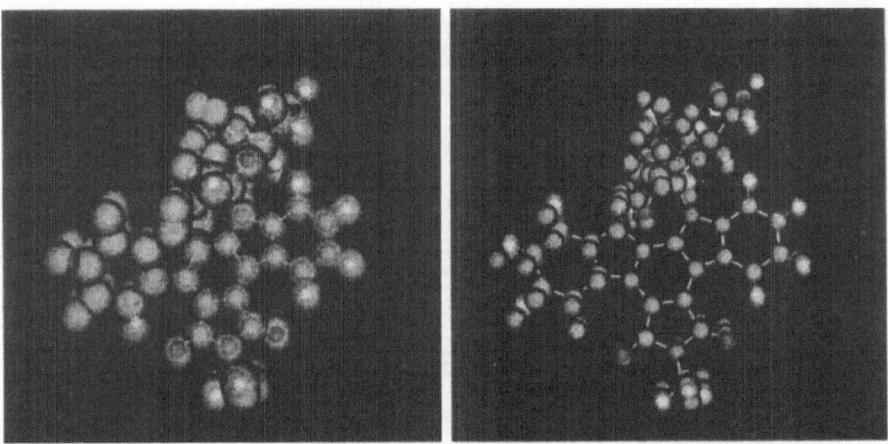

Fig. 3. Gaussian splatting and isosurface extraction. a) isolevel value of 0.9, b) isolevel value of 0.95.

can be used to sample unstructured points into a structured point set. Vtk's GaussianSplatter composites 3D footprints into a 3D volume by accumulating the maximum splat value at each voxel. The splatting function used here is the uniform Gaussian distribution. The function can be cast into the form

$$SF(x, y, z) = se^{-f(r/R)^2} \tag{1}$$

where s is a scale factor, f is the exponent factor controlling decay rate, r is the distance between the point and the Gaussian center point, and R is the radius of influence of the Gaussian.

Using this technique, an overall function that predicts the position of the drug molecule can be constructed. Each atom from each of the molecule family members is splatted into a 3D structured points object to produce aggregate information. This resultant aggregate is the value of the function that indicates the likelihood of an atom occupying that location.

An isosurface can be extracted from this volume, using the marching cubes algorithm [8]. This isosurface contains the volume in which atoms are most expected to be located. See Figure 3a. The blobby portion in the top center part of the image is the area having high uncertainty (the dynamic side chain of the molecule). The choice of isolevel employed to extract the isosurface has a great influence on the characteristics of the final image. A good isolevel is one that can extract the structure and geometry of the static part of the molecule, keeping the blobby representation of the dynamic part of the molecule unchanged. Alternatively, the isolevel can be chosen to give atoms isosurfaces which approximately match the van der Waals radius of the atom. Atoms in stable sections appear to be spheres. Atoms in more dynamic sections become less regular isosurfaces which reflect the likely range of positions. Figure 3b shows this representation.

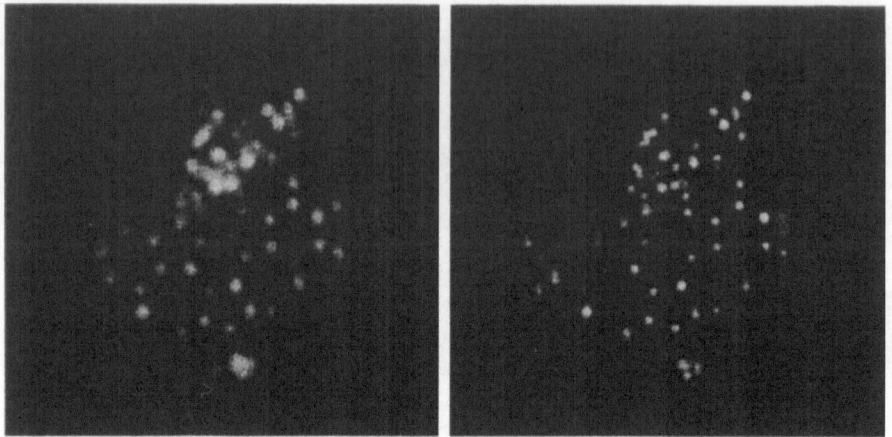

Fig. 4. Volume rendering of likelihood volume. a) emphasizing areas of greatest likeli-hood, b) greater contribution from less likely areas.

A different approach is to render the likelihood volume using direct volume rendering. This approach can show the whole distribution of the volume, rather than just those portions exceeding a threshold likelihood. Techniques to volume render molecular properties expressed as a structured point dataset have been previously used by Goodsell [3]. By volume rendering, we can see the interior of the volume associated with the object. This can give a better insight as to the distribution of the position and orientation of the molecule along with the uncertain region in which the side chain may exist. The choice of transfer function to convert voxel values to resultant color and opacity values can greatly change the characteristics of the image produced. Figure 4 shows volume renderings of the likelihood volume using different transfer functions. Figure 4a emphasizes the areas where atom occupation is probable. Figure 4b includes a greater contribution of areas where atom occupation is much less likely.

4   Implementation

The techniques described here are implemented using the Visualization Toolkit (vtk) [9]. Vtk is a freely available 3D graphics and data visualization toolkit which can be used for a wide variety of computer graphics and data visualization problems. This software employs an object-oriented approach using C++ with an optional Tcl/Tk interface. The package can be extended with the construction of additional classes. To facilitate the exchange of information between Tcl and C++ methods, automatically generated wrapper code written in C is used. Applications using this toolkit are easy to implement and portable across UNIX and PC platforms.

5 Summary and Conclusions

The field of data visualization offers graphical solutions to interpreting large amounts of data. Molecular graphics is one such area where a few good images can greatly enhance insight into the problem. The goal of this project was to create these few good images. The wire frame and the ball and stick model give a good starting point for visualizing the complex. By putting all the copies of the drug molecule into one image and controlling the opacity, the resulting image can be divided into stable and dynamic regions. Splatting provides an straightforward and useful way of converting the unstructured point dataset of atom positions into a structured point set describing a likelihood volume. The choice of isolevel for the isosurface from the splatted data can be altered to create visualizations with slightly different goals. Alternatively, the likelihood volume can be directly volume rendered, creating a more holistic representation of the entire likelihood distribution.

6 Acknowledgements

This work was conducted while the authors were at the University of Mississippi. Data was provided by Dr. David Graves, Department of Chemistry, University of Mississippi. This material based on work supported by the National Science Foundation under Grant No. ACIR 9996043.

References

1. Kate Beard and William Mackaness: Visual Access to Data Quality in Geographic Information Systems. Cartographica, vol. 30, no. 2-3. (1993) 37-45
2. Michael Burnett and Carroll Johnson: ORTEP-III: Oak Ridge Thermal Ellipsoid Plot Programs for Crystal Structure Illustrations. Oak Ridge National Laboratories Report ORNL-5138. (1996)
3. S. D. Goodsell, S. Mian, and A. J. Olson: Rendering volumetric data in molecular systems. Journal of Molecular Graphics, vol. 7. (1989) 41-47
4. S. D. Goodsell: Visualizing Biological Molecules. Scientific American, vol. 267, no. 5. (1992)
5. Adrian Herzog: Modeling reliability on statistical surfaces by polygon filtering. In The Accuracy of Spatial Databases, Michael Goodchild and Sucharita Gopal, eds.. Taylor&Francis, London, (1989) 209-218
6. Suresh Lodha, Alex Pang, and Robert Sheehan, and Craig Wittenbrink: UFLOW: Visualizing Uncertainty in Fluid Flow. Proceedings of IEEE Visualization '96. IEEE Computer Society Press (1996) 249-254
7. Suresh Lodha, C.M. Wilson, and Robert Sheehan: LISTEN: sounding uncertainty visualization. Proceedings of IEEE Visualization '96. IEEE Computer Society Press, (1996) 189-196
8. W. E. Lorensen and H. E. Cline: Marching Cubes: A High Resolution 3D Surface Construction Algorithm. Computer Graphics, vol. 21. (1987) 163-169

9. W. Schroeder, K. Martin,. and W. E. Lorensen: The Design and Implementation of an Object-oriented Toolkit for 3D Graphics and Visualization. Proceedings of IEEE Visualization '96 IEEE Computer Society Press. (1996) 93-100

10. David Theobald: Accuracy and bias in surface representation. In The Accuracy of Spatial Databases, Michael Goodchild and Sucharita Gopal, eds. Taylor&Francis, London, (1989) 99-106

11. R. M. Wadkins and D. E. Graves: Interactions of Anilinoacridines with Nucleic Acids: Effects of Substituent Modifications of DNA-Binding properties. Biochemistry, vol. 30, no. 17. (1991) 4277-4283

12. Matthew Ward and Junwen Zheng: Visualization of Spatio-Temporal Data Quality. Proceedings of GIS/LIS '93. (1993) 727-737.

13. L. Westover: Footprint Evaluation for Volume Rendering. Computer Graphics, vol. 24. (1990) 367-376

14. Craig Wittenbrink: IFS Fractal Interpolation of 2D and 3D Visualization. Proceedings of IEEE Visualization '95. IEEE Computer Society Press, (1995) 77-84

Editors' Note: see Appendix, p. 340 for colored figures of this paper

Authors Index

Color Plates

Ebert et al. (pp. 3–12)

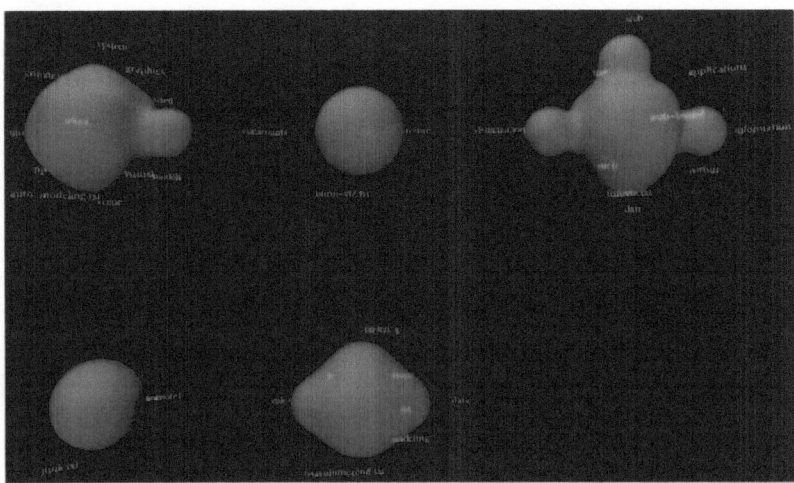

Fig. 5. Multiple documents term frequency visualized as implicit surface shapes. The documents in the upper left and the upper right both have a high frequency of the term "information" (bulge to the right). The upper right blob is the same as that in Fig. 3

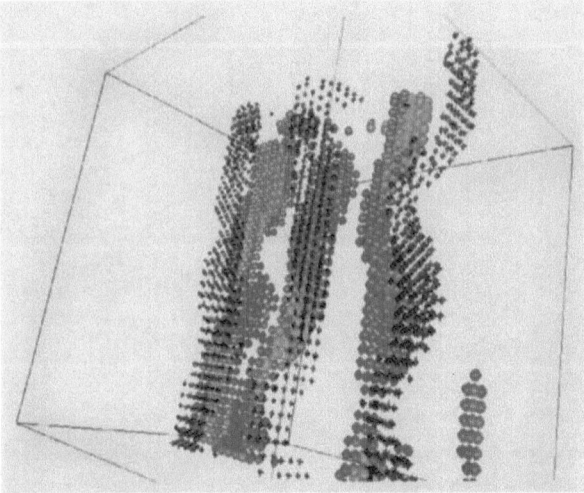

Fig. 7. Visualization of a magnetohydrodynamics simulation of the solar wind in the distant heliosphere displaying 3 vortex tubes with positive j vorticity (cuboids and ellipsoids) and 3 vortex tubes with negative j vorticity (stars)

Herman et al. (pp. 13–22)

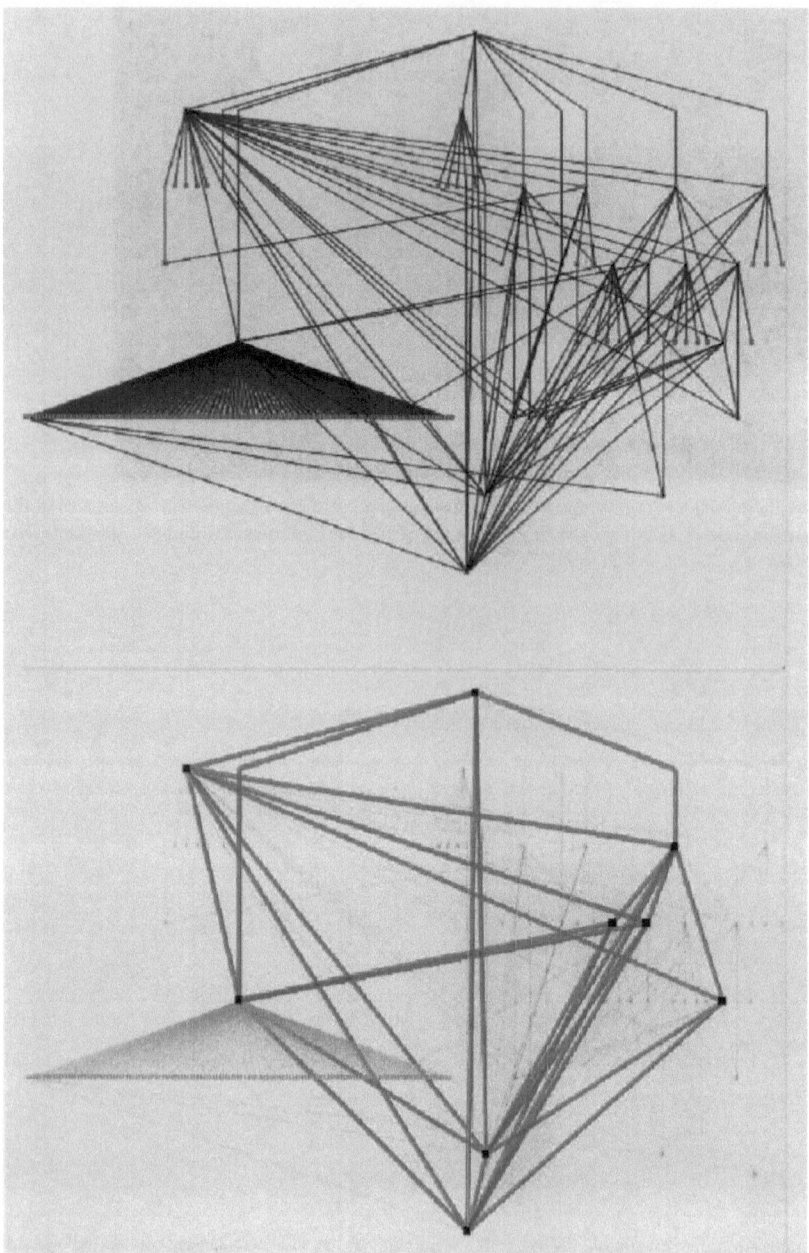

Fig. 7. Underlying DAG extracted from a web site (approx. 200 nodes) and skeleton based on average of modified Flow metric and its dual

Alexa and Müller (pp. 23–32)

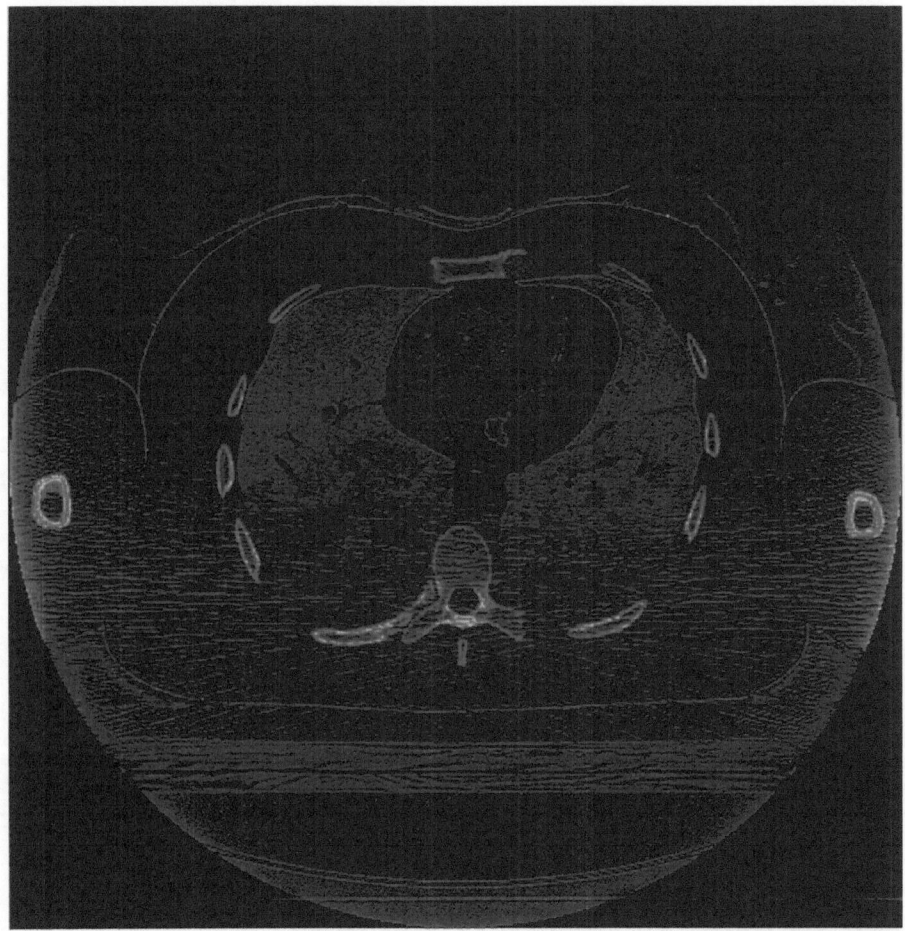

Fig. 4. This image demonstrates the benefits of displaying scalar data with multidimensional visual representations. In addtion to an already defined grayscale, the soft structures of the bones were mapped to red color

Khouas et al. (pp. 35–44)

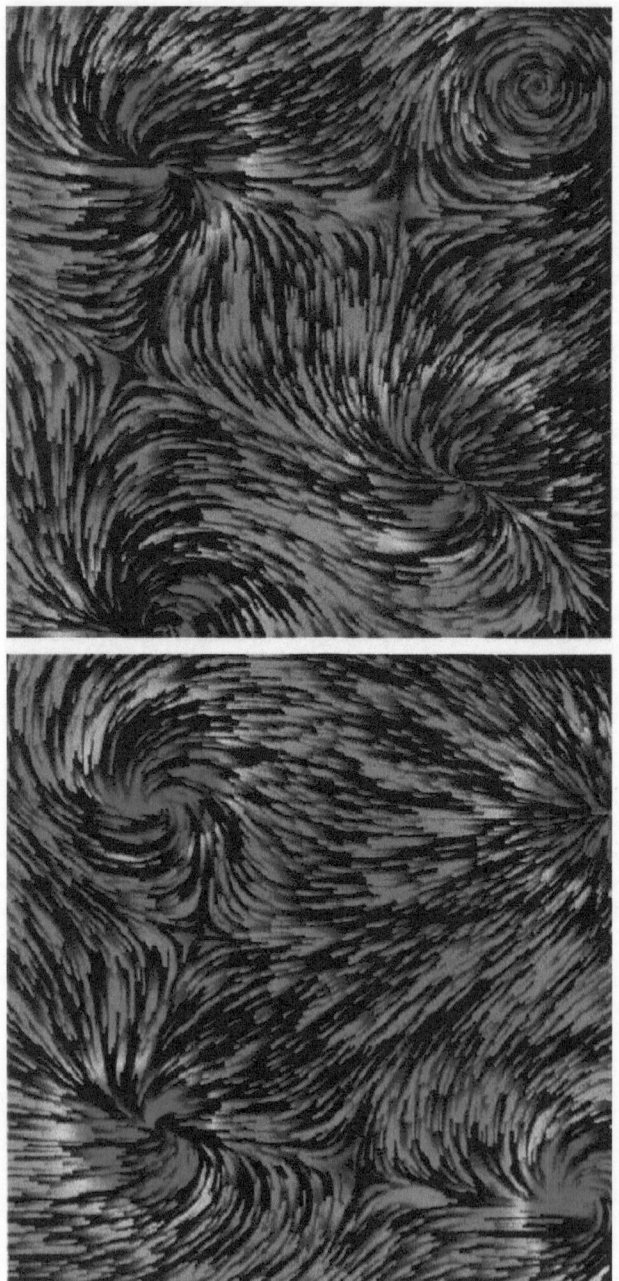

2D Vector field visualization using an autoregressive synthesis of furlike texture with color encoding of the magnitude

de Leeuw and van Liere (pp. 45–52)

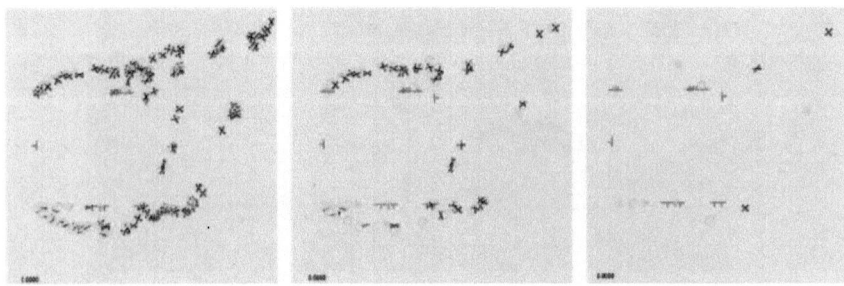

Fig. 4. Three views of the implicit method using a box filter. Left: original data set (322 critical points). Middle: 2×2 box filter (179 critical points). Right: 8×8 box filter (40 critical points)

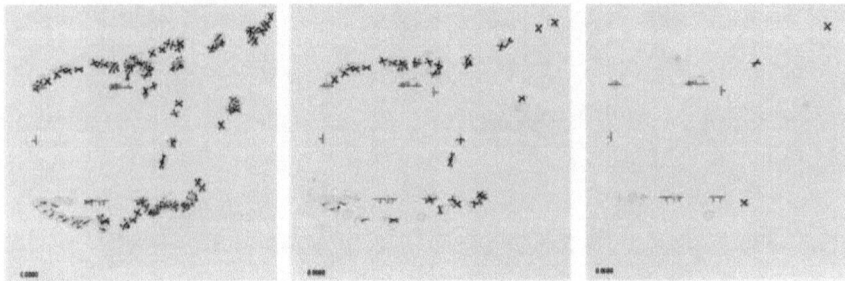

Fig. 6. Three views of the explicit method using the pair filter. Left: original data set (322 critical points). Middle: $0.0001 * H$ distance (114 critical points). Right: $0.001 * H$ filter (34 critical points)

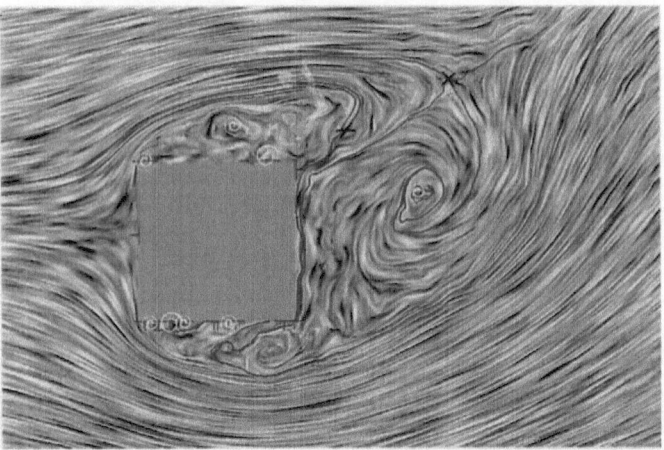

Fig. 7. A view of global flow structure around a square cylinder. The pair distance filter is used with a distance of $0.001 * H$

Sadarjoen and Post (pp. 53–62)

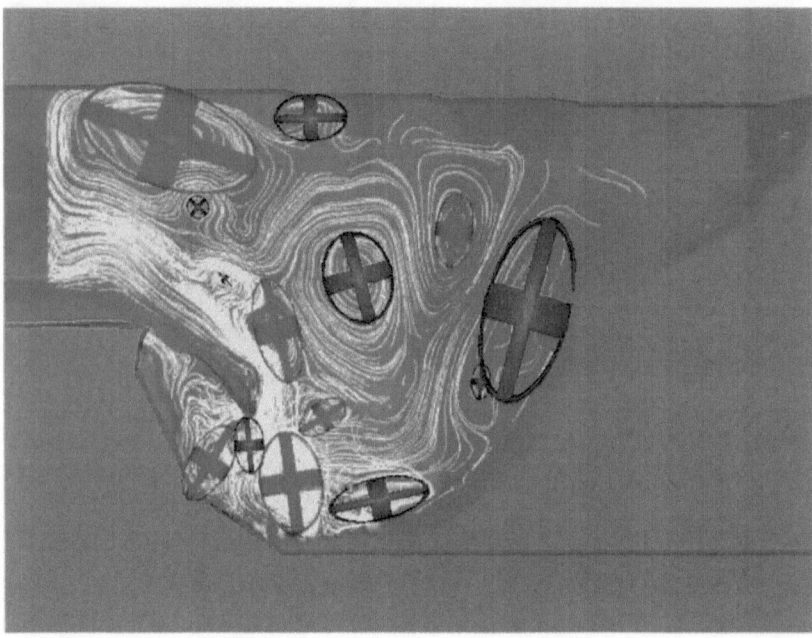

Fig. 8. Bay of Gdańsk with streamlines and vortices approximated by ellipses. Red and green indicate rotation in opposite directions

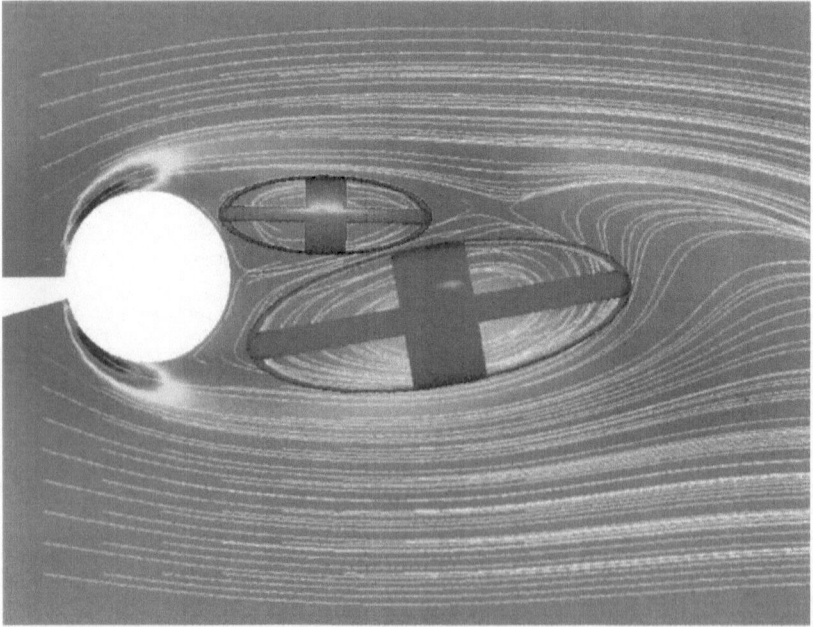

Fig. 9. Flow past a tapered cylinder with vortices approximated by ellipses, and streamlines released in a slice coloured with λ_2

Reinders et al. (pp. 63–72)

Fig. 1. A visualization of a path (red objects), its prediction (transparent object) and one of the candidates (yellow object)

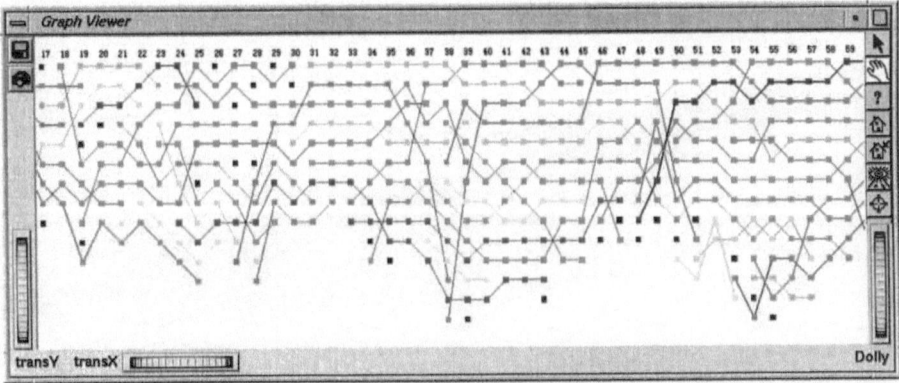

Fig. 2. The feature graph: a node represents a feature in a certain frame and an edge between two nodes represents a positive correspondence between the two features

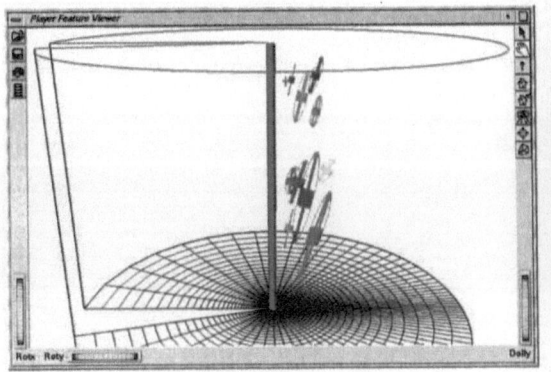

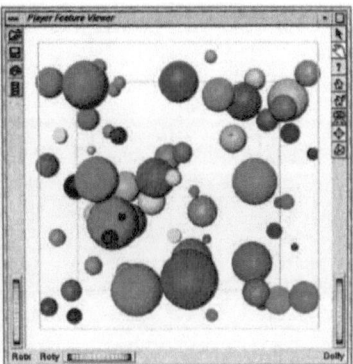

Fig. 3. Flow past a tapered cylinder

Fig. 4. Turbulent vortex structures

318

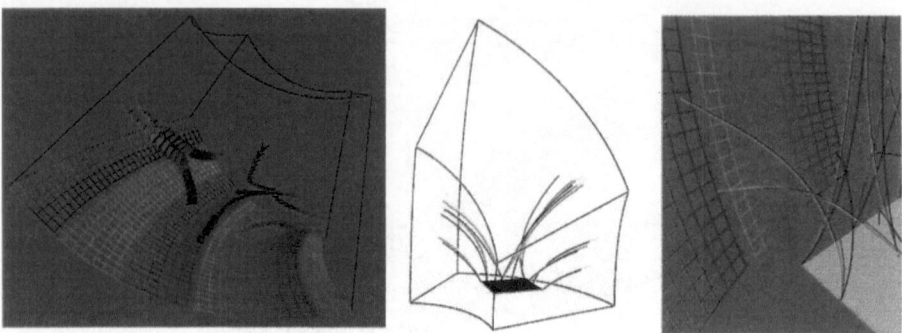

Fig. 3. In the image on the left hand side, streak tetrahedra in an analytically given flow on a curvilinear sparse grid of level 5 are shown. In the next two pictures streak lines depict a vortex flow field; yellow lines display the flow on a full grid, blue, green, and red lines on a curvilinear sparse grid of level 2, 3, and 4 respectively. In the closeup on the right hand side, it can be seen that the traces computed on the full grid and the sparse grid of level 4 are almost identical

Fig. 4. Streak balls display the flow in the blunt fin data set; the red balls are computed on a curvilinear sparse grid of level 4, the yellow ones on a grid of level 3, and the green ones on a grid of level 2

Kim and Pang (pp. 87–98)

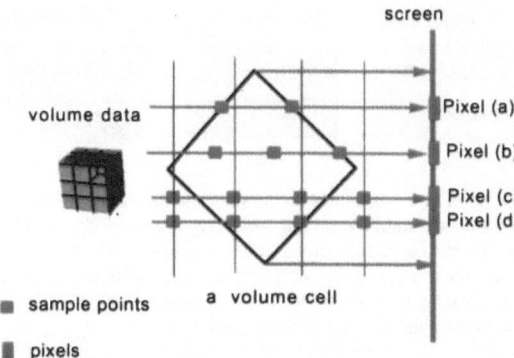

Fig. 2. Simulations of image based algorithms using our projection base algorithms. The black square and the green vertical line illustrate a volume cell and its projection to the screen. For each pixel, the color contributions are computed by taking sample points and compositing their values along the viewing direction. The blue vertical lines indicate a volume slicing plane shared by pixels (c) and (d)

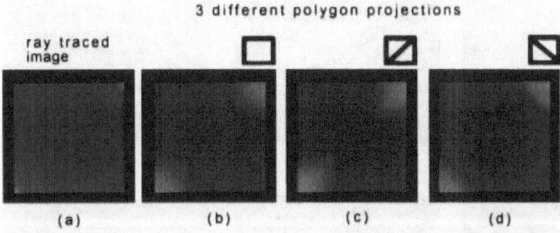

Fig. 3. Ray casting and projection algorithms with different polygonalization. The volume data is 4^3 with uniform values except at the two corners. The viewing direction is orthogonal to the front face. Above image **b**, **c** and **d**, we show the three types of polygonalization: **b** a square, **c** and **d** two different triangulations of a square. The images show that different triangulations have different effects on Gouraud shading

Fig. 4. Volume rendering and image level comparisons of Hipip. Images **a** and **b** are generated using our simulations of **a** ray casting with data samplings and reconstructions at cell faces, and **b** polygon projection such as coherent projection. Image **c** shows the absolute differences between **a** and **b**. MSE (Mean Square Errors) and RMSE (Root Mean Square Errors) of actual difference intensities are (4.882487, 1.574969, 2.983297) and (2.209635, 1.254978, 1.727222) for each red, green and blue channel respectively. All image sizes are 256 × 256

320

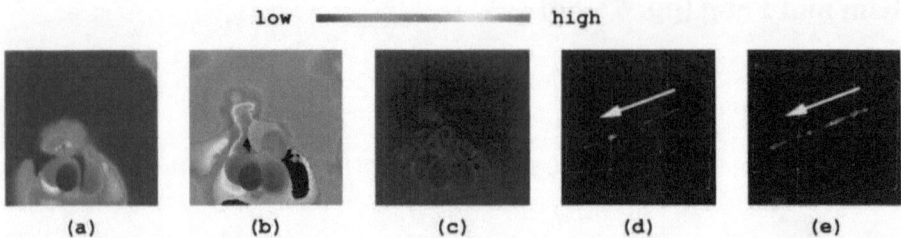

(a) (b) (c) (d) (e)

Fig. 5. Data level comparisons of algorithms used in Fig. 4. **a** shows the number of cells needed to reach opacity of 0.11 with ray casting simulation, **b** shows the number of cells needed to reach opacity of 0.21 with the polygon projection algorithm, **c** shows differences in the number of cells needed to reach opacity of 0.15 for both algorithms. Colors are used to indicate relative values of the metric. Black indicates regions that did not reach the threshold. The pixel probe visualizations shows the absolute **(d)** and additive **(e)** red intensity contributions of data from the volume to the pixel marked by the cross hair

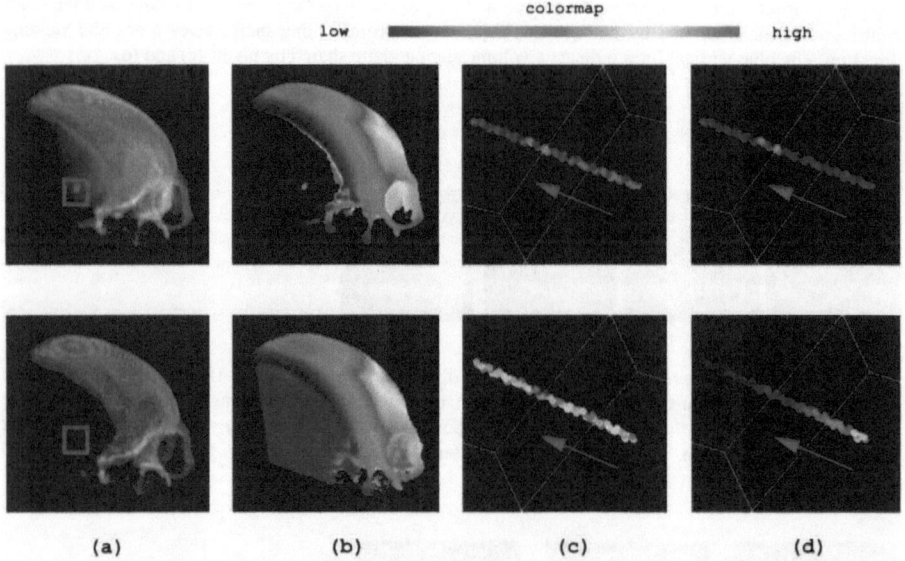

(a) (b) (c) (d)

Fig. 6. Case study illustrating a hypothetical tumor and how the *volume distance* metric (**b**) and the *pixel probe* (**c**) and (**d**) can shed more insight. Visualizations of column **c** and **d** show differences in absolute and additive color contributions from volume cells

Udeshi and Hansen (pp. 99–108)

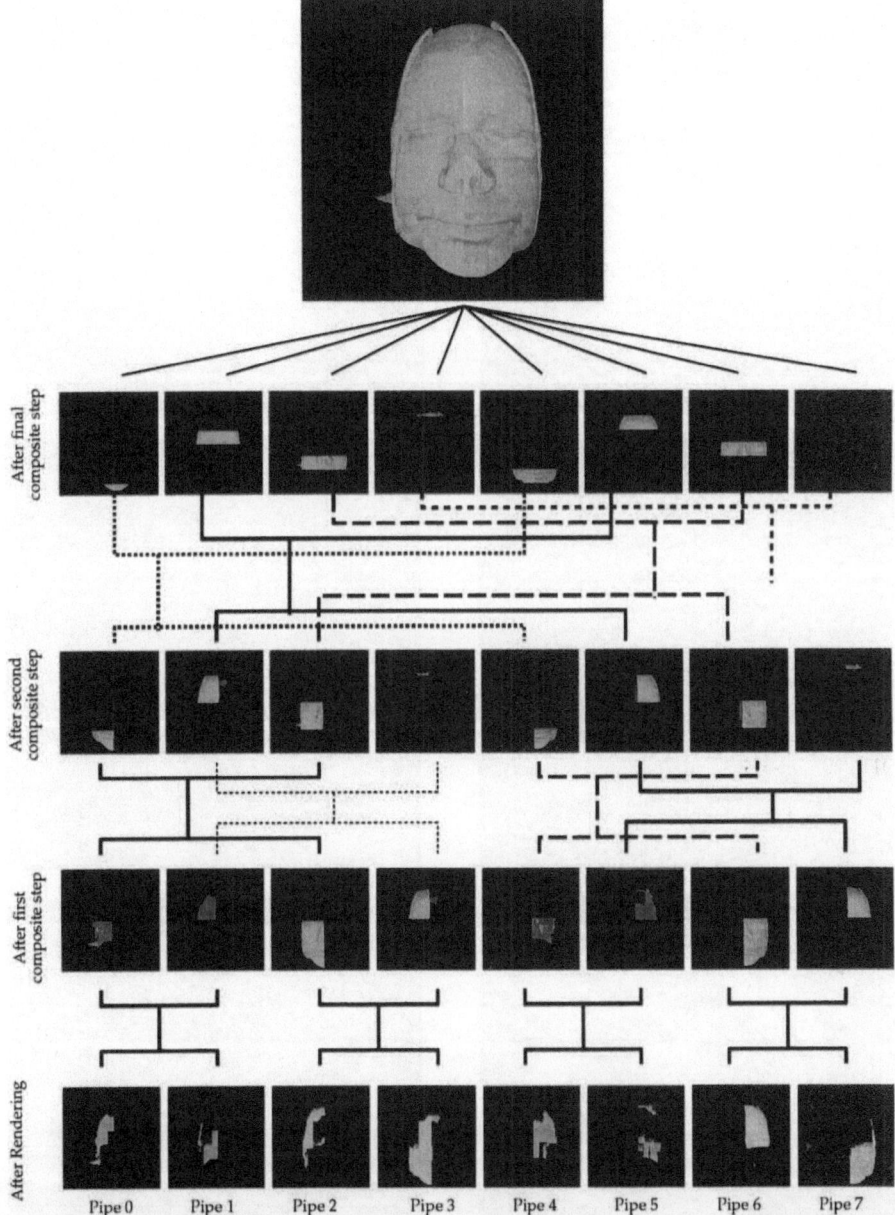

Fig. 1. The compositing levels are along the verticle axis and the graphics adapters are along the horizontal axis. After the final step (the top most level), the final image is composed from each of the partial images

Clyne and Dennis (pp. 109–120)

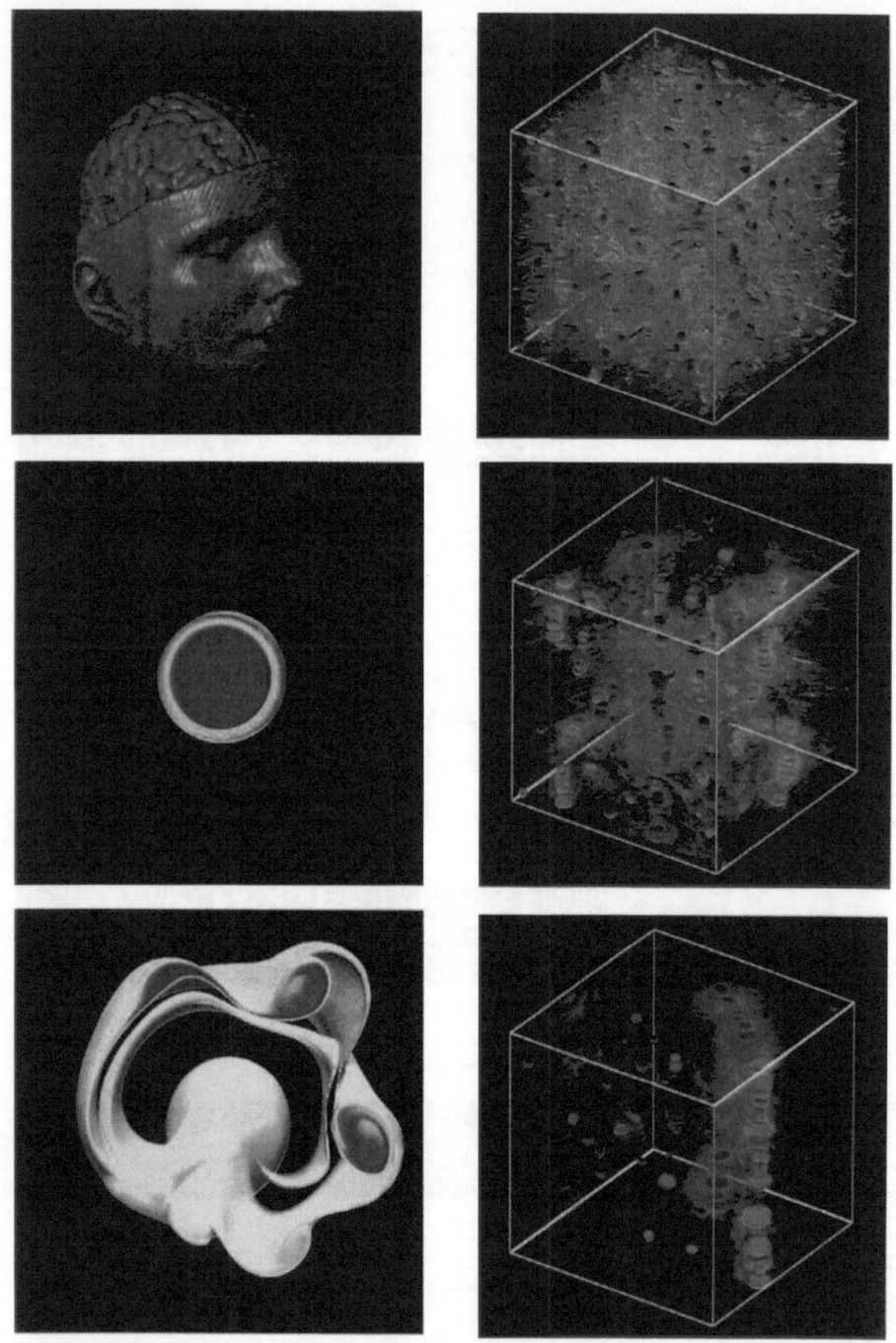

Plate 1. MRBrain (top), Polar Vortex data at time 50 (middle), and 400 (top)

Plate 2. QG data at time 100 (top), 499 (middle), and 1492 (bottom)

Tong et al. (pp. 121–132)

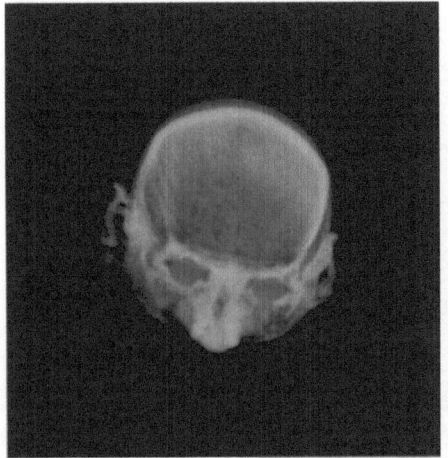

Fig. 8. Image of the Head I data set rendered by conventional method

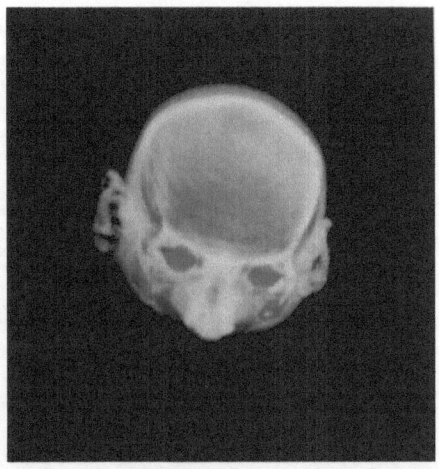

Fig. 9. Image of the Head I data set rendered by new method

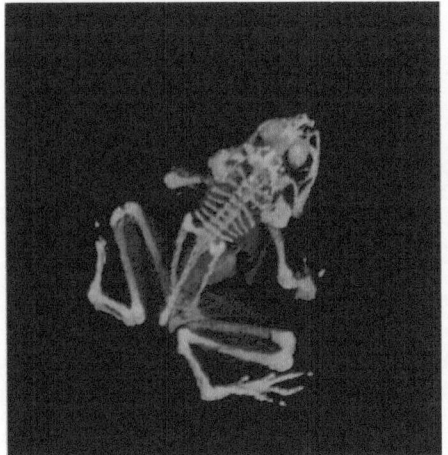

Fig. 10. Image of the Frog data set rendered by new method

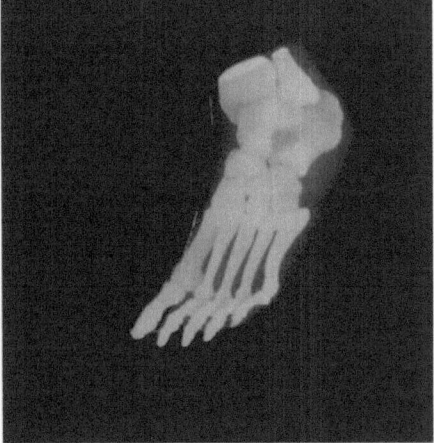

Fig. 11. Image of the Foot data set rendered by new method

Bartz and Skalej (pp. 155–166)

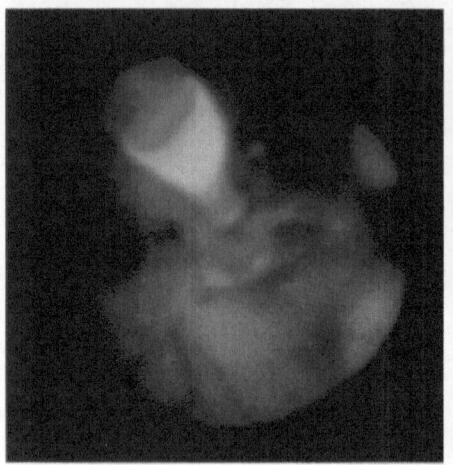

Fig. 2. View through optical endoscope

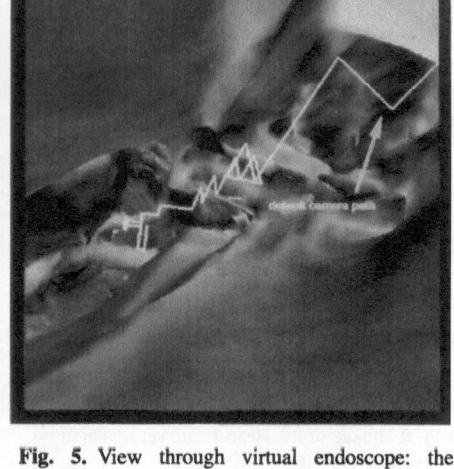

Fig. 5. View through virtual endoscope: the voxel-based default camera path is rendered as white line

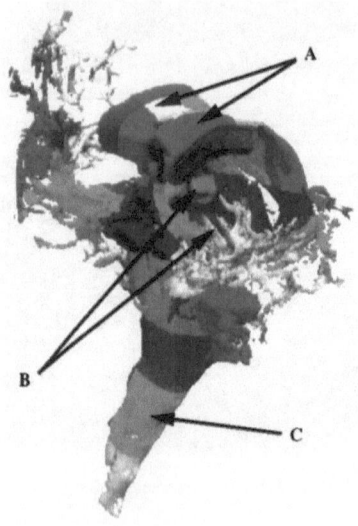

Fig. 6. Octree Subdivision of Ventricular system – different blocks are marked with different colors. A: Lateral Ventricles. B: 3rd and 4th Ventricles. C: Central Canal

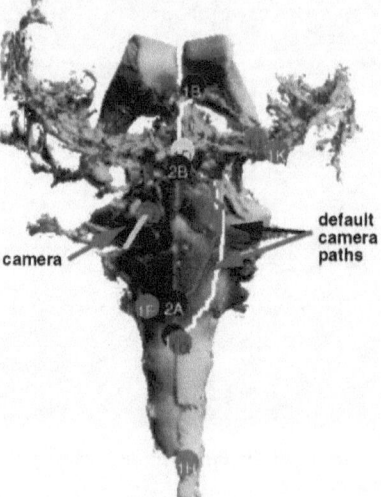

Fig. 7. Snapshot from Scout Panel with frontview of ventricular system: two default camera paths are shown (white and red): the pre-defined markers are colored in blue

Bender et al. (pp. 167–176)

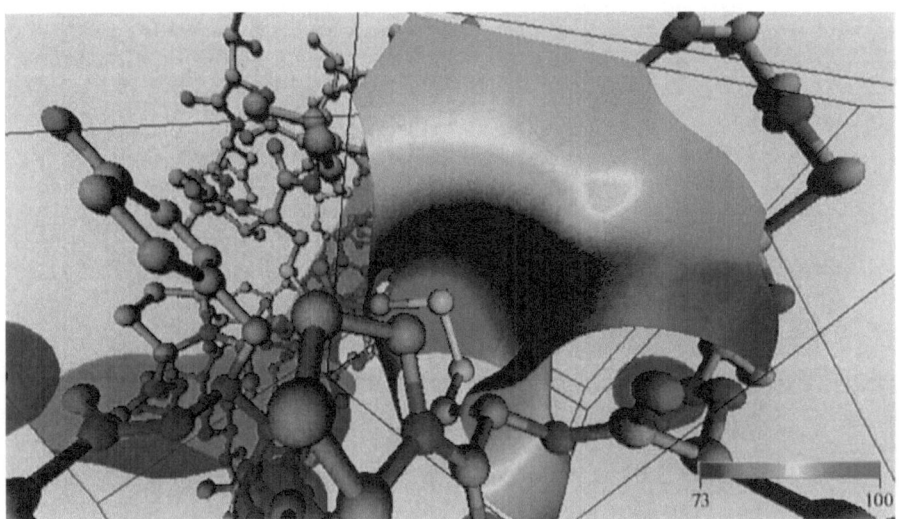

Fig. 1. Plant seed protein (*Ball-and-Stick Model, different Isosurfaces*)

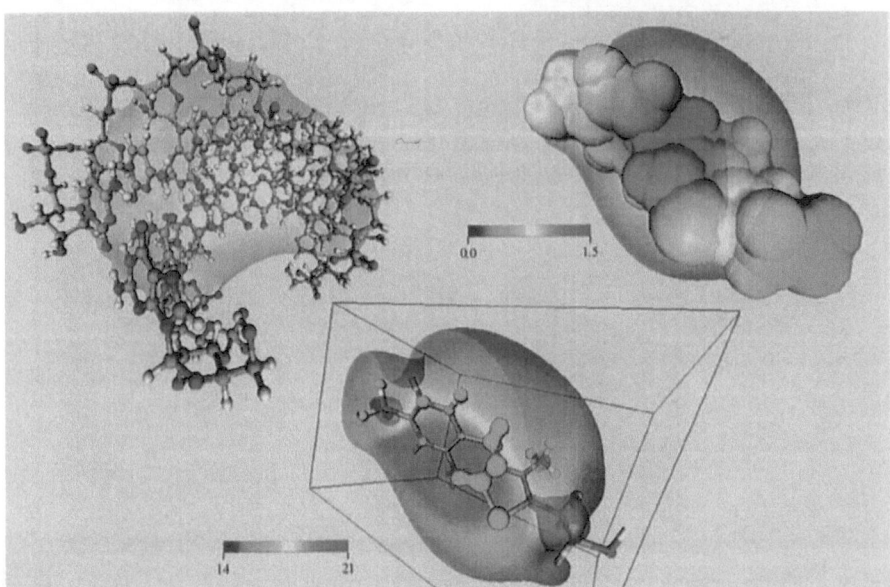

Fig. 5. Ribonucleic acid (*Ball-and-Stick Model, Isosurface*), vitamin-b1 (*Richards' Contact Surface, Isosurface*), vitamin-b1 (*Stick Model, different Isosurfaces*)

Polthier and Schmies (pp. 179–188)

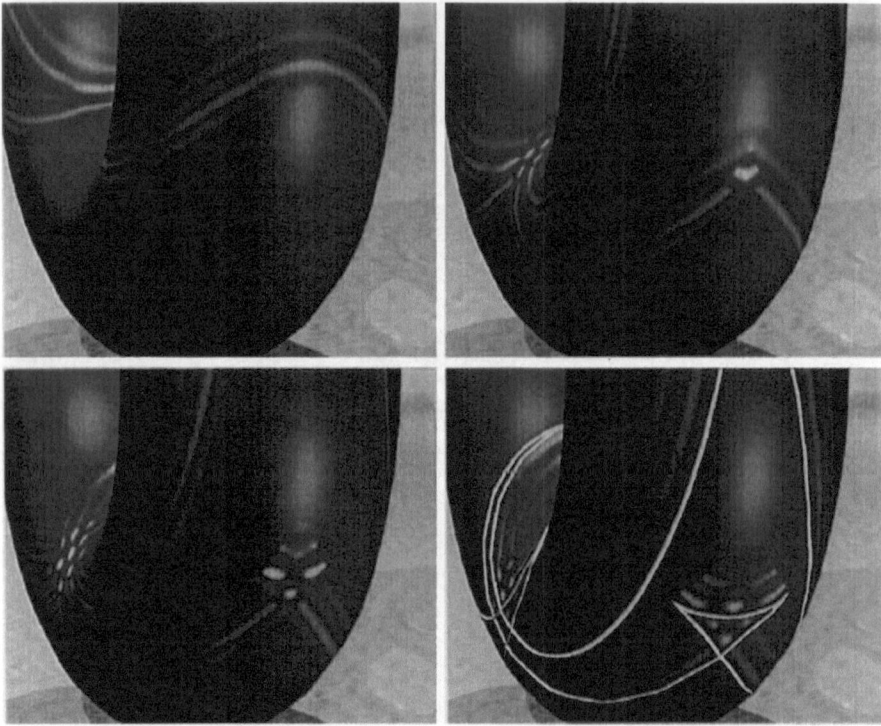

Fig. 7. A point wave on a polyhedrally approximated torus initiated at top branches at the conjugate point in the form of a swallow's tail. Visualization of the interference uses branched texture maps

Fig. 8. Straightest geodesics are defined to have equal angle on both sides at each point. On planar faces they are straight lines, and across edges they have equal angles on opposite sides. Straightest geodesics can be extended through polyhedral vertices, a property not available for shortest geodesics

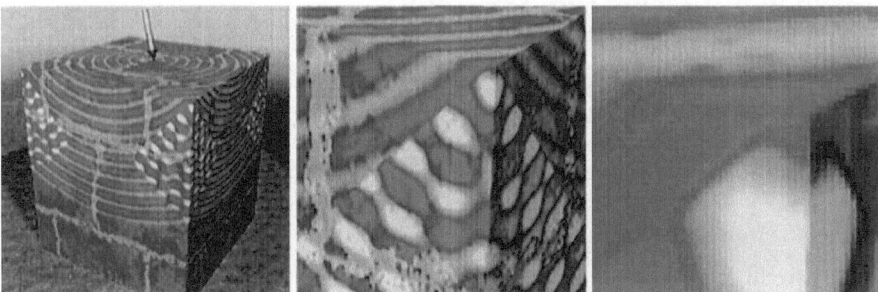

Fig. 9. A point wave branches at the vertices of a cube which are equivalents of conjugate points on smooth surfaces. The section behind a vertex is covered by three interfering texture layers while other parts are covered once. The zoom shows the triangles and texels on the surface

 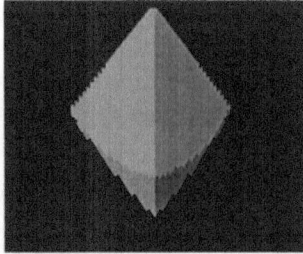

Fig. 10. Interference of a point wave at the vertex of a cube and texture layers. The left picture shows the interference behavior of the wave at the vertex of a cube. The wave is stored as a branched texture map, where to each point on the surface a stack of texels is associated. In the right picture the surface texels are colored according to the height of the texture stack at each point

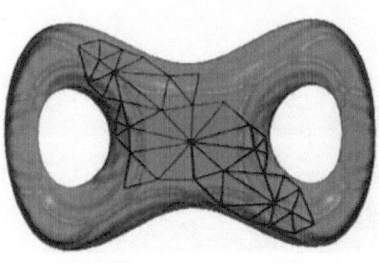 

Fig. 11. Evolution of distance circles under the geodesic flow on the (highly discretized) polyhedral model of a pretzel and a torus

Kreylos and Hamann (pp. 189–198)

Fig. 12. Seventh experiment. Original RGB image, 329 × 222 pixels

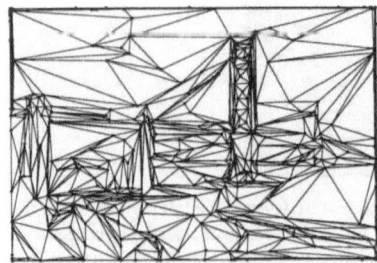 

Fig. 13. Seventh experiment. Left: final configuration; right: final linear spline approximation

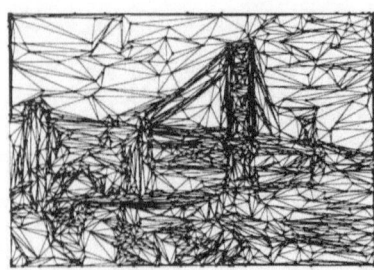 

Fig. 14. Eighth experiment. Left: final configuration; right: final linear spline approximation

Gerstner et al. (pp. 199–211)

a Adaptive projection of the geographical map   **b** Original data

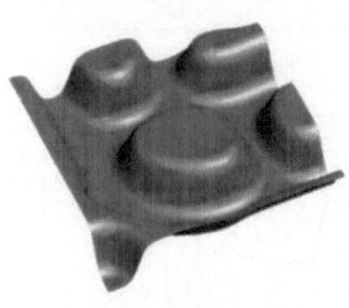 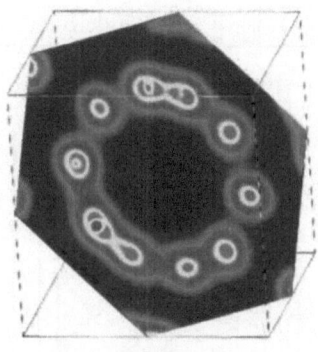

c Timestep of Cahn-Hilliard Equation   **d** Color shaded slice of the bucky ball

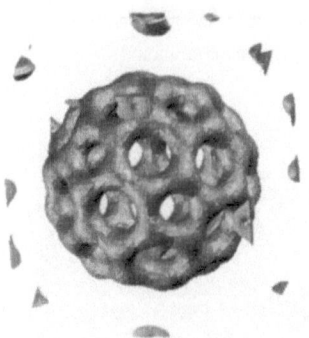 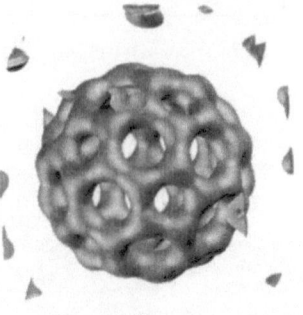

e Adaptive projection of the isosurface   **f** Original data

Fig. 4. Above the graph of a geographic height field, its adaptive projection and a timestep of the Cahn-Hilliard-Equation are shown. Of the bucky ball data set we show a color shaded diagonal slice, an adaptive projection and a full resolution isosurface

Wartell et al. (pp. 213–223)

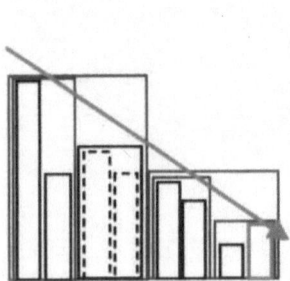

Fig. 3. General algorithm

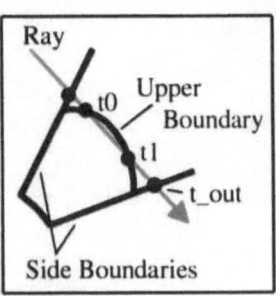

Fig. 4. Intersection test

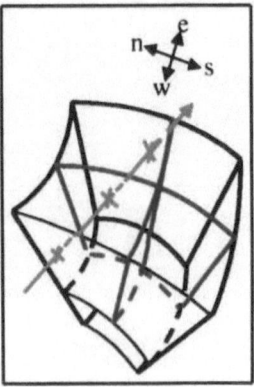

Fig. 5. Quad traversal

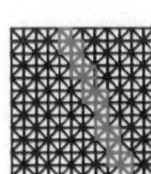

Fig. 6. Triangle grid

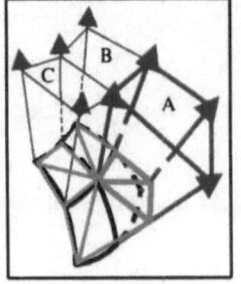

Fig. 7. Triangle traversal

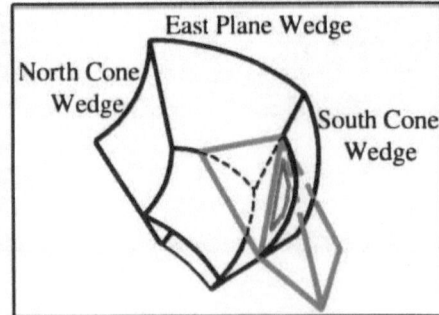

Fig. 8. Triangle containment problem

a

b

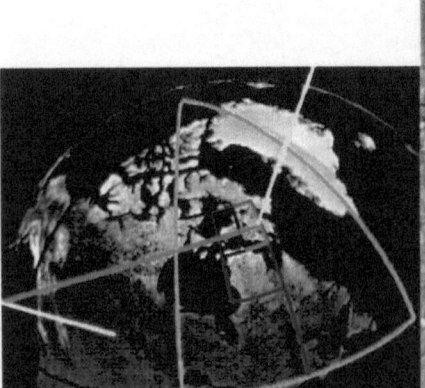

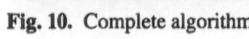

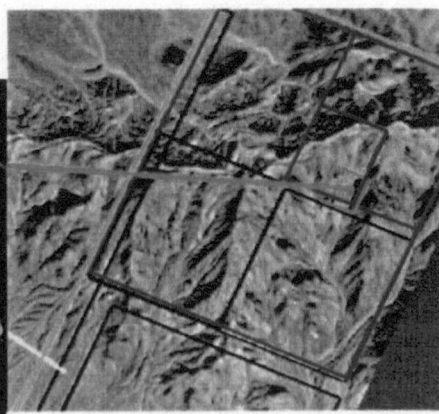

Fig. 10. Complete algorithm

Telea and van Wijk (pp. 225–234)

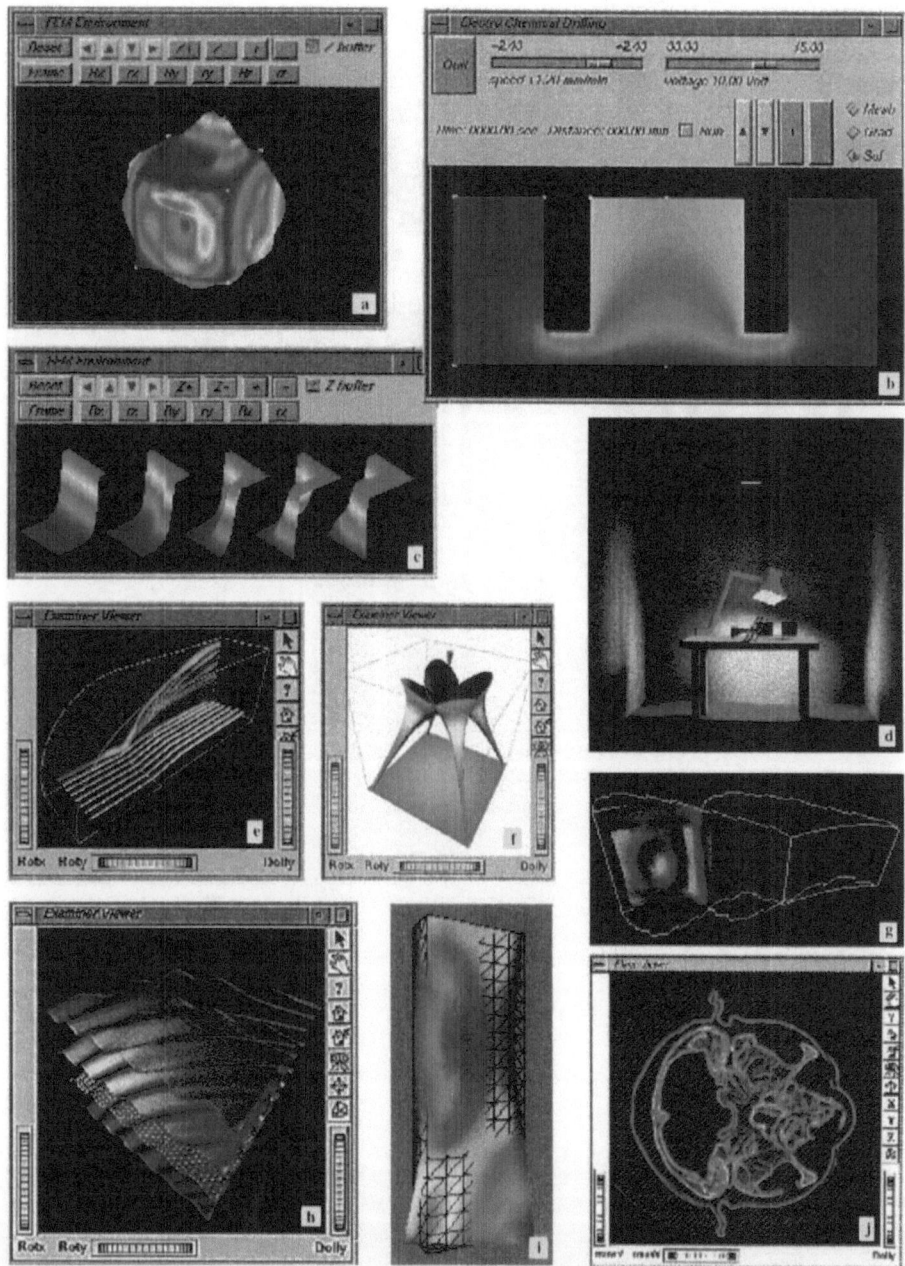

Fig. 7. Visualizations and simulations performed in the VISSION environment

Gell et al. (pp. 237–246)

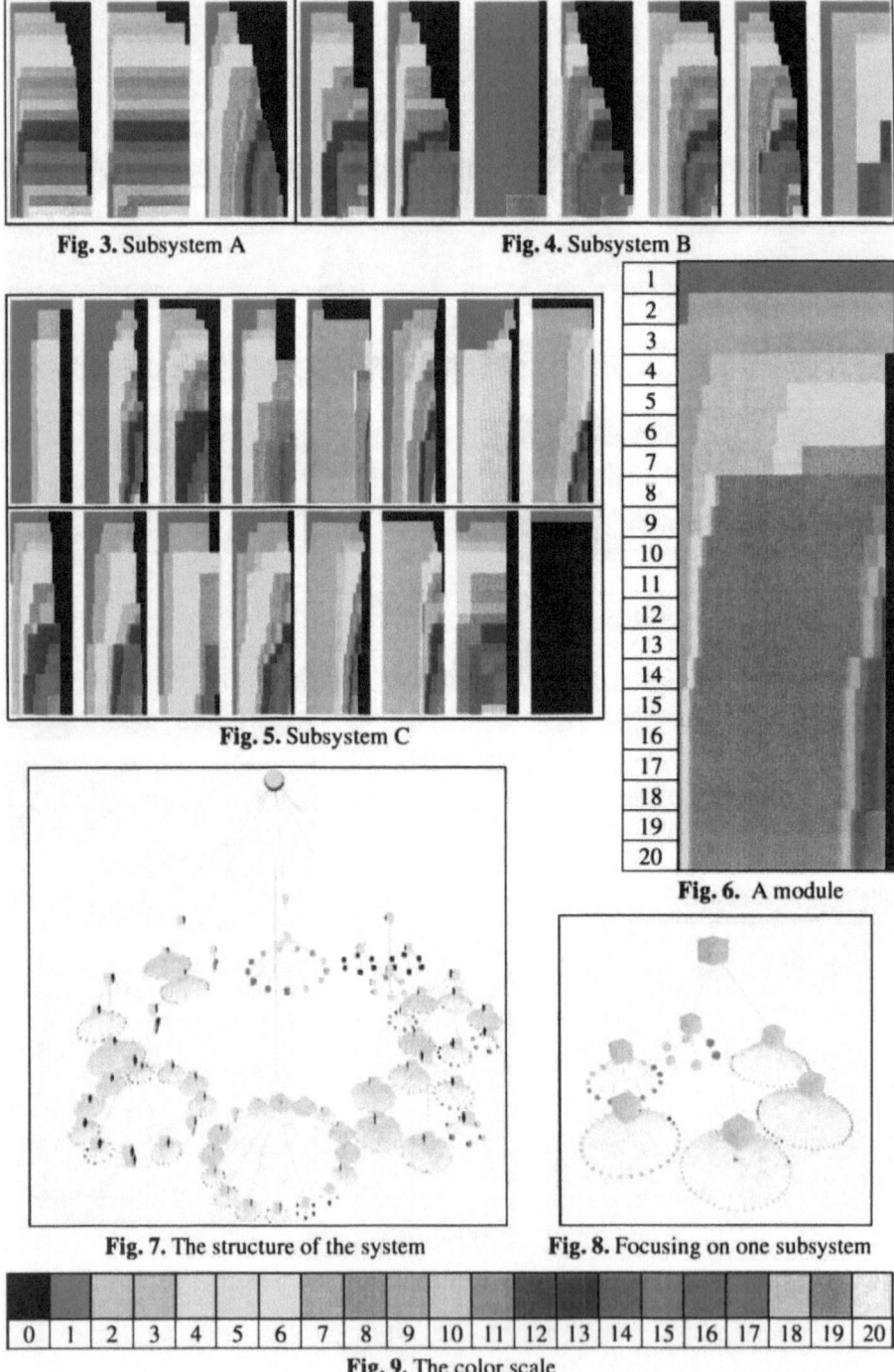

Fig. 3. Subsystem A

Fig. 4. Subsystem B

Fig. 5. Subsystem C

Fig. 6. A module

Fig. 7. The structure of the system

Fig. 8. Focusing on one subsystem

Fig. 9. The color scale

He (pp. 247–252)

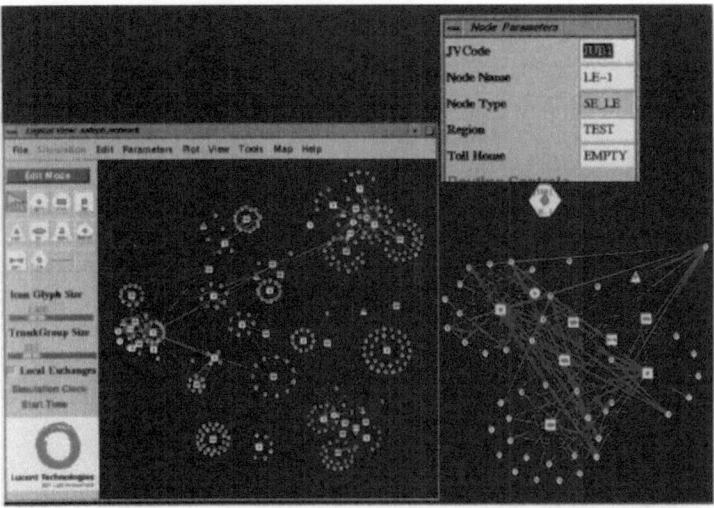

Fig. 1. Logic network view. Switches are placed according to their homing information. Only the trunks with above-threshold capacities are displayed in the full network view (bottom left), and the highlighted LCA is focused in the zoom view (bottom right)

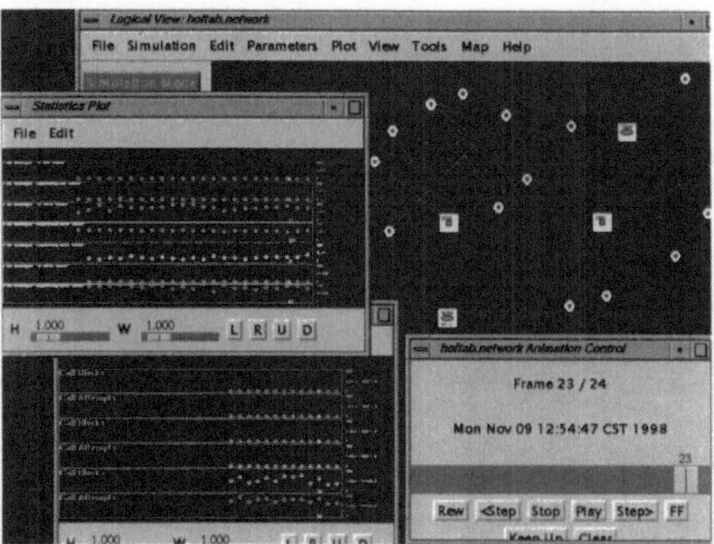

Fig. 2. Real time monitoring of network performance and statistics

334

Haase et al. (pp. 261–266)

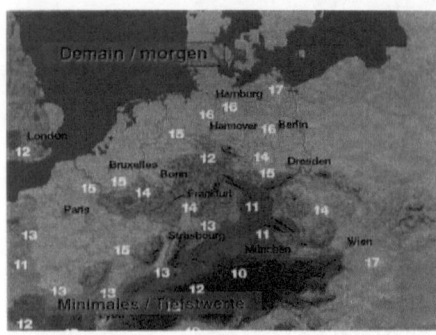

Fig. 1. TriVis: 2D visualisation showing temperatures

Fig. 2. TriVis 3D visualisation with snowfall

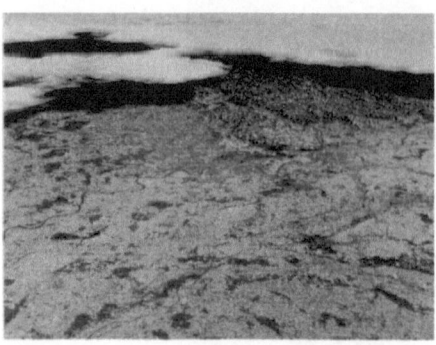

Fig. 3. TriVis: high-resolution terrain with satellite texture and model clouds

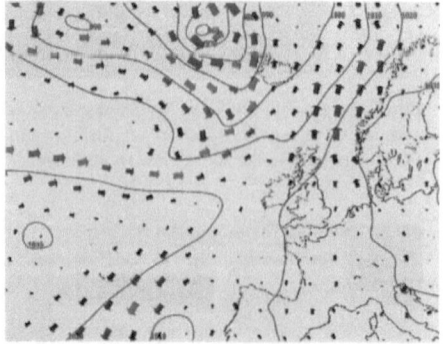

Fig. 4. VISUAL: visualising wind, temperature, liquid water content, and pressure

Fig. 5. VISUAL: treatment of the model's vertical coordinates (grid over Greenland)

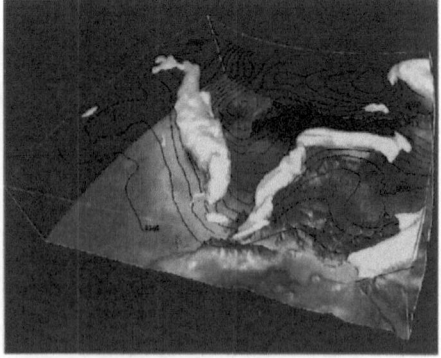

Fig. 6. VISUAL showing isosurfaces of wind speed

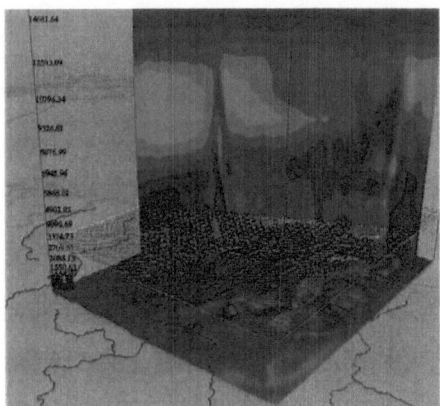

Fig. 7. VISUAL: vertical slice of wind speed generated by solid contour algorithm

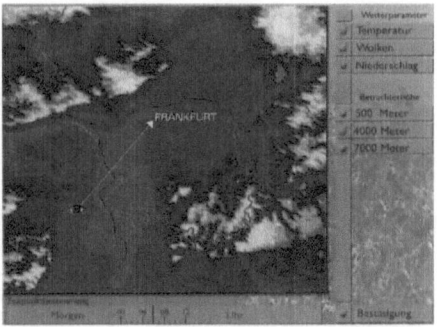

Fig. 8. WxoD: CityWeather for visualisation on server

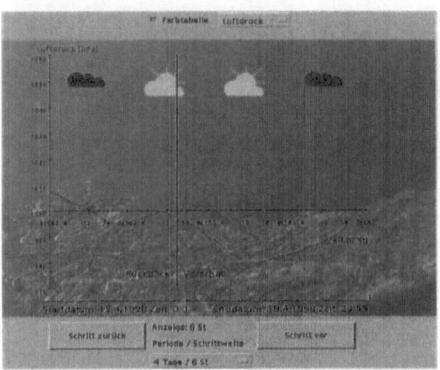

Fig. 9. WxoD: meteogram visualisation locally on client

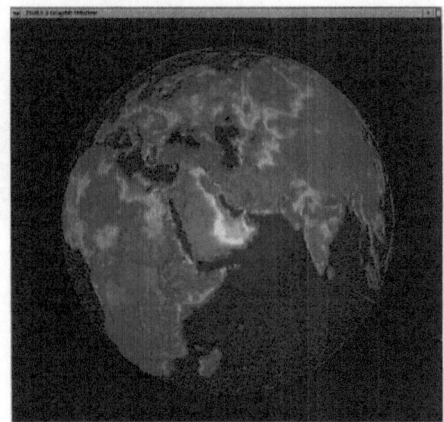

Fig. 10. DWD's new Global Model visualised with IGD's ISVAS system

Fig. 11. Virtual Studio: TriVis 3D weather geometries can fully be integrated into virtual studio systems

Fig. 12. Augmented Reality: vertical red line cuts the terrain to better show the position of the arrows

Bajaj et al. (pp. 269–276)

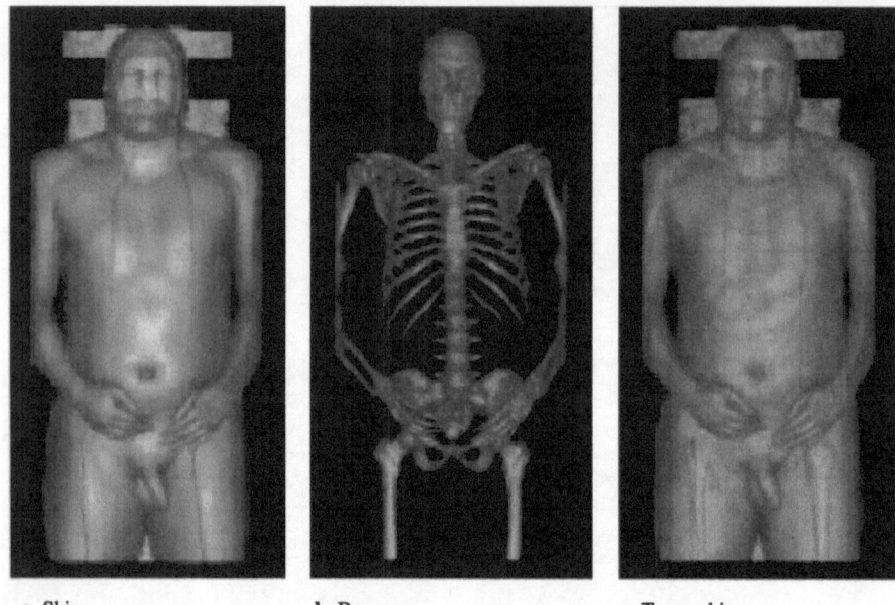

a Skin   b Bone   c Trans. skin

Fig. 1. Parallel ray-cast visible man

Glau (pp. 277–283)

Fig. 3. Volume rendered temperature dataset inside a car cabin. LUT merging was applied for feature highlighting

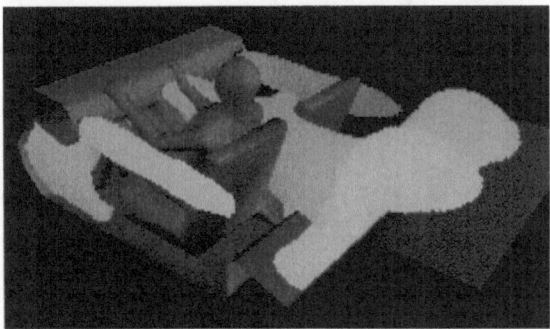

Fig. 4. Iso-surface generated using a depth shading approach

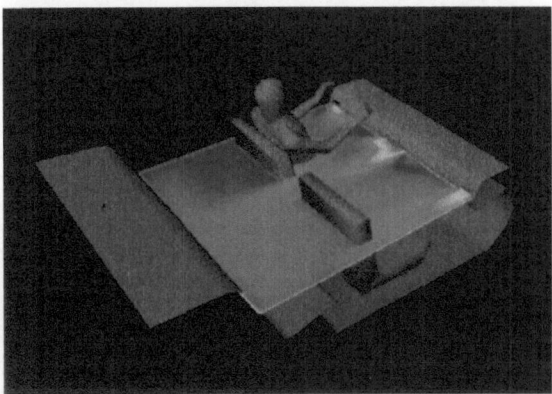

Fig. 5. Clipping plane

Lürig et al. (pp. 285–289)

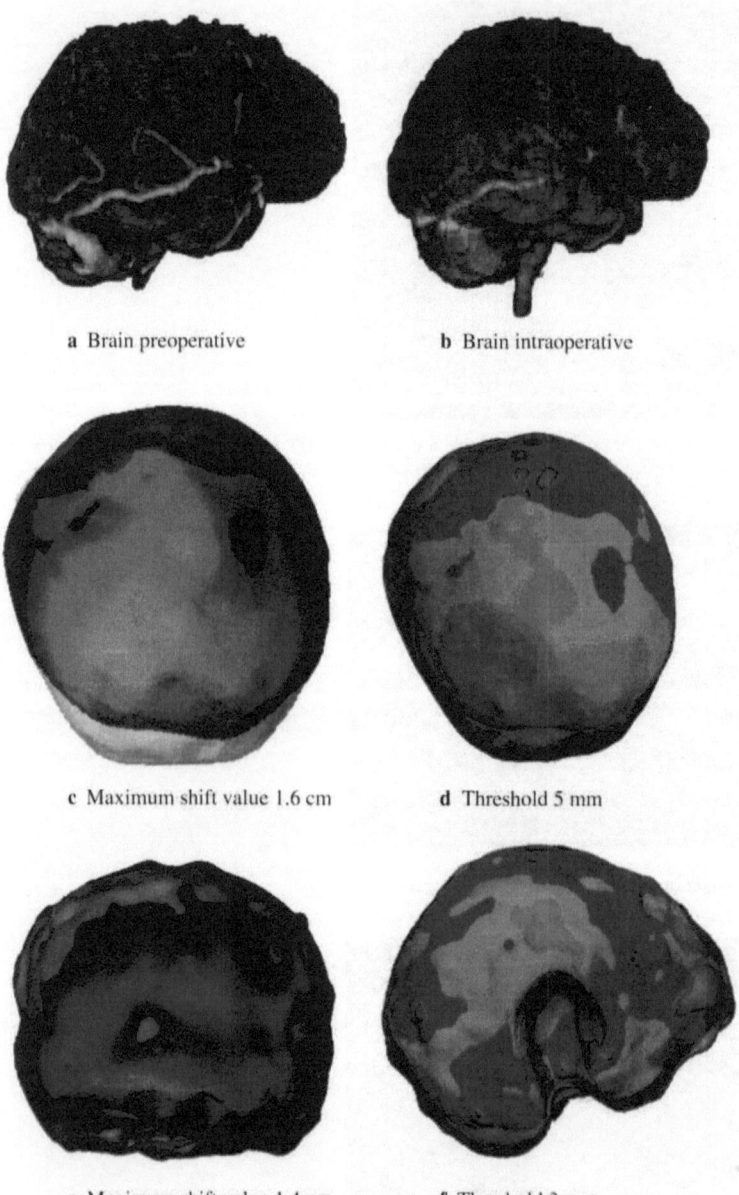

a Brain preoperative   b Brain intraoperative

c Maximum shift value 1.6 cm   d Threshold 5 mm

e Maximum shift value 1.4 cm   f Threshold 3 mm

Fig. 1. Examples for the brain shift visualization

Walter et al. (pp. 291–298)

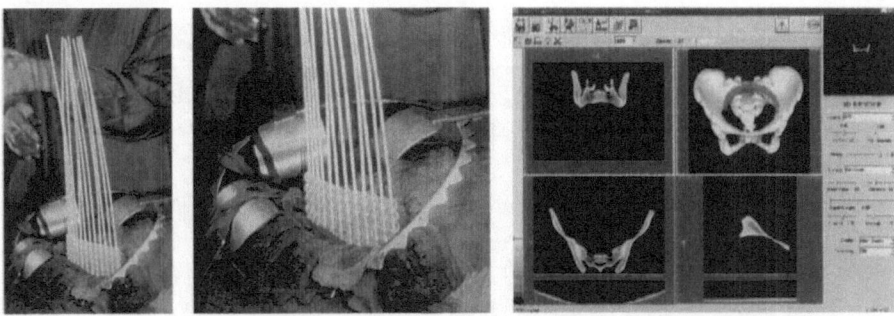

Fig. 6. Left: Flab with rubber pellets and flab in situ with applicator pipes; right: InViVo-IORT user interface: three windows with orthogonal cutting planes through the CT volume data and a 3D surface reconstruction (upper right window)

Fig. 7. Display of CT scans and intersected pellets

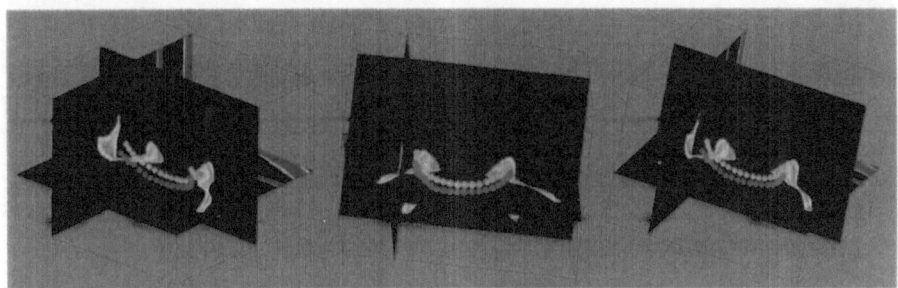

Fig. 8. 3D visualisation as 3 orthogonal cutting planes with overlaid flab pellets of two applicators

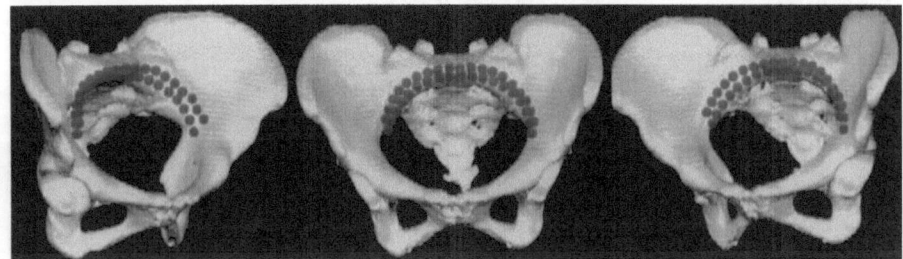

Fig. 9. 3D visualisation (surface volume rendering) with overlaid flab pellets of two applicators

Rheingans and Joshi (pp. 299–306)

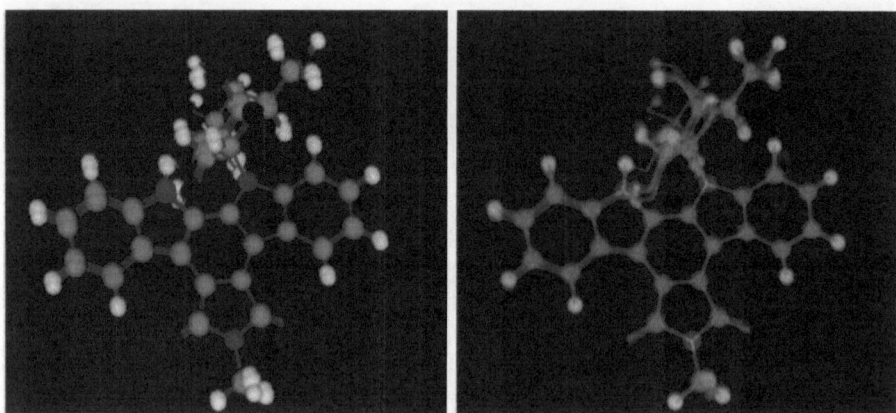

Fig. 2. Ball and Stick. Eight members, **a** each full opacity, **b** each with opacity of 0.125. Carbon atoms are grey, nitrogen red, oxygen blue, and hydrogen white

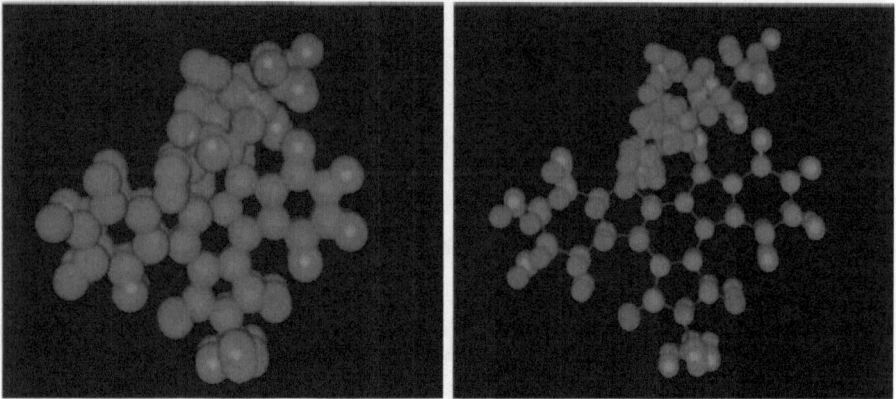

Fig. 3. Isosurface of likelihood volume. **a** Isolevel value of 0.9, **b** isolevel value of 0.95

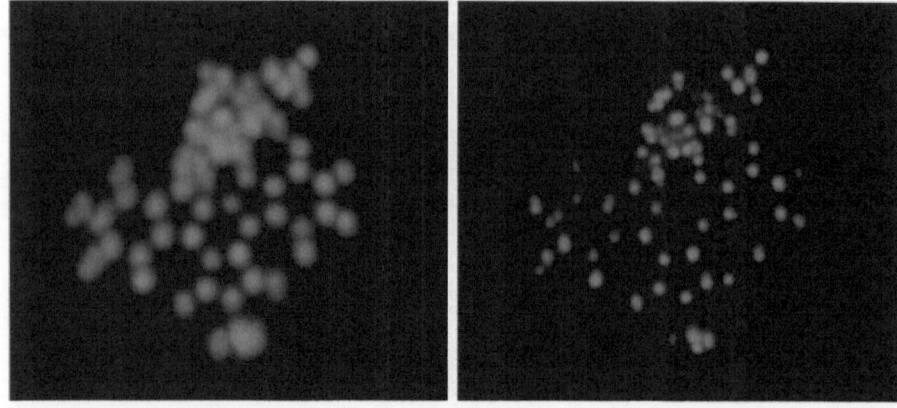

Fig. 4. Volume rendering of likelihood volume, **a** emphasizing areas of greatest likelihood, **b** greater contribution from less likely areas

SpringerEurographics

Bruno Arnaldi,

Gérard Hégron (eds.)

Computer Animation and Simulation '98

Proceedings of the Eurographics
Workshop in Lisbon, Portugal,
August 31–September 1, 1998

1999. VII, 126 pages. 82 figures.
Softcover DM 85,–, öS 595,–, sFr 77,50
ISBN 3-211-83257-2. Eurographics

Contents:
- J.-D. Gascuel et al.: Simulating Landslides for Natural Disaster Prevention
- G. Besuievsky, X. Pueyo: A Dynamic Light Sources Algorithm for Radiosity Environments
- G. Moreau, S. Donikian: From Psychological and Real-Time Interaction Requirements to Behavioural Simulation
- N. Pazat, J.-L. Nougaret: Identification of Motion Models for Living Beings
- F. Faure: Interactive Solid Animation Using Linearized Displacement Constraints
- M. Kallmann, D. Thalmann: Modeling Objects for Interaction Tasks
- M. Teichmann, S. Teller: Assisted Articulation of Closed Polygonal Models
- S. Brandel, D. Bechmann, Y. Bertrand: STIGMA: a 4-dimensional Modeller for Animation

Michael Gervautz,

Axel Hildebrand,

Dieter Schmalstieg (eds.)

Virtual Environments '99

Proceedings of the Eurographics
Workshop in Vienna, Austria,
May 31–June 1, 1999

1999. X, 191 pages. 78 figures.
Softcover DM 85,–, öS 595,–, sFr 77,50
ISBN 3-211-83347-1. Eurographics

The special focus of this volume lies on augmented reality. Problems like real-time rendering, tracking, registration and occlusion of real and virtual objects, shading and lighting interaction and interaction techniques in augmented environments are addressed. The papers collected in this book also address levels of detail, distributed environments, systems and applications and interaction techniques.

All prices are recommended retail prices

 SpringerWienNewYork

Sachsenplatz 4–6, P.O.Box 89, A-1201 Wien, Fax +43-1-330 24 26, e-mail: books@springer.at, Internet: http://www.springer.at
New York, NY 10010, 175 Fifth Avenue • D-14197 Berlin, Heidelberger Platz 3 • Tokyo 113, 3-13, Hongo 3-chome, Bunkyo-ku

SpringerEurographics

Martin Göbel,

Jürgen Landauer, Ulrich Lang,

Matthias Wapler (eds.)

Virtual Environments '98

Proceedings of the Eurographics Workshop
in Stuttgart, Germany, June 16–18, 1998

1998. VIII, 335 pages. 206 partly coloured figures.
Softcover DM 128,–, öS 896,–, sFr 116,50
ISBN 3-211-83233-5. Eurographics

Ten years after Virtual Environment research
started with NASA's VIEW project, these
techniques are now exploited in industry to
speed up product development cycles, to
ensure higher product quality, and to encour-
age early training on and for new products.
Especially the automotive industry, but also
the oil and gas industry are driving the use of
these techniques in their works.
The papers in this volume reflect all the different tracks of the workshop: reviewed technical papers as research contributions, sum-
maries on panels of VE applications in the
automotive, the medical, the telecommunication and the geoscience field, a panel dis-
cussing VEs as the future workspace, invited
papers from experts reporting from VEs for
entertainment industry, for media arts, for
supercomputing and productivity enhancement. Short industrial case studies, reporting
very briefly from ongoing industrial activities
complete this state of the art snapshot.

Panos Markopoulos,

Peter Johnson (eds.)

Design, Specification

and Verification

of Interactive Systems '98

Proceedings of the Eurographics Workshop
in Abingdon, U.K., June 3–5, 1998

1998. IX, 325 pages. 119 figures.
Softcover DM 118,–, öS 826,–, sFr 107,50
ISBN 3-211-83212-2. Eurographics

Does modelling, formal or otherwise, have a
role to play in designing interactive systems?
A proliferation of interactive devices and
technologies are used in an ever increasing
diversity of contexts and combinations
in professional and every-day life. This
development poses a significant challenge
to modelling approaches used for the design
of interactive systems. The papers in this
volume discuss a range of modelling approaches, the representations they use, the
strengths and weaknesses of their associated specification and analysis techniques
and their role in supporting the design of
interactive systems.

All prices are recommended retail prices

SpringerWienNewYork

Sachsenplatz 4–6, P.O.Box 89, A-1201 Wien, Fax +43-1-330 24 26, e-mail: books@springer.at, Internet: http://www.springer.at
New York, NY 10010, 175 Fifth Avenue • D-14197 Berlin, Heidelberger Platz 3 • Tokyo 113, 3-13, Hongo 3-chome, Bunkyo-ku